Seaweeds of the British Isles

Seaweeds of the British Isles

FRONT COVER: *Corallina officinalis* with epiphytic *Mesophyllum lichenoides*

Seaweeds of the British Isles

A collaborative project of the British Phycological Society
and the Natural History Museum

Volume 1 Rhodophyta

Part 2B Corallinales, Hildenbrandiales

Linda M Irvine & Yvonne M Chamberlain

Natural History Museum, London

First published by the Natural History Museum,
Cromwell Road, London SW7 5BD
© Natural History Museum, London, 1994

This edition printed and published by Pelagic Publishing, 2011,
in association with the Natural History Museum, London

The Authors have asserted their right to be identified as the Authors of this work under the Copyright, Designs and Patents Act 1988.

ISBN 978-1-907807-10-7

This book is a reprint edition of 0-11-310016-7.

All rights reserved. No part of this publication may be transmitted in any form or by any means without prior permission from the British Publisher.

A catalogue record for this book is available from the British Library.

Contents

Foreword

There is an increasing demand for taxonomic expertise and knowledge, including the development of sound species concepts and the provision of authoritatively written identification guides. The need for the accurate identification of British seaweeds is met by the series *Seaweeds of the British Isles*. To date, six parts have been published, each involving close collaboration between The Natural History Museum and the British Phycological Society. Authors who have contributed to the series include Dr E M Burrows (University of Liverpool, UK), Professor T Christensen (Sporeplanter Institute, Denmark), Professor P S Dixon (University of California, USA), Professor M D Guiry (University of Galway, Ireland), Drs W F Farnham & R L Fletcher (University of Portsmouth, UK), Professor M H Hommersand (University of North Carolina, USA) and Dr C A Maggs (Queens University, Belfast, UK). The present volume completes the Florideophycideae, the major and more advanced subclass of Rhodophyta. The remaining part of the Rhodophyta volume, dealing with the subclass Bangiophycideae, is in preparation.

The close association of The Natural History Museum with the *Seaweeds of the British Isles* goes back to 1967 when Linda Irvine was appointed to assist with the production of the Rhodophyta volume. She contributed to two of the earlier parts, the first published in 1977 and the second in 1983. The original intention was that the 'Coralline Algae' and *Hildenbrandia*, now the Corallinales and Hildenbrandiales, would be included in the second part dealing with the Cryptonemiales, Palmariales and Rhodymeniales. However, information of these poorly understood and much neglected groups was so incomplete that it was decided to publish them separately. Linda Irvine and Yvonne Chamberlain have spent the past ten years resolving many of the taxonomic and biological problems. An undertaking of this magnitude required collaboration with specialists in other parts of the world and often involved spending extended periods working in overseas laboratories. Two of these specialists have contributed directly to the volume, namely Dr Curt Pueschel (State University of New York, USA) and Dr Bill Johansen (Clark University, Massachusetts, USA).

Of all photosynthetic organisms, algae are perhaps the most poorly known. Algal research continues to be a high priority for The Natural History Museum, with specialists in four of the Botany Department's Divisions. *Seaweeds of the British Isles* is an extremely important project that meets a fundamental need for sound taxonomic work and accessible guides to identification.

Dr Stephen Blackmore, Keeper of Botany, Associate Director – Life Sciences, The Natural History Museum, London.

Acknowledgements

We would like to record our gratitude to our collaborators, Dr H.W.Johansen (Corallinoideae) and Dr C.M.Pueschel (Hildenbrandiales), and also to both Dr W.H.Adey, who generously provided expertise and encouragement at the beginning of the project, and Dr W.J.Woelkerling, whose continued interest and cooperation has helped us to bring it successfully to a conclusion.

We would also like to thank Dr F.Ardré, the late Dr H.Blackler, Dr G.Boalch, Dr D.Bosence, Dr J.Brodie, Mr F. and Mrs A.Bunker, Dr J.Cabioch, Mr S.Campbell, Dr T.Christensen, Mr J.Clokie, Miss P.Cook, Mr P.Cooke, Dr P.Cornelius, the late Prof. M.De Valera, the late Dr P.Dixon, Mr A.Eddy, Dr R.Edyvean, Dr G.Elliott, Dr W.Farnham, Dr R.Fletcher, Dr D.George, Mr M.Gill, Dr M.Guiry, Mr J.Hall-Spencer, Miss C.Hayek, Mr J.Hepburn, Mr S.Honey, Mr R.Hooper, Dr D. and Mr A.Irvine, Dr D.John, Mr C.Jermy, Dr E.B.G.Jones, Dr W.E.Jones, Dr J.Kain, Mr J.Laundon, Dr P.Leukart, Dr C.Maggs, Dr T.Masaki, Mr D.Minchin, Mr O.Morton, Dr A.Pentecost, Dr D.Penrose, Mrs J.Pope, Dr H.Powell, Mr J.Price, Dr W.Prud'homme van Reine, Mrs S.Scott (formerly Hiscock), Mr C.Spurrier, Mrs M.Steentoft, Miss S.Stone, Dr L.Terry, Mr I.Tittley, Col. P. & Mrs V.Trueman, Dr R.Walker and Miss A.Webster for helpful discussions, assistance with collecting, the loan or gift of specimens, records etc., and Mr R.Ross who prepared the latin diagnosis and, with Dr C.Jarvis and Dr P.C.Silva, helped with nomenclature. Dr G.Lawson gave valuable editorial help.

Additional to the lists in Vol. 1 parts 1, 2A and 3A, we are indebted to the directors and curators of the following institutions for providing loans, working facilities and assistance:

Herbarium, Naturwet. Museum, Antwerp (Belgium) AHW
Herbarium, National Botanic Gardens, Glasnevin, Dublin DUB
Herbarium, Forschungsinstitut Senckenberg, Frankfurt (Germany) FR
Herbarium, University of Michigan, Ann Arbor, Michigan (USA) MICH

We would like to express our appreciation of the financial support provided by the British Council (to LMI), the Natural Environment Research Council (Grant no.GR3/4818A to YMC) and the British Phycological Society (Frontispiece subsidy). We are grateful for the support and congenial working conditions provided by both the Natural History Museum and the University of Portsmouth.

Most of the illustrations were prepared by YMC, those for the Corallinoideae and Hildenbrandiales were drawn by HWJ, Mrs V.C.Gordon Friis and CMP, and the cover drawing of *Mesophyllum lichenoides* on *Corallina officinalis* is by Mr L. Ellis. Help with photography was provided by Mr G.Bremer, Mr F.Bunker, Mr C.Derrick, Mr J.Hall-Spencer, Mr P.Hurst, Mr P.Mullock, Mrs S.Scott, and Dr W.Woelkerling; the *British Phycological Journal* and the *Bulletin of The Natural History Museum* kindly allowed us to re-use some previously published plates.

Linda Irvine would like to add special thanks to all those who made the completion of the book possible by providing support and sharing the care of her husband during his long, difficult and continuing illness.

Yvonne Chamberlain thanks Dick, Mary, Sarah and her late father for all their encouragement and tolerance over the last 18 years.

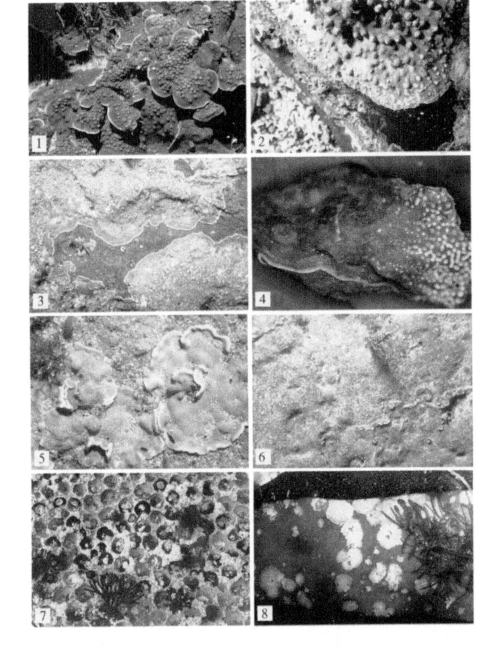

INTRODUCTION

by Linda M. Irvine, Yvonne M. Chamberlain and H. William Johansen

This part of Volume 1 deals with the orders Corallinales and Hildenbrandiales. For the taxonomic treatment of the Hildenbrandiales see p.235. A concise account of the structure and classification of the Corallinales is given here, features of taxonomic significance and practical aid in identification being emphasised. This is followed by a guide to the layout and conventions used in the descriptions. Finally, illustrated descriptions are given of 14 genera and 45 species, one with two varieties and another with four; keys to genera and species are provided. One species *(Phymatolithon brunneum)* is newly described and three new combinations *(Hydrolithon boreale, H. cruciatum* and *H. sargassi)* are made.

1 Vegetative characteristics of the Corallinales

Algae of the order Corallinales are almost entirely calcified and comprise both nongeniculate genera (Frontispiece) that are, with few exceptions, encrusting, and geniculate genera (Fig.14) that are erect and have successive, uncalcified joints (genicula) separating calcified segments (intergenicula).

Woelkerling & Irvine (1988a) have recently summarised the main characteristics of nongeniculate genera and the following account is partly based on the information provided there. However, only information relevant to taxa found in the British Isles is included here, and a simplified terminology is used in line with that in the rest of the Volume (see Chamberlain, 1990).

Information on geniculate genera can be found in Johansen (1976, 1981).

1.1 Thallus form
1.1a Nongeniculate genera
Most corallines are epilithic and occur as encrustations on rocks and stones, but some are epiphytic on algae and seagrasses, epizoic on shells, hydroids etc., unattached or, as in *Choreonema,* semi-endophytic in geniculate corallines. There are no representatives of erect

---

FRONTISPIECE: (1) *Mesophyllum lichenoides* ×1, brownish (Methuen reddish brown) (photo S. Scott); (2) *Lithothamnion glaciale* ×1, reddish pink (Methuen dull/greyish red) (photo J. Hall-Spencer); (3) *Phymatolithon lenormandii* ×1, reddish grey (Methuen dull red at margin, reddish grey in centre) (photo S. Scott); (4) *Phymatolithon purpureum* shedding cortical discs ×1, dried specimen with greenish tinge, note thick rolling margin (photo C. Derrick); (5) *Lithophyllum incrustans* ×0.8, chalky lavender (Methuen brownish grey) (photo F. Bunker); (6) *Phymatolithon lamii* ×1, milky pink (Methuen greyish rose/ruby) (photo F. Bunker); (7) *Lithophyllum incrustans* surrounding urchin cups, Fanore, W Ireland ×0.1, note habit and bleached greyish thalli as seen under water (photo P. Cooke); (8) *Pneophyllum limitatum* growing on *Laminaria* ×1, note habit of pale thalli seen in strong light.

nongeniculate genera (e.g. *Mastophora*) in the British Isles. In all British Isles taxa rhizoids are unknown and, except in *Choreonema*, attachment of the lower surface of the thallus is presumably by the secretion of adhesive mucilage (Jones & Moorjani, 1973; Chamberlain, 1984).

Most species are fundamentally dorsiventral. The thallus may remain flat and smooth, as in *Phymatolithon laevigatum*, or it may develop excrescences. These vary from warty, to lumpy, to fruticose (i.e. having branches with radial construction), as in various forms of *Lithothamnion glaciale*. Excrescences can also be discoid, layered or foliose, as in the leaf-like lamellae of *Mesophyllum lichenoides* or forms of *Lithophyllum incrustans*. All these growth forms commonly intergrade with each other, limiting their value for the identification of species. Unattached specimens (maerl, rhodoliths), such as occur in *Lithothamnion corallioides* and *Phymatolithon calcareum*, vary from relatively smooth, compact balls to warty or fruticose thalli. It is believed that they always originate from attached encrusting thalli, but, apart from *Lithothamnion glaciale*, the attached forms have been found very rarely in the British Isles and the bulk of the populations comprise thalli that have presumably continued to grow and fragment after becoming detached.

1.1b Geniculate genera
Geniculate corallines consist of firmly attached holdfasts and erect branching systems of intergenicula and genicula. The holdfasts are attached to hard substrata such as rocks or shells or to various other algae. In all species in the British Isles erect fronds (Fig.14) are initially produced from encrusting holdfasts (Fig.15) which, in *Corallina*, become extensive crusts that closely resemble thalli of *Lithothamnion*. In *Jania* and *Haliptilon*, on the other hand, encrusting portions are short-lived after initiating erect axes and become essentially replaced by stolons that arise from the lower intergenicula (Cabioch, 1966b). These stolons are modified branches in which the intergenicula tend to be cylindrical or barrel-shaped (Figs 16B; 17D).

1.2 Thallus anatomy
1.2a Nongeniculate genera
In common with the majority of red algae, coralline thalli are composed of filaments each of which arises from a separate initial (meristematic cell). The filaments are nearly always united into a pseudoparenchyma; exceptions are the semi-endophytic filaments of *Choreonema* (*q.v.*) and unconsolidated, creeping filaments that sometimes occur in *Pneophyllum* (Fig.62A) and the *Fosliella*-state of *Hydrolithon* (*q.v.*).

Two distinct types of organisation, termed dimerous and monomerous, are found in dorsiventral thalli. In the *dimerous* thallus a spore germinates (Fig.1) to produce a unistratose layer of radiating, repeatedly branched filaments, here termed basal filaments (Fig.1). These filaments grow radially by the activity of terminal initials (Fig.1) and branch pseudo-dichotomously (Fig.1). Their growth contributes to areal expansion of the thallus. Each cell cuts off a mainly uncalcified, small epithallial cell from its dorsal surface (Fig.1). In a few taxa (e.g. *Pneophyllum confervicola*) no further vegetative cell types are formed except in relation to the development of conceptacles; in most species, however, cells of basal filaments divide periclinally (Fig.2A), the upper cell, the subepithallial initial, remaining meristematic. This initial cuts off successive cells from its ventral surface to form erect filaments (Fig.2A) which contribute to increase in thickness of the thallus. The subepithallial initial also cuts off further cells, known as epithallial cells, dorsally; usually old epithallial cells are shed but in some cases they are retained and short filaments are formed. The dimerous type is similar to the construction of some noncalcified crustose red (e.g. *Rhodophysema*) and brown (e.g. *Pseudolithoderma*) algae (Irvine, 1983; Fletcher, 1987) where the filaments have similarly been termed basal and erect. Dimerous thalli occur in *Lithophyllum*, *Titanoderma*, *Fosliella*-state of *Hydrolithon*, *Pneophyllum*, *Exilicrusta* and *Melobesia*.

In a dorsiventral *monomerous* thallus several layers of pseudodichotomously dividing filaments arise peripherally from a germinating spore (Fig.2B) producing a multistratose ventral system of filaments that grow by the activity of terminal initials (Fig.2B). Although such filaments branch repeatedly, the ventral layer remains parallel to the substratum and more or less constant in thickness (Fig.2C) while the upper branches curve towards the dorsal surface. Downward curving filaments are usually suppressed and occur only vestigially (Fig.2C). The upward curving portions (Fig.2C) terminate in epithallial cells subtended by subepithallial initials cutting off cells ventrally and renewing epithallial cells dorsally. Apart from the presence of epithallial cells, a feature unique to the Corallinales, the multistratose ventral system of filaments and their peripheral derivatives are considered to be analogous to the medulla and cortex, respectively, in other pseudoparenchymatous red algae (see, for example, *Furcellaria* in Dixon, 1973, fig. 16E) and the same terms are used in brown algae for the central and peripheral thallus areas respectively (Fletcher, 1987, fig. VIII). Dorsal filaments may go on to produce radially constructed branching excrescences (e.g. *Lithothamnion corallioides*), or dorsiventral lamellae unattached to the substratum, in which

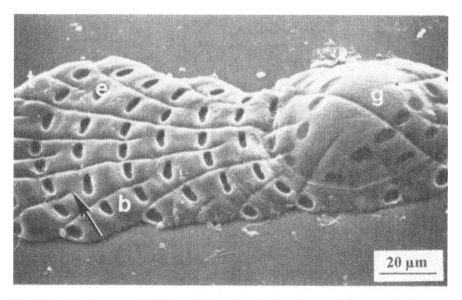

Fig. 1 Germinating spore (g) of *Pneophyllum* showing developmental features of a dimerous thallus, b = basal filament cell, e = epithallial cell, t = terminal initial, arrow = pseudodichotomous division.

cortex develops both ventrally and dorsally (e.g. *Mesophyllum*). Monomerous thalli occur in *Leptophytum*, *Lithothamnion*, *Phymatolithon* and *Mesophyllum*. In *Lithophyllum* the primary thallus is dimerous, but monomerous regions often develop secondarily from cells of erect filaments (e.g. *L.incrustans*).

In monomerous construction the multistratose system of filaments is produced at a very early stage and the unistratose basal layer is absent or confined to the few-celled sporeling stage. The difference between mono- and dimerous construction is not always clear cut. In some plants there is an apparent change from dimerous to monomerous, sometimes with reversions as described by Cabioch (1972). The differences are not of fundamental taxonomic importance and most subfamilies contain examples of both types. In fact, the presence of a unistratose basal layer appears to be correlated with the epiphytic habit (see Woelkerling & Irvine, 1988a) and possibly with a faster growth rate; some monomerous

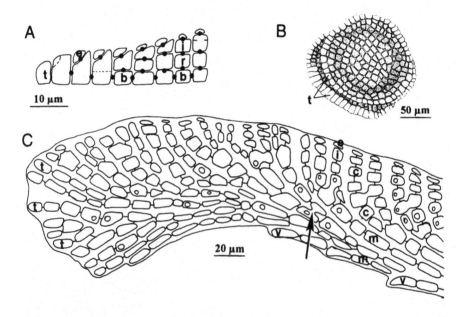

Fig. 2 Developmental features of nongeniculate thallus structure. (A) Diagrammatic VS of dimerous thallus showing position of primary pit connections (dot), terminal initial (t), basal filament cell (b), epithallial cell (e), subepithallial initial (i) and erect filament cell (r) (redrawn from Cabioch 1972, fig. 12A). (B) Surface view of germinating spore of *Phymatolithon lenormandii* showing two layers of initials (t) (from Cabioch, 1972, fig. 6A, as *Lithothamnion*). (C) Vertical section of monomerous thallus of *Leptophytum* showing medullary filaments (m), cortical filaments (c), terminal initials (t), subepithallial initials (i), epithallial cells (e), vestigial downwards-orientated filaments (v) and a fused cell (arrow) in which a fusion in face view is also present.

epiphytes are known to be fast-growing, however (YMC obs.). Genera such as *Lithophyllum* and *Spongites*, in which plants may have both mono- and dimerous construction, appear to have retained a degree of flexibility in construction which could be of adaptive significance in these large and successful genera. Elliott (1970) pointed out that a young developing plant must occupy territory, secure anchorage and sometimes smother organisms before turning to a thickening of vegetative tissues and eventual formation of reproductive structures; the differences in structure are a response to its different needs. Plants are well adapted to returning to a primary type of growth in order rapidly to encase objects in their path or recover from damage (Cabioch, 1972).

Traditionally the lower, uni- or multistratose layers have been termed the hypothallium and the upper, erect filaments the perithallium (see Johansen, 1981). These specialised terms not only fail to discriminate between the two different developmental modes described above (a concept proposed by Cabioch, 1972; see also Woelkerling & Irvine, 1988a,b), but also falsely imply that the anatomy of the coralline thallus differs fundamentally from that of other red algae. The terms dimerous and monomerous proposed by Woelkerling & Irvine (1988b) are accepted here. Woelkerling & Irvine (1988b) also coined the terms primigenous and postigenous for the basal and erect filaments respectively of the dimerous thallus, and core filaments and peripheral derivatives of core filaments for the medulla and cortex respectively of the monomerous thallus. These terms have their use in specialised studies, but the more general terms for algae have been used in previous parts of the Flora and seem more appropriate here.

An interesting and conspicuous feature of internal structure is that cells of contiguous filaments are sometimes aligned laterally in rows (see *Lithophyllum crouanii*). These rows can be so pronounced as to obscure the actual filamentous structure and give the thallus a tiered appearance in vertical section (Fig.12). Rows and tiers can be curved (Figs 98A-C), in which case the structure is referred to as coaxial (see *Mesophyllum lichenoides*).

1.2b Geniculate genera

In British Isles species a more or less extensive basal encrusting thallus develops as described above. Erect fronds are multiaxial and are initiated from the encrusting base when a group of subepithallial initials produces a tier of cells that become a geniculum. Genicular cells are uncalcified, do not branch and lack fusions. The subepithallial initials above the geniculum then elongate and divide synchronously to form a core of intergenicular medullary filaments (Fig.12) composed of several arching tiers of calcified cells (single tiers in a few non-British species of *Jania*) always surmounted by subepithallial initials. Some cells at the periphery of the medulla divide pseudodichotomously and form short-celled filaments growing towards the lateral surfaces of the intergeniculum; these filaments contain plastids and constitute a cortex. Intercellular fusions are common between contiguous filaments in both medulla and cortex in all British Isles taxa. Cortical filaments terminate in short epithallial cells which form a layer over all calcified surfaces. Subepithallial initials produce new cortical cells inwardly, adding to the diameter of an intergeniculum, and new epithallial cells outwardly. The frond is formed of alternate genicula and intergenicula and usually becomes pinnately and/or dichotomously branched.

1.3 Spore germination

Newly released coralline spores (i.e. carpospores, tetraspores or bispores) are approximately globular, deep red cells encased in a wall and surrounded by colourless mucilage (Jones &

Moorjani, 1973; Chamberlain, 1984). On settling they become attached to the substratum by adhesion and the cell contents divide into as many as 56 cells within the confines of the original spore wall (Figs 3, 94G). This type of segmentation was described by Chemin (1937) as the *Dumontia*-type and occurs almost throughout the Corallinales when spores germinate under favourable conditions. Spore segmentation was further studied by Cabioch (1972), Chihara (1973a,b,c, 1974a,b), Notoya (1974, 1976a,b), Bressan (1980), Chamberlain (1983, 1984), Jones & Woelkerling (1984), and Harlin *et al.* (1985) all of whom recognized that particular sequences could generally be related to taxonomic groups. Germination discs may remain visible even in quite mature plants in the thin, dimerous genera *Hydrolithon* (*Fosliella*-state), *Pneophyllum* and *Melobesia* (*q.v.*) which have four- (Figs 3A-C), eight- (Fig.3D) and up to fiftysix- (Fig.94G) celled centres to the germination disc respectively. In

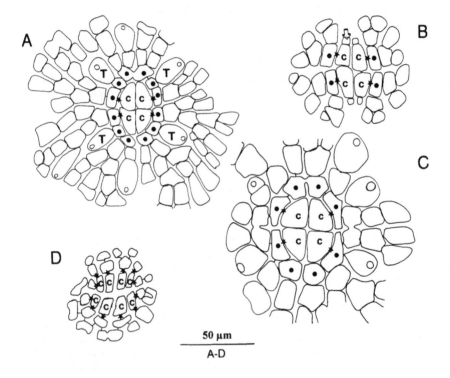

50 μm

A-D

Fig. 3 Germination disc features in *Pneophyllum* and the *Fosliella*-state of *Hydrolithon*. (A) *Hydrolithon farinosum* with 12 surrounding cells; note characteristic position of trichocytes (T) (type specimen, CN). (B) *Hydrolithon boreale* with 4 surrounding cells; note small cells (arrow) that develop no further (cultured from YMC 80/93). (C) *Hydrolithon cruciatum* with 8 surrounding cells (YMC 88/81). (D) *Pneophyllum fragile* (YMC 77/262) with 8 central cells. c = central cells; black dot = surrounding cells; star = primary pit connections linked to central cells.

addition, the pattern of cells surrounding the four-celled centre of *Hydrolithon* (*Fosliella-* state) is diagnostic for species recognition.

1.4 Cytological features

The most notable feature of coralline cells is the presence of a more or less heavy deposition of calcite crystals at right angles to the lumen (Fig.4) within the wall of most vegetative cells, including those forming conceptacles (see Giraud & Cabioch, 1979). This property has allowed corallines to be one of the few algal groups to fossilise.

It has been assumed that calcium carbonate is deposited on or within marine algae as a result of the extraction of carbon dioxide from water during photosynthesis, since seawater is often saturated or supersaturated with calcium bicarbonate, especially in the upper layers of tropical waters. Whilst this may be to some extent true in groups where the calcium carbonate is deposited in the orthorhombic form as aragonite (e.g. *Peyssonnelia*, Cryptonemiales), experiments indicate that in groups such as the Corallinales, where the calcium carbonate occurs almost entirely in the rhombohedral (hexagonal) calcite form, the situation is more complex and other metabolic processes are involved. In contrast to the aragonite-forming groups, the calcite-forming groups are much more widely distributed in colder as well as warmer waters (Lewin, 1962; Moberly, 1968). Bohm *et al.* (1978) found that the maximum rate of calcification for *Phymatolithon calcareum* in the Baltic occurred at 10°C. Okazaki *et al.* (1982) suggested that the substance involved in the process of calcium uptake and/or as a template for calcite nucleation may be alginic acid (or a close relative) otherwise known only in the brown algae; these authors reported alginic acid as 2% of the organic matter in *Serraticardia maxima* (Corallinoideae), a taxonomically as well as physiologically interesting situation. For further information see Johansen (1981), Borowitzka (1977) and Cabioch & Giraud (1986).

Calcification requires an organic matrix within which crystals are deposited (Giraud &

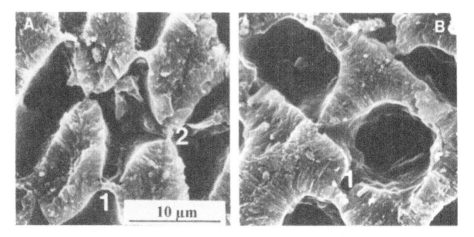

Fig. 4 (A & B) Vertical fractures (to same scale) of *Lithophyllum crouanii* and *L. nitorum* respectively showing calcite crystals orientated at right angles to the cell lumen; note primary (1) and secondary (2) pit connections.

Cabioch, 1979) and begins soon after spore germination (Cabioch, 1972), even as early as the four-celled stage in *Corallina* (Jones & Moorjani, 1973). It is most intense in the meristems, is energy demanding and light dependent and in *Corallina officinalis* calcium carbonate comprises 70–90% of the dry weight (Pentecost, 1978). A layer of calcite crystals 100µm thick may attenuate light by up to 70% (Pentecost pers. comm.). Littler (1976) presented several hypotheses for the adaptive significance of calcification, including mechanical support, resistance to abrasion, wave shock and grazing, and protection against fouling epiphytes and intense light.

Until recently most methods of examining coralline algae required decalcification, but the advent of the scanning electron microscope has enabled observation of calcification in situ, considerably improving our understanding of anatomical and morphogenetic features.

It is characteristic of coralline algae that adjacent (as opposed to successive) cells of contiguous filaments may be joined by cell fusions (Fig.2C) or direct secondary pit connections (Fig.4A). The latter are formed by dissolution of the walls at the point of contact and so differ in development from the indirect connections seen in other red algae (Dixon, 1973). Pueschel (1987, 1989) has found that coralline primary pit connections have plugs with two cap layers, the outer one being domed, and that they lack cap membranes: they share these characters with members of the Batrachospermales. The deeply staining pit bodies referred to by Adey & Adey (1973) have been shown to be equivalent to the outer of the two cap layers (Pueschel & Trick, 1991).

Starch grains (Figs 85C,G) are common, especially in lower parts of encrusting thalli and medullary filaments of geniculate thalli. Cabioch (1971a) and Chamberlain (1991b) found that the size of mature grains has diagnostic significance in some groups. Associated with starch accumulation there are often extensive cell fusions whilst plastids and other organelles decrease in number.

Adey & Adey (1973) found the presence or absence of staining bodies to be of diagnostic significance in melobesioid species. These possibly proteinaceous intracellular bodies stain very intensely with phosphotungstic haematoxylin. Giraud & Cabioch (1983) found various types of intensely staining cell inclusions, but their presence varied geographically and seasonally and was not sufficiently consistent to be diagnostically reliable. Ultrastructural studies have also been reported by Borowitzka & Vesk (1978, 1979), Millson & Moss (1985) and Pueschel (1992).

1.5 Trichocytes

Trichocytes (hair-bearing cells) are produced in many species. In the *Fosliella*-state of *Hydrolithon* (*q.v.*), trichocytes terminate basal filaments (Fig.9B) whereas in *Pneophyllum* (*q.v.*) they are intercalary (Fig.9A) and this provides a useful diagnostic character. Chamberlain (1985b) showed that trichocytes are present only at times of relatively high temperature and high light intensity (photon flux density) and are absent in winter and/or at great depth. Trichocytes are rare and of little taxonomic significance in the larger encrusting species (see Walker, 1984). They also occur in some of the geniculate genera, e.g. *Jania* (Johansen, 1981), but their taxonomic value has not been evaluated.

2 Reproductive characteristics of the Corallinales

2.1 Life History

A *Polysiphonia*-type life history (Fig.5) has been shown to occur in cultured plants of various species of *Pneophyllum* (Chamberlain, 1987; Fujita, 1988) and a self-perpetuating, bisporangial life history (Fig.5) is present in *Hydrolithon* (Chamberlain, 1977b, as *Fosliella*), *Pneophyllum* (Chamberlain, 1987), and *Titanoderma* (Suneson, 1982, as *Lithophyllum*). Both types of life history probably occur widely throughout the Corallinales. Further, it is not uncommon to find individual plants bearing, for example, both spermatangial and bisporangial conceptacles, while other populations consist entirely of tetrasporangial plants which appear to be self-perpetuating, suggesting considerable flexibility in reproductive strategy and ploidy level.

2.2 Gametangial Plants

All reproductive bodies are borne in conceptacles comprising a chamber surrounded by walls and a roof; conceptacles are produced by the vegetative thallus either superficially (Fig.52) or immersed within it (Fig.20C). Haploid gametangial plants and diploid tetra/ bisporangial plants are morphologically similar, but the carposporophyte (see Vol.1(1), p. 19) is microscopic and develops within the original carpogonial conceptacle. Most taxa are dioecious but some species of *Jania*, for example, are monoecious while both dioecious and monoecious plants occur in, for example, *Melobesia membranacea*. All gametangia are borne in uniporate conceptacles. Male gametes (spermatia) are differentiated within

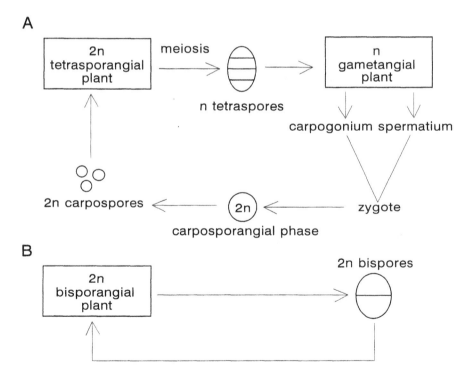

Fig. 5 Diagrams to show *Polysiphonia*-type (A) and self-perpetuating bisporangial type (B) life-histories.

specialised cells (spermatangia) (Fig.22E) which in turn arise from spermatangial initials. They may be restricted to the conceptacle floor or occur on the wall as well. The filaments bearing spermatangia may be simple structures (e.g. *Lithophyllum*) or dendroid systems (e.g. *Phymatolithon*) and these differences appear to be taxonomically significant (Lebednik, 1978). Many spermatangial conceptacles have a spout (e.g. *Hydrolithon*) which may facilitate delivery to the carpogonial plant.

Female gametes (carpogonia) develop terminally on carpogonial branches that arise only from the conceptacle floor. Mature carpogonia (Fig.33F) are more or less bottle-shaped and comprise a swollen basal portion (the egg) and a whip-like upper portion (trichogyne) which may elongate to protrude through the pore in the conceptacle roof. Spermatia may become attached to the trichogyne and presumably the male nucleus travels downwards to fuse with the female nucleus to form a diploid zygote.

The zygote undergoes a complicated series of nuclear and cytological changes to form a microscopic, diploid carpospore-producing plant (carposporophyte) within the carpogonial (female) conceptacle. The carposporophyte (Fig.33D) usually comprises: 1] a central, more or less discoid, multinucleate fusion cell that remains anchored to the conceptacle floor; 2] short gonimoblast filaments each terminating in a single carposporangium. In the Melobesioideae the central fusion cell is reduced and sometimes absent and a discontinuous fusion cell (Lebednik, 1977b, table 3) bearing scattered gonimoblast filaments occurs across the conceptacle floor. In some genera (e.g. *Phymatolithon*) no central fusion cell develops and each gonimoblast filament develops individually (Fig.110D), or the fusion cell is branched (e.g. *Leptophytum*) rather than discoid.

2.3 Tetrasporangial Plants

When diploid carpospores are released they germinate and develop into diploid tetra-sporangial plants. Tetrasporangial initials develop on the floor of the conceptacle chamber and each then divides (presumably meiotically) into four simultaneously produced zonately arranged tetraspores. Because the division into four spores occurs simultaneously, the final spore number is obvious as soon as it begins and there is no possibility of confusion with bisporangia. The (presumably haploid) spores are usually released through the pore as a tetrad which quickly separates, each spore potentially giving rise to a gametangial plant.

Two principal types of tetrasporangial conceptacle occur: 1] uniporate (Fig.21F), in which tetrasporangia may either be scattered across the conceptacle floor (e.g. *Lithophyllum crouanii*) or confined to the periphery, leaving a central group of sterile filaments (columella) (e.g. *Titanoderma*); 2] multiporate (Fig.116E), in which each tetrasporangium terminates in a plug occupying a pore in the conceptacle roof through which the tetrad is released individually (e.g. *Phymatolithon*). *Choreonema* is unusual because although each tetrasporangium has a plug they are enclosed within a uniporate conceptacle (Fig.11C).

2.4 Bisporangial Plants

In the British Isles possibly the predominant type of reproduction is by bisporangia. The self-perpetuating bisporangial plant is externally nearly always identical with tetrasporangial plants of the same species and bisporangia are borne in the same range of uniporate (Fig.29B) and multiporate (Fig.79C) conceptacles. Uninucleate diploid bisporangia are released and germinate to produce further diploid, bisporangial plants in a form of apomixis. Occasionally bisporangia containing uninucleate or irregularly nucleated bispores are seen in otherwise tetrasporangial conceptacles and it may have been by this means that bisporangial populations arose.

References to further publications on vegetative and reproductive structure are provided in Woelkerling & Irvine (1988a).

3 History of Classification

The first coralline alga to be recognized as a living organism was probably *Corallina*, by Pliny in the first century AD, and it became well-known in the Mediterranean as a vermifuge. The words coralline and *Corallina* are diminutives of the italian *corallo* = coral and were given to coral-like organisms of smaller size. Later, a range of organisms were described as corallines and began to appear in herbaria, for example those of Morison, Ray and Dillenius in Oxford; they included not only algae but also corals, hydroids, sponges, bryozoa, etc. In the eighteenth century all corallines were transferred to the animal kingdom because of the discovery of polyps in some of them (Ellis, 1755). At this stage coralline algae were classified as *Corallina* (which also included the green jointed alga *Halimeda*) and *Corallium*, in which two species were distinguished: crustose plants and maerl, later named by Pallas (1766) *Millepora agariciformis* and *M. calcarea*. There followed a descriptive phase during which many genera and species were distinguished on the basis of external appearance and the linking of their names to modern concepts has proved a laborious and time-consuming procedure. Philippi (1837) finally recognized that coralline algae were not animals and he proposed the two generic names *Lithophyllum* and *Lithothamnion* as *Lithothamnium*, which together with *Melobesia* Lamouroux (1812) and *Mastophora* Decaisne (1842) are probably the most noteworthy names among the nongeniculate entities.

A hundred years ago Foslie began working with tremendous energy and published 69 papers between 1887–1909 (Woelkerling, 1984) during which time he increased the number of species from 175 to 650 species and forms. At first he thought that conceptacles provided useful taxonomic characters but later found they were very variable and one of his most important contributions was the recognition of the extreme variability possible in both vegetative and reproductive features. Heydrich, a contemporary of Foslie, published many detailed observations between 1897–1911 (see Heydrich references in Woelkerling, 1988); these are often difficult to interpret, but he did emphasise the significance of multiporate tetra/ bisporangial conceptacles, a feature which distinguishes the subfamily Melobesioideae from the others, where they are uniporate. Rosenvinge (1917) described secondary pit connections in Danish *Lithophyllum* spp., thus setting the stage for the distinction of the subfamily Lithophylloideae. Lemoine (Ardré & Cabioch, 1985; Chamberlain, 1985c), working in France, published continuously for most of this century, concentrating her studies on vegetative anatomy. The results she obtained used features which could also be seen in fossils, in which she was interested, and palaeontologists have continued to use her schemes. She drew attention to the difference between species with unistratose and multistratose basal layers and distinguished between plumose and coaxial types of multistratose construction. This led to the description of the genus *Mesophyllum* (Lemoine, 1928) with a vegetative anatomy similar to her concept of *Lithophyllum* combined with the multiporate tetra/ bisporangial conceptacles of the Melobesioideae, to which subfamily it has been assigned. Suneson (1945), working in Sweden on *Mastophora* from South Africa, realised the importance of the character combination of cell fusions and uniporate tetra/ bisporangial conceptacles and was the first to suggest that *Lithophyllum*, *Mastophora* and *Melobesia* represented three distinct evolutionary lines. Working with geniculate corallines, Johansen (1969) recognized that three lines of evolution led to these plants and segregated three subfamilies: Corallinoideae, Amphiroideae and Metagoniolithoideae. Only members of the first occur in the British Isles. On the basis of anatomical studies, Cabioch (1972) suggested that some genera such as *Corallina* are related to the Mastophoroideae, whereas others such as *Amphiroa* are closer to the Lithophylloideae.

Many generic and suprageneric classification schemes have been proposed for coralline algae (Cabioch, 1972, fig.40; Johansen, 1981; Chamberlain, 1983, p.330; Woelkerling, 1988, chapters IV, V). For many years they were included in the order Cryptonemiales as the family Corallinaceae until Silva & Johansen (1986) erected the order Corallinales, and gave the main features as: 1) walls of most vegetative cells impregnated with calcite; 2) in addition to terminal meristems characteristic of other red algae, intercalary meristems are common; 3) plugs of primary pit connections have two-layered, dome-shaped caps; 4) reproductive structures are produced in roofed conceptacles (except *Sporolithon*); 5) tetrasporocytes usually undergo simultaneous zonate division (cruciate in *Sporolithon*); 6) post-fertilization events involve a cluster of procarpial filament systems. Characters of secondary importance include consistently two-celled carpogonial filaments; the prevalence of secondary lateral anastomoses (pit connections and cell fusions); the frequent occurrence of a self-perpetuating, presumably diploid, bisporangial phase; and the presence of a stalk cell subtending tetra/ bisporangia. As they remark, many of these features are present in other orders, but collectively they are unique to the Corallinales. Woelkerling (1988, p.83) accepted this definition of the order, with some modifications.

The Corallinales at present includes two families with living representatives, the Sporolithaceae and Corallinaceae. The Sporolithaceae was described by Verheij (1993) to accommodate the genus *Sporolithon* (not present in the British Isles) which has simultaneously cruciately divided tetrasporangia borne in sori rather than conceptacles. The Corallinaceae has simultaneously zonately divided tetrasporangia borne in conceptacles and contains several subfamilies; the principal characters upon which their classification depends are: 1) the presence of cell fusions and/or secondary pit connections; 2) the presence or absence of genicula; and 3) the presence of either uniporate or multiporate tetra/ bisporangial conceptacles. Two principal schemes of classification are at present in use. Cabioch (1972, 1988) considered that the presence or absence of genicula is of secondary significance: she classified geniculate taxa together with nongeniculate taxa in subfamilies that share the same cell anastomosis/ conceptacle roof characters. Johansen (1981) and Woelkerling (1988), on the other hand, have opted for a scheme that segregates geniculate genera into separate subfamilies. The latter scheme is adopted here as being more practical, but with the reservation that Cabioch's scheme probably represents a more acceptable evolutionary hypothesis. Johansen (1981) and Woelkerling (1988, pp.62–75) both discussed features diagnostic of genera and the characteristics are presented here in the form of keys and generic descriptions.

4 Environmental interactions

4.1 Habitat
Until relatively recently a lack of basic taxonomic information resulted in many ecological studies referring to the entire group of crustose corallines as 'lithothamnia'. It is now known that some species show quite narrow tolerances for shore level etc. and more accurate recording will provide information useful for a wide range of future surveys.

Large encrusting coralline species form quite distinct zones on the shores of the British Isles. They occur mainly on open, rocky shores where there is plenty of water movement and are rare in sheltered harbours and estuaries or on sandy or muddy shores. *Phymatolithon lenormandii* is the one species that can grow in these otherwise unpromising localities. In some places the littoral zone is dominated by *Lithophyllum* spp. (e.g. Shetland Isles, Alderney, LMI obs.) but elsewhere *Phymatolithon* spp. predominate. The reasons for this difference are unknown.

On suitable rocky shores the highest zone usually colonized by corallines is at about midlittoral and the principal species inhabiting littoral zones are as follows:

Mid littoral: *Phymatolithon lenormandii* (Frontispiece; Fig.111A) and/or *Lithophyllum orbiculatum* (Fig.32) may be common in pools. The latter is characteristically bleached to a pale putty colour at high shore levels and is confined to pools and gullies. *P.lenormandii* is usually pinkish to mauvish and often extends across rock surfaces under cover of fucoids such as *Fucus serratus*. *Corallina officinalis* also occurs in pools and crevices at this level.

Mid to lower littoral: *Phymatolithon lenormandii* and *C. officinalis* continue to be common and are often accompanied by *P. laevigatum* (Fig.105) which is particularly luxuriant on Scottish shores and may be locally common in the south. Another species that occurs somewhat sporadically in the south but may form extensive, low littoral rock cover in Scotland is *Lithophyllum crouanii*. The most vigorous species is *Lithophyllum incrustans* (Figs 6; 27A,B) which forms thick encrustations at several shore levels; the plants are smooth in shady situations or deeper water but become extravagantly crested in more open habitats such as midlittoral rocks and shallow pools obviously thriving under high light intensity. Other species, such as *Phymatolithon lenormandii* and *P. purpureum*, were shown by Figueiredo *et al.* (1992) to be shade-loving, becoming bleached when fucoid cover was removed and failing to recover when grazing animals prevented the growth of epiphytes. In the north, as far south as Yorkshire and the Isle of Man, *Lithothamnion glaciale* (Frontispiece) forms characteristic bright pink, knobbly encrustations in protected places. More locally one of the prettiest corallines, *Mesophyllum lichenoides* (Frontispiece, front cover), occurs as leafy plants attached to *Corallina*. Where seagrass populations are found the epiphytic species *Pneophyllum fragile* (Fig.64A) is often abundant and *Hydrolithon cruciatum* occurs locally. Two other epiphytic species are abundant. Thin, translucent thalli of *Melobesia membranacea* (Fig.93A), with dark-centred conceptacles, occur on virtually all shores where corallines are present and are particularly common on *Furcellaria* and *Laminaria*. Thicker, bright pink plants of *Titanoderma pustulatum* var. *pustulatum* (Fig.39A), with conspicuously raised conceptacles, are found in profusion on *Chondrus* and *Mastocarpus*, particularly on wave-exposed shores. Lower littoral: lower littoral and upper sublittoral rocks throughout the British Isles are covered with *Phymatolithon purpureum* (Frontispiece) which forms gleaming deep brownish-pink expanses particularly noticeable in winter. At the same level thick, smooth, rather pale pink plants of *P. lamii* (Frontispiece) occur sporadically. *Corallina officinalis* and *C. elongata* occur in pools and also hang in festoons from the rocks as the tide retreats. *Jania rubens* is a common epiphyte on *Cladostephus* and *Cystoseira* at this level. These species are joined by *Haliptilon squamatum* on both rock and *Cystoseira* in protected places in well-aerated water. Many species commonly grow in pools; Kooistra *et al.* (1989) discussed factors affecting species distribution in pools at Roscoff, Brittany, where the species composition is similar to that in pools in the British Isles.

Sublittoral: Adey & Adey (1973) found the small, bright pink plants of *Lithothamnion sonderi* (Fig.90A) to be the commonest species round the British Isles accompanied by *L. glaciale* in the north. *Phymatolithon purpureum* and *P. lamii* (syn. *P. rugulosum*) also grow commonly in the sublittoral and are thought to be the principal components of vast coralline populations covering bedrock in the north Atlantic ocean. *P. lamii* has been recorded at a depth of 90m off Rockall (Clokie *et al.*, 1981). Maerl/rhodolith deposits are quite common round our coasts, comprising mainly *Phymatolithon calcareum* (Fig.103) and *Lithothamnion corallioides* (Fig.84) in the south with the latter being replaced by *L. glaciale* and, to a lesser extent, *L. lemoineae* in the north. *Lithophyllum incrustans* also occurs as a maerl component; in the west of Ireland particularly, there are spectacularly branched *Lithophyllum* plants of various forms that may all pertain to this species although they have been given the names *L. dentatum*, *L. duckeri* (=*L. racemus f. crassum*), *L. fasciculatum* and *L. hibernicum* (Figs 23–26).

According to Bosence (1983b) unattached plants occur in a wide range of non-muddy marine environments normally with depths less than 100m. Attempts have been made to distinguish between branched forms (maerl) and nodules (rhodoliths) which may or may not be defined as having a nonalgal core. Whilst these are useful general terms, there are intergrades between the two in both appearance and mode of formation and strict definitions are probably not practical.

Living unattached corallines occur from the lowermost intertidal down to 27m but are most abundant between depths of 1–10m. The environmental conditions required for luxuriant growth are provided by shallow, clear waters sheltered from excessive wave action and tidal currents where branching forms can build up to 10–20cm above the surrounding sea floor (see Bosence, 1983b; Farrow *et al.*, 1979), sometimes occupying thousands of square metres. The beds create an important habitat for a wide variety of animals and other algae. A matrix of trapped mud and algal mucilage gradually buries the algae and the beds are broken locally by physical and biological erosion to form gravel-sized sediment often rippled by oscillatory currents. Bosence (1976, 1983b) showed that the nature and density of branching is a sensitive indicator of hydraulic conditions. He adopted three classes of shape (spheroidal, ellipsoidal, discoidal) and four classes of branching density (single main branch, few branches, many branches, dense nodule), showing this scheme could be used to classify samples from different habitats. From wave tank experiments he found that branching increases with increasing exposure to water movement, with ellipsoidal forms being least, and discoidal forms most, stable.

According to Tittley *et al.* (1989) the rhodolith assemblage in eastern Canada originates from attached plants probably bieroded by undercutting and burrowing animals; they may cover large areas but unlike maerl beds are not usually associated with currents. For further information see Davis & Wilce (1987).

Species contributing to maerl-rhodolith beds seem to be those capable of growing on lightweight, mobile substrata and/or of continuing to grow when portions of a thallus become detached and themselves become mobile (LMI obs.). Species which grow only on fixed or heavier, rarely mobile substrata are rarely found in such beds (see Adey, 1970d). In the British Isles the species most commonly found in maerl/rhodolith beds are *Lithothamnion corallioides*, *L. glaciale* and *Phymatolithon calcareum* (Table 4). Other species which occur occasionally are *Lithophyllum* spp. *Leptophytum laeve*, *Lithothamnion lemoineae*, *L.sonderi*, *Phymatolithon lenormandii* and *P. purpureum*.

As a consequence of the correlation between appearance and environmental factors, different species can look deceptively similar. Examples are seen in specimens of the three common maerl species which can be very difficult to identify (Table 4), and species of the *Lithophyllum duckeri-racemus* complex.

4.2 Distribution
Adey & Adey (1973, figs 68–70) summarised the nongeniculate coralline flora of the British Isles as primarily boreal with a strong element of Lusitanian species extending throughout and accounting for half of the species in the southwest. There is also a weak element of subarctic species (*Lithothamnion glaciale, L. lemoineae* and *Leptophytum laeve*) which does not extend as far as the south coast.

Norton (1985) included some coralline species in his preliminary atlas of British Isles algae.

4.3 Growth and seasonal variation
Early experiments in Brittany (Lemoine 1913c, 1940) showed that growth is much slower in encrusting algae than in the erect forms where elongation can be up to 50mm per month. Her observations showed an average growth in extent of 2–3mm per annum, the highest recorded growth being 6–7mm per annum in *Lithophyllum incrustans*. Figures for increase in thickness are much lower, e.g. 50–200μm per annum in *Phymatolithon laevigatum* (Adey, 1964, 1970b; see also Adey & Johansen, 1972; Steneck & Adey, 1976). Adey (1970b) found that in several North Atlantic species of Melobesioideae growth rates were highest, and strongly light-dependent, between 9–15°C. Adey (1964, 1965) also found that conceptacles take between two months (*Phymatolithon*) and three months (*Clathromorphum*) to develop from primordia to maturity. As first noted by Foslie (1896), many northern species are winter fertile, especially species in the Melobesioideae, e.g. *Phymatolithon purpureum*, although others such as *P. lenormandii* are fertile throughout the year. Jones & Moorjani (1973) found a difference in the growth rates of tetraspores of *Jania rubens* (epiphytic) and *Corallina officinalis* (epilithic). In the former, secure attachment occurred in 4.5h and an erect axis developed from a limited basal disc in 8 days whereas in the latter secure attachment took 2 days and axes were produced from expanded bases only after 12 weeks. Chamberlain (1984) found that *Pneophyllum* spores showed amoeboid changes in shape in suspension, that the mean diameter of tetraspores within a species was consistently smaller than that of carpospores and bispores, and that adhesive mucilage was secreted on attachment. In some cases spores may be viable only at specific times of the year even though they are produced throughout (Jones & Woelkerling, 1983; Chamberlain, 1987); this may account for the lack of success with germination and culture experiments (Woelkerling *et al.*, 1983). We have found no evidence to suggest that conceptacles develop more than once a year in perennial species, or that the conceptacles produce sporangia more than once.

Chamberlain (1977b) found that in *Hydrolithon boreale* (as *Fosliella farinosa*) conceptacles appeared 22.5 weeks after inoculation of bispores, and that mature bisporangia developed after a further 4 weeks (at 10°C, 8h light:16h dark). Jones & Woelkerling (1983) found that Australian material of *Hydrolithon cruciatum* (as *Fosliella cruciata*) reached reproductive maturity in up to 72 days after spore germination over a range of light regimes provided the water temperature was 15–22°C. They found a mean growth rate of up to 500μm²/day. Such a short lifespan enables species to succeed as opportunistic ephemerals both as epiphytes and as primary colonizers on bare rock surfaces. In contrast, the age of nodules 3cm in diameter from deep water has been estimated as 75 years (Bosellini & Ginsburg, 1971) and 20–30cm in diameter as 500–800 years (Adey & MacIntyre, 1973). For comparison, *Corallina officinalis* was found to produce an intergeniculum 1mm long every 12 days in eastern USA, the growth rate being affected by both light and temperature (Colthart & Johansen, 1973).

Steneck & Adey (1976) confirmed experimentally that both water motion and light controlled growth and branch formation in *Lithophyllum congestum* in the Caribbean (see comments under *L. incrustans*), whilst grazing by parrot fish also had a marked influence. Desiccation and competition were important at higher levels on the shore. *Lithophyllum incrustans* shows a similar range of form in the British Isles, probably under the influence of the same environmental factors. For further information on growth and reproduction, with particular reference to *Lithophyllum incrustans*, see the summary of the work of Edyvean and collaborators under the entry for that species.

Brown *et al.* (1977) found that the growth of geniculate corallines is significantly inhibited by a medium enriched with orthophosphate at a concentration normally used in culturing other groups of marine algae; apparently the phosphate interferes with the process of calcification. The presence or absence of coralline algae in certain types of habitat may prove to be a reliable biological indicator of relative levels of man-made phosphate pollution in seawater.

4.4 Form variation

Many nongeniculate species develop excrescences (Frontispiece: *L. glaciale*) from their surfaces. Whilst the propensity for this seems to be species-related, Adey (1966b, pp.326–8) provided conclusive evidence that degree and form of excrescence development are highly variable within a species and even within a single plant, due to a variety of ecological conditions. Similarly, intergenicula in geniculate corallines exhibit variation in form (Johansen & Colthart, 1975) and it is necessary to make allowance for this when identifying plants.

One of the interesting characteristics of some of the larger encrusting species is their ability to form crests when two thalli, or parts of the same thallus, meet (Fig.6). This can result in very elaborately-shaped plants and accounts for the high degree of form variation, particularly in the genus *Lithophyllum*. In other cases thallus growth ceases when margins meet, or one grows over the other. Plants of a particular species seem consistent in their behaviour pattern in these respects, thus providing an aid to identification.

4.5 Animal interactions

Coralline algal crusts have been a prominent feature in photic marine environments since the Cambrian Period and herbivory is the principal means by which they remain free of potentially lethal epiphytes (Steneck, 1983). The main animal groups involved are molluscs, urchins and, in warmer waters, fish. Regardless of their taxonomic affinity, encrusting corallines from heavily grazed environments are frequently thick and have sunken or flush (i.e. protected) conceptacles. Although thicker crusts grow more slowly, they are better protected from excavating herbivores and are more capable of recovering from injury. Thin, poorly attached forms are found only where herbivory is low, e.g. in the littoral, in deep water or in cryptic habitats. Plants often show evidence of herbivore-induced injuries in the form of grazing trails on their surfaces and healed wounds internally and the morphology can be greatly affected by herbivory. There seems to be a delicate biological balance in these species between being simultaneously herbivore-dependent, resistant and susceptible. Steneck (1985) regarded the inter-relationships with grazers as species specific and possibly co-evolved. Feeding experiments (Keats *et al.*, 1983) showed that the grazing rate for animals eating *Lithothamnion glaciale* and *Phymatolithon lamii* (as *P. rugulosum*) is low compared with that for large, fleshy algae such as *Laminaria*. Hagen (1983) described overgrazing of *L. hyperborea* beds in northern Norway, resulting in 'barren' sea floor dominated by sea urchins and crustose corallines bleached by the increased light intensity after the kelp canopy disappeared. The term Isoyake is used by the Japanese for similar barren areas, which appear to occur periodically as a result of raised sea temperatures (Noro *et al.*, 1983). Older, thicker crusts can be invaded by polychaete worms such as *Dodecaceria* and *Polydora*; these leave holes which are sometimes mistaken for conceptacles (pers. obs.). Geniculate corallines are also frequently grazed by invertebrate herbivores in coastal zones (Littler & Kauker, 1984). Refuges are provided by the fronds of some species (e.g. *Corallina officinalis*) and Dommasnes (1968, 1969) reported on the animals inhabiting this niche in Norway. For further information on this fascinating subject, see Chamberlain & Cooke (1984), John *et al.* (1992), Larson *et al.* (1980) and Paine (1984).

Work throughout this century in many disciplines has confirmed the preponderance of calcareous algae as both bulk producers and consolidators in the majority of reef limestone deposits, and substantiated the predominant role of coralline algae in cementing sediments produced by calcareous green algae, molluscs, foraminifera and hermatypic corals. For

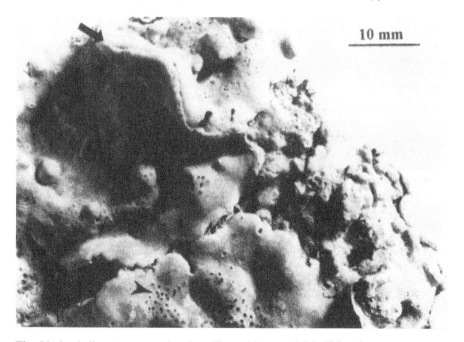

Fig. 6 *Lithophyllum incrustans* showing adjacent bisporangial thalli forming a crest (arrow); note spores being released (arrowhead) (YMC 84/354).

further information See Littler & Littler (1984) and Bosence (1984). Johnson & Mann (1986) showed that, after removing grazing animals, the biomass of recruited fleshy algae on crustose *Phymatolithon* spp. was less than half that on granite; this is probably due to epithallial sloughing although chemical inhibition cannot be ruled out. This discovery contrasts with the hypothesis mentioned above (Steneck, 1983) that crustose corallines depend on grazers for maintenance. Johnson & Mann (1986) concluded that 'crustose corallines amount to a guild of prominent ecological significance in shaping the structure of shallow hard-bottom communities.'

Research reviewed in Morse & Morse (1984) has shown that larvae of certain benthic invertebrates (e.g. species of *Acmaea* [limpets], *Haliotis* [abalone] and *Tonicella* [chitons]) preferentially settle and metamorphose on crustose corallines. The result of this coevolution is possibly to reduce epibionts on the host algae and to provide inducing chemicals and nutrients for the animals.

5 Fossils

Because coralline algae are heavily calcified, they occur extensively as fossils. Adey (1979) suggested that they evolved in high latitudes during the Mesozoic, moved into cryptic situations in low latitudes during the early Tertiary, with genera such as *Porolithon* and *Neogoniolithon* evolving later as 'sun forms' tolerant of bright light conditions. Further comments on the evolution of the group were given by Hoek (1984) but studies have been handicapped by the difficulty of observing key features in fossil material. Recently, however, Braga *et al.* (1993) have been successful in using the SEM to differentiate between cell fusions and secondary pit connections in fossil material, thus paving the way to a better understanding of the evolution of the group and the palaeoenvironmental conditions under which they lived.

A late cretaceous *Pneophyllum* described by Voigt (1981, as *Fosliella*) is discussed under *Pneophyllum*.

6 Practical uses of coralline algae

For many years maerl has been harvested in Cornwall, Ireland and Brittany for use as an agricultural and horticultural fertilizer. It has mainly been used as a source of lime, but it is popular today as a natural, organic fertilizer and is believed to contain trace elements and organic compounds that enhance its value. Because of its increasing use, greater understanding of the component species and their biology is desirable in relation to cropping policy.

As coralline algae are one of the few macroscopic algal groups to fossilise successfully, they are used as stratigraphical markers by the oil industry, and a considerable literature exists on the subject (Wray 1977). They also provide a beautiful building stone as seen, for example, in Vienna.

Corallina officinalis,q.v., was traditionally used as a vermifuge and an intriguing technique for orthodontists has recently been pioneered by Kasperk *et al.*, 1988). Fragments of *Corallina* thallus are used as a framework round and through which bone cells can grow to form an excellent repair medium for damaged or diseased jaw bones; the fact that they induce no immune response makes them particularly valuable in this context.

7 Previous publications on British Isles coralline algae.

In an early reference to coralline algae Ray (1690) described branched corallines occurring near Falmouth in southern Cornwall where maerl beds exist to this day (Woelkerling &

Irvine, 1986a). Two of the earliest and most important authors to describe species in the British Isles were Ellis (1755, 1768; Ellis & Solander, 1786) and Johnston (1842). Their works were quoted and amplified by Harvey (1848, 1849a,b, 1850) who figured a number of coralline algae. Unfortunately all their specimens seem to be lost and it is difficult to evaluate their descriptions and records, particularly those of small, epiphytic species. Batters (1890, 1902) industriously gathered records for his compendia and corresponded with the Norwegian coralline specialist, Foslie (see Foslie, 1905a). These lists formed the basis on which Newton (1931) founded her descriptions in *A Handbook of British Seaweeds*. Knight & Parke (1931) included a number of corallines in *Manx Algae* published at the same time and Lemoine (1913a) described a number of species from Clew Bay in western Ireland. There are further Irish records in Morton & Chamberlain (1985, 1989) and Guiry (1978a). A few species were included in *British Seaweeds* by Dickinson (1963), fourteen in a recent field key (Hiscock, 1986) and several in a list of Berwickshire red algae (Hardy, 1993). A catalogue of coralline type specimens in the BM was published by Tittley *et al.* (1984). Price (1978) lists natural history books many of which have references to coralline algae.

The present account relies heavily on the extensive study made by Adey & Adey (1973), but their work was based mainly on sublittoral collections and excluded epiphytic species. Since then studies have been made of the mainly epiphytic genera *Fosliella* (now *Hydrolithon*) and *Pneophyllum* (Chamberlain, 1983) and *Titanoderma* (Chamberlain, 1991b).

8 Methodology

8.1 Collecting

Geniculate and epiphytic coralline algae can be collected in the same ways as other marine algae but epilithic crustose plants require special treatment. The most important tools are a hammer and chisel, the choice of which is largely a matter of experience and personal preference. Rock samples bearing whole plants or at least large pieces should be collected whenever possible, and/or several chips from a large plant placed together in a polybag; precut waterproof paper is very useful for labels. Loose encrusted cobbles and maerl can be collected into mesh bags. Although cobbles are easy to collect, the plants growing on them tend to be atypical and often sterile, due to overturning and abrasion, and are difficult to identify. It is important to collect fertile specimens, so a handlens is useful for checking for conceptacles. Many species, especially Melobesioideae, are fertile in winter, so serious attention must be paid to tide times, weather forecasts, daylength and air pressure; under such circumstances warm waterproof clothing and footwear are essential.

8.2 Preservation

The samples should be washed in seawater after removing epiphytes, and then brushed with a soft brush to remove debris, which inhibits drying and obscures surface detail. Labelling with waterproof ink can be done directly or on a patch of quick-drying white paint. It is important to record the colour when fresh. Specimens can then be dried at room temperature or preserved in 5% formalin/seawater, the latter being better for subsequent microscopical examination but the colour is lost unless kept cool and dark. We have found that one of the most satisfactory storage methods for dry material is to place a *fully dried* specimen in a plastic bag together with a large index card, which provides both support and space for information. The bags can be stored vertically in shallow drawers or on shelves, with each specimen visible and accessible.

8.3 Preparation

Surface features, including conceptacles, can be examined with reflected light using a dissecting microscope and a magnification of between ×10 and ×50. A sandbag is useful for levelling to focus on details in an irregularly-shaped sample. Conceptacles can be crushed or picked apart to disclose contents, although this is difficult in practice. Chips removed with a scalpel or small hammer and chisel can be supported by plasticine: the fractured surfaces often reveal conceptacle shape or cell alignment. Similar chips can be decalcified in a weak acid such as 10% nitric and sectioned with a freezing microtome. For preparing microscope slides, we have used glycerine jelly or Karo (a US proprietary brand of maize syrup) with a stain such as crystal violet or aniline blue incorporated in the mountant to avoid handling the sections. A range of stains and mountants have been used by others (see Adey & Adey, 1973). A simple embedding technique using gelatine has been successful, but for critical studies paraffin wax or resin embedding methods such as described by Johansen (1981) and Woelkerling (1988) are preferable. Very satisactory SEM micrographs can be obtained of air-dried surfaces, including carefully orientated fractures, providing three-dimensional information on anatomy and conceptacle structure with the calcification in situ. Details of fine structure can be observed using TEM methods. Techniques for geniculate plants are similar and an additional simple method involves mounting whole portions of decalcified plants; these are transparent enough for many features to be observed after staining with, for example, fast green.

8.4 Identification

One of the most bewildering apects of the study of red algae is that a similarity in external form of two individuals is frequently no indication of their taxonomic affinity. Plants that are morphologically similar can be totally unrelated to each other and the converse, that related plants can be dissimilar in appearance, is also true. Nowhere is this phenomenon more apparent than in the Corallinaceae. This factor, perhaps even more than the obvious one of technical difficulty, has repeatedly resulted in confusion of identity and has been instrumental in delaying progress towards an understanding of the taxonomy of the group. As pointed out strongly by Huvé (1962), it is folly to use only the external aspect of the thallus to identify coralline algae. Since surface features can also vary according to season and environmental conditions, these characterisitcs are best used after some experience has been gained in other ways. Although decalcifying and sectioning is time-consuming, the insight gained rewards the effort. Some species under some circumstances are unmistakable, e.g. well-grown *Mesophyllum lichenoides* on *Corallina* or bright pink, protuberant *Lithothamnion glaciale* in Scotland, but even these species can appear quite different under other conditions and many species are never distinctive enough for positive identification on sight. Some species are more likely to form crests where two thalli, or parts of the same thallus, meet (Fig.6) and we have also noticed that certain species regularly overgrow, or are overgrown by, others (Figs 20A; 39A) in a consistent hierarchy.

Thallus surface features under SEM have been proposed by several authors as useful for identification for both geniculate species (Garbary & Veltkamp, 1980; Garbary & Johansen, 1982) (see Figs 7; 8) and nongeniculate species (Garbary, 1978; Chamberlain, 1990) (see Fig.18). Chamberlain (1990) recognized three surface patterns that she named *Pneophyllum*-type (Fig.7A), *Phymatolithon*-type (Figs 7C; 8C) and *Leptophytum*-type (Fig.8D) after genera in which they were seen in the British Isles: the names were intended as a concise means of labelling rather than implying strict taxonomic significance. *Melobesia membranacea* (Fig.7B) and *Lithothamnion corallioides* (Fig.8A) have further distinct appearances. Surface pattern can sometimes help in solving precise taxonomic problems. For example, sterile maerl branches of *Lithothamnion corallioides* and *L. glaciale* are difficult to distinguish (Table 4), but their surface patterns (Figs 8A,B) are different. Also, all three species of *Leptophytum* (*q.v.*) in the British Isles have a distinct surface pattern (Fig.8D) that apparently derives from the relatively unusual presence of epithallial cells with calcified outer walls (Fig.8E). A similar surface is seen in the taxonomically remote *Lithophyllum nitorum* (Fig.7D) but in no other species of *Lithophyllum* in the British Isles. It is useful to record surface type of mature thalli (young, relatively lightly calcified, areas

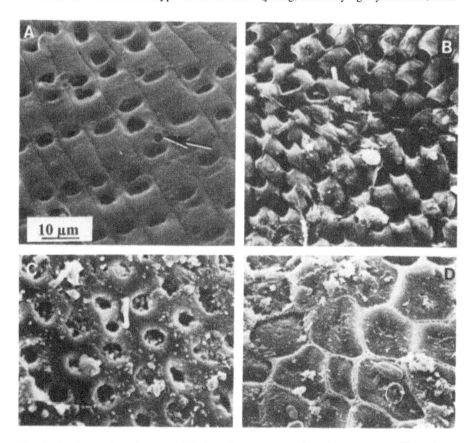

Fig. 7 Thallus surface features (SEM, A-D to same scale) of crustose coralline algae. (A) *Pneophyllum myriocarpum*; note intercalary trichocyte (arrow). (B) *Melobesia membranacea*. (C) *Lithophyllum incrustans*. (D) *Lithophyllum nitorum* showing *Leptophytum*-type surface.

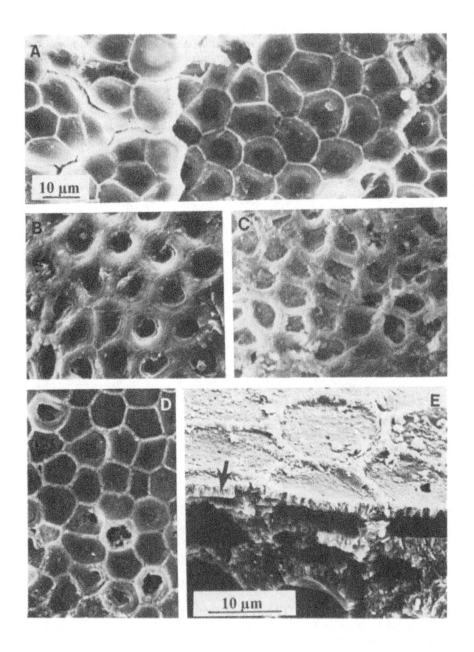

of thallus may differ from mature areas in any one species) as it might provide distinguishing characters in specimens where other features are absent.

Reference is made (e.g. Figs 77C; 87A) to thallus zonation. This feature is clearly visible in sections of some species and comprises lateral lines of cells that appear distinct from surrounding cells. This may be due to differences in cell size and/or shape, or to the cells themselves (or the middle lamella between them) staining more heavily than surrounding ones. Such zonation or banding has been observed by many authors (e.g. Suneson, 1943a, p. 9, fig. 4C; Hamel & Lemoine, 1953, p. 96, fig. 60; Mason, 1953, p. 341, pl. 45, fig. c). The feature may be a consequence of periodicity in growth, but this has not been confirmed.

Conceptacles provide several features which aid identification. Multiporate roofs can usually be detected with a handlens or by scraping off a roof and viewing under a compound microscope. They indicate tetra/ bisporangial material of a melobesioid species, though it should be remembered that gametangial conceptacles in this subfamily are uniporate, in common with all conceptacles in all other subfamilies. Size, shape and degree of prominence or immersion of conceptacles are other useful features. Thallus thickness is less useful than might be supposed, especially as some species, in the Mastophoroideae in particular, can reproduce when very young but at other times can become massive, though they rarely do in the British Isles. In common with encrusting species in other algal groups, many species are fertile only in winter, others are fertile principally in summer and a third group produces conceptacles throughout the year, although the spores are probably not always viable. The above characteristics can be used in combination to generate Keys for identification.

It is useful to refer to shore zonation (Section 4.1) for a guide as to which species are common in each zone.

The seasonal factor is very important in the identification of crustose corallines. It is crucial to collect the large crustose species in winter when they are intensely coloured and very conspicuous until about March. The majority are fertile in winter and in November and December rocky shores are often covered with colourful expanses of *Phymatolithon purpureum, P. laevigatum* and *P. lamii* with white cortical discs popping off the conceptacles (Fig.115B). In summer they become bleached and buried under a mass of annual algae, and are much more difficult to identify. Epiphytic species occur throughout most of the year but one of the commonest associations, *Pneophyllum fragile* growing on *Zostera* (Fig.64), is present mainly from about July to December.

9 Arrangement of the work.

In general the style adopted by Dixon & Irvine (1977) and Irvine (1983) has been followed. Irvine (1983, pp.1–5) discussed such matters as the restriction of synonyms etc. which also apply to this work. Five subfamilies of Corallinaceae are dealt with alphabetically, and, within each, the genera are similarly treated. Herbarium abbreviations, which follow Holmgren *et al.* (1990), are listed at the end of this Introduction.

Fig. 8 Thallus surface features (SEM, A-D to same scale). (A) *Lithothamnion corallioides*, characteristic surface with conspicuous middle lamellae (white) on right, old epithallial cells being shed on left. (B) *Lithothamnion glaciale*. (C) *Phymatolithon calcareum*. (D) *Leptophytum elatum*. (E) *Leptophytum laeve* in vertical fracture/surface view showing calcified epithallial cell outer wall (arrow) (photo J. Hall-Spencer).

The present authors had the opportunity to work closely with Dr Woelkerling during the preparation of his book on nongeniculate genera (Woelkerling, 1988) and are fully conversant with his generic circumscriptions, which are followed here unless otherwise stated. For geniculate genera we are following Johansen (1981).

The notes below are additional to those given under this heading in Vol. 1 (1, 2A), to which the reader is referred for background information.

9.1 Typification

Generic typification follows that given in Woelkerling (1988) and the type of each species is listed whenever possible, including the original locality, present location and appropriate references. We endorse the selection of neotypes as necessary (e.g. *Phymatolithon calcareum*, *Mesophyllum lichenoides*) especially when based on a new collection of fertile material chosen to preserve current usage of the name wherever possible. Dixon & Irvine (1977) stated in Vol.1(1), p.70 that permanent collections in many herbaria were examined in detail in connection with typification. In both this and the subsequent part 2A this statement was intended to imply that type specimens had been examined unless otherwise stated. In order to make this clearer in the present work we have adopted the symbol !, as used in the nineteenth century, to indicate specimens seen by the authors.

9.2 Keys

Keys to species and varieties are given under each genus and a comprehensive Key to genera of the Corallinaceae found in the British Isles is given on p.30. Key characters are given under each species: these are features we have found easy to recognize (e.g. strong orbital ridges at the margin of most plants of *Phymatolithon lenormandii*) and are not necessarily diagnostic in the taxonomic sense. Section 4.1 of this Introduction indicates the species most likely to be seen in particular shore zones.

9.3 Notes accompanying descriptions

Short notes are given on habitat, distribution, and growth, including both seasonal occurrence and information on the development of reproductive structures. The distribution of each species is given on the basis of the county system (see map p. 276) in use before 1974 with a few subdivisions as necessary. A brief list of easily observed, distinguishing characters is given at the beginning of each species description. Comments are included on the range of form we have encountered together with notes and references on any known or surmised correlations with environmental factors. Lastly, we have included notes of general interest for each species concerning aspects not covered elsewhere, such as inter-relationships with other organisms, practical uses, problems of identification etc.

9.4 Transfers, Additions and Deletions

There have been a number of amendments since the most recent checklist (Parke & Dixon, 1976), as follows:

9.4a Transfers

Corallina granifera Ellis & Solander to *Haliptilon virgatum* (Zanardini) Garbary & Johansen
Dermatolithon adplicitum Foslie to *Titanoderma pustulatum* var. *macrocarpum* (Rosanoff) Y.Chamberlain
D. confinis Adey et Adey to *T. pustulatum* var. *confine* (P.& H.Crouan) Y.Chamberlain

D. corallinae (P.& H. Crouan) Foslie to *T. corallinae* (P.& H.Crouan) Woelkerling, Y.Chamberlain & P.Silva

D. crouanii (Foslie) Lemoine pro parte, excl.typ. to *T. laminariae* (P.& H.Crouan) Y.Chamberlain

D. hapalidioides (P.& H.Crouan) Foslie to *T. pustulatum* var. *macrocarpum*

D. litorale (Suneson) Lemoine to *T. pustulatum* var. *canellatum* (Kützing) Y.Chamberlain

D. pustulatum (Lamouroux) Foslie to *T. pustulatum* (Lamouroux) Nägeli var. *pustulatum*

Fosliella farinosa (Lamouroux) Howe to *Hydrolithon farinosum* (Lamouroux) D.Penrose & Y.Chamberlain; pro parte excl. type to *Hydrolithon boreale* (Foslie) Y.Chamberlain (*Fosliella*-state)

F. lejolisii (Rosanoff) Howe to *Pneophyllum fragile* Kützing

F. limitata (Foslie) Ganesan to *P. limitatum* (Foslie) Y.Chamberlain

F. minutula (Foslie) Ganesan to *P. confervicola* (Kützing) Y.Chamberlain

F. tenuis Adey & Adey pro parte to *P. myriocarpum* (P.& H.Crouan) Y.Chamberlain

F. zonalis (P.& H.Crouan) Ganesan to *P. caulerpae* (Foslie) P.Jones & Woelkerling

Jania corniculata (Linnaeus) Lamouroux to *Jania rubens* var. *corniculata* (Linnaeus) Yendo

Phymatolithon bornetii (Foslie) Foslie to *Leptophytum bornetii* (Foslie) Adey

P. polymorphum (Linnaeus) Foslie to *P. purpureum* (P.& H.Crouan) Woelkerling & L.Irvine

P. rugulosum Adey to *P. lamii* (Lemoine) Y.Chamberlain

9.4b Additions
Exilicrusta parva Y.Chamberlain
Hydrolithon boreale (Foslie) Y.Chamberlain comb.nov.
H. cruciatum (Bressan) Y.Chamberlain comb.nov.
H. samoënse (Foslie) D.Keats & Y.Chamberlain
H. sargassi (Foslie) Y.Chamberlain comb.nov.
Pneophyllum lobescens Y.Chamberlain
P. myriocarpum (P.& H.Crouan) Y.Chamberlain
Leptophytum elatum Y.Chamberlain
Lithophyllum crouanii Foslie
L. dentatum (Kützing) Foslie
L. duckeri Woelkerling
L. hibernicum Foslie
Phymatolithon brunneum Y.Chamberlain n.sp.

9.4c Deletions
Clathromorphum compactum (Kjellman) Foslie
Dermatolithon cystoseirae (Hauck) Huvé
Fosliella valida Adey & Adey
Lithothamnion fruticulosum Foslie excluded, see under *Lithothamnion*
L. norvegicum (Areschoug in J.Agardh) Kjellman excluded, see under *Lithothamnion*
Schmitziella endophloea Bornet & Batters in Batters excluded from the Corallinales by Woelkerling & Irvine (1982).

9.5 Measurements and Illustrations
Measurements (Fig.9), particularly such values as thallus diameter, are rarely diagnostic but nevertheless are useful in indicating limits. Only maxima and minima are included

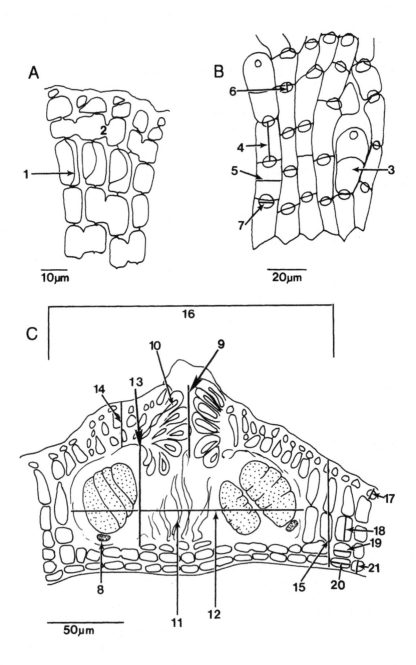

for all dimensions; it is probable that mean plus standard deviation would be a more useful value but it was not usually possible to calculate this. The numbers outside brackets are the maxima and minima measured by the present authors, additional values in brackets are those given by Adey & Adey (1973) when they extend the range we have found.

Cell measurements (Fig.9B,C) refer to cell lumens and length is always measured between primary pit connections as shown. Conceptacle measurements (Fig.9C) follow the convention in Adey & Adey (1973).

It is frequently impossible to identify coralline algae without recourse to microscopic examination. Photographs are provided of each species at the level of magnification seen with a good hand lens or dissecting microscope. Drawings of vegetative and reproductive features are also provided and where possible comparable features are drawn to the same scale as indicated by the symbol * on anatomical drawings, or to similar scale as indicated by • on diagrammatic drawings. This convention will enable users to assess, for example, relative conceptacle sizes at a glance. It may seem superfluous to include in a Flora features that can only be seen under a scanning electron microscope (indicated by SEM in brackets). However, coralline algae are very easy to prepare for the SEM as they only need to be air dried in contrast to the tedious process of decalcifying and sectioning. Scanning electron micrographs, for example, are particularly useful for confirming, or otherwise, the presence of the *Leptophytum*-type cell surface and three-dimensional features can be conveyed more readily than by other methods of illustration.

9.6 Descriptive Terminology
The terms used for anatomical features of coralline algae have been discussed under *Thallus anatomy* and *Reproduction* and these are used in synoptic descriptions. As it is difficult to select terms to describe the external features of crustose plants, a representative description, together with explanation, follows: '1) *Thallus* encrusting, closely adherent, 2) to at least 8cm diameter, to 250μm thick, 3) thallus entirely flat or with mosaic of small mounds, or partly or entirely squamulose, 4) thallus orbicular becoming confluent, 5) confluences flat or slightly crested, 6) *margin* entire, thin, with or without conspicuous orbital ridges, 7) *surface* smooth, 8) cell surface *Phymatolithon*-type (SEM), 9) brownish, greyish, mauvish, pinkish, or reddish (Methuen dark ruby, greyish ruby, dull violet, dull red etc.), often young parts reddish and old parts brownish on same plant, 10) glossy, matt or rough;'
1) General thallus habit, e.g. encrusting, unattached etc.
2) General size limits

Fig. 9 Position of features and measurements in dimerous thallus cells and a conceptacle. (A) Surface view of *Pneophyllum* thallus showing an intercalary trichocyte (1) and a cell fusion (2). (B) Surface view of *Hydrolithon* (*Fosliella*-state) thallus showing a terminal trichocyte (3), basal cell length (4), basal cell diameter (5), epithallial cell length (6), epithallial cell diameter (7). (C) VS of *Pneophyllum* tetrasporangial conceptacle showing a tetrasporangial stalk cell (8), length of pore canal (9), pore filaments (10), columella (11), conceptacle chamber diameter (12), conceptacle chamber height (13), roof thickness (14), thallus thickness (15), conceptacle external diameter (16), epithallial cell length (17), erect filament cell length (18), erect filament cell diameter (19), basal filament cell length (20), basal filament cell diameter (21).

3) Thallus configuration varies from being entirely flat to having low mounds, larger excrescences etc.

4) In general shape most encrusting thalli are orbicular when young, later becoming irregular and usually confluent with adjacent thalli

5) Confluences between thalli may remain flat, become crested, or overlap: these features are sometimes diagnostic

6) Margin may be entire, lobed etc.; it may vary from thin and bistratose to thick and chunky; its surface may vary from completely flat to having pronounced orbital ridges

7) In microscopic features the surface may be smooth, rough, flaky etc.

8) Cell surface (SEM) may be *Pneophyllum-*, *Leptophytum-*, or *Phymatolithon-*type etc. (see Figs 7–8).

9) Generalised colours of living plants are given, together with detailed matching of fresh plants, when available, with Methuen colour charts (Kornerup & Wanscher, 1978). Information about dried material is sometimes given.

10) Texture may be glossy, matt, grainy etc.

Woelkerling *et al.* (1993) have published a useful guide to external form in which they suggest a standardized terminology. This should help to unify future systematic work on nongeniculate taxa.

Further terms are defined in the Glossary.

9.7 Abbreviations
Herbaria:

AHW	—	Naturwet. Museum, Antwerp (Belgium)
BM	—	British Museum (Natural History) = The Natural History Museum, London.
C	—	Botanical Museum, Copenhagen (Denmark)
CHE	—	Société des Sciences Naturelles et Mathématiques de Cherbourg (France)
CN	—	Université de Caen (France)
CO	—	Laboratoire Maritime, Concarneau (France)
FR	—	Forschungsinstitut Senckenberg, Frankfurt (Germany)
L	—	Rijksherbarium, Leiden (Netherlands)
LD	—	Botanical Museum, The University, Lund (Sweden)
MICH	—	University Herbarium, Ann Arbor, Michigan (USA)
PC	—	Museum National d'Histoire Naturelle, Paris (Cryptogams) (France)
TRH	—	Kongelige Norske Videnskabers Selskab Museet, Trondheim (Norway)
TSB	—	Trieste
UC	—	University Herbarium, Berkeley, California (USA)
USNC	—	United States National Coralline Collection, Washington, DC (USA)

c.	—	circa
h.	—	hours
ICBN	—	International Code of Botanical Nomenclature, (Greuter *et al.*, 1988)
LS	—	longitudinal section
q.v.	—	which see
RVS	—	radial vertical section
SEM	—	scanning electron microscopy
TEM	—	transmission electron microscopy

TVS — tangential vertical section
VS — vertical section
h — height
d — diameter
l — length
! — specimen seen
* — anatomical drawings to same scale
• — diagrammatic drawings to similar scale

YMC 90/236 etc. — herbarium numbers
Collectors' names indicated by italics

KEY TO CORALLINACEAE

1 Plant with erect branches appearing jointed, composed of calcified intergenicula separated by non-calcified intergenicula..................................... Corallinoideae p.37

   Plant entirely encrusting or with erect branches not jointed..................................... 2

2 Thallus semi-endophytic in *Jania* and *Haliptilon*.............................. *Choreonema* p.33

   Thallus not semi-endophytic 3

3 Cells in contiguous filaments joined by secondary pit connections (Fig.4)[+]............. 4

   Cells in contiguous filaments joined by cell fusions (Fig.2C)[+] 5

4 Thallus with single, basal palisade layer (Fig.37), margin usually bistratose (Fig.40B), mainly epiphytic.. *Titanoderma* p.88

   Thallus with single basal layer of mainly squarish, non-palisade cells, thickening immediately behind margin (Fig.33A), nearly always epilithic ..... *Lithophyllum* p.59

5 Tetra/bisporangial conceptacles uniporate* (Figs 51D,E)... 6

   Tetra/bisporangial conceptacles multiporate* (Fig.87B).. 7

6 Pore of tetra/ bisporangial conceptacle surrounded in VS by enlarged, vertically orientated cells (Fig.53A), germination disc with 4-celled centre (Figs 3A-C), basal filament trichocytes terminal (Fig.50C)...................................... *Hydrolithon* p.114[†]

   Pore of tetra/bisporangial conceptacle surrounded in VS by tiny (Figs 65E,G) to elongate (Fig.61H) cells orientated horizontally at least initially; germination disc with 8-celled centre (Fig.3D), basal filament trichocytes intercalary (Fig.60B)..*Pneophyllum* p.131[†]

7 Epithallial cells flared (Figs 74 *Lithothamnion*; 85E).. 8

   Epithallial cells domed (Fig.74 *Phymatolithon*) to flattened (Fig.74 *Leptophytum*) or rectangular (Fig.99B), not flared.. 9

8 Lower filaments unistratose... *Exilicrusta* p.162

   Lower filaments multistratose .... *Lithothamnion* p.176

9 Thallus dimerous, not exceeding 10 cells thick except near conceptacles ....
   .... *Melobesia* p.195

   Thallus monomerous, exceeding 10 cells thick 10

10 Medulla coaxial (Figs 98A-C) comprising *c*.80% thallus thickness . *Mesophyllum* p.201

   Medulla not coaxial 11

11 Outermost epithallial cells flattened; conceptacle pore plugs surrounded by differentiated cells; cell surface (SEM) *Leptophytum* type (Figs 8D,E)............. *Leptophytum* p.165

   Outermost epithallial cells domed or rectangular; conceptacle pore plugs not surrounded by differentiated cells; cell surface (SEM) *Phymatolithon*-type (Fig.8C)
   .... *Phymatolithon* p.206

For Abbreviations see end of Introduction.

* All gametangial/ carposporangial conceptacles are uniporate

\+ See section 8.4 of Introduction for guide to finding this character

† If no listed characters are evident, go straight to species keys.

CORALLINALES Silva & Johansen

CORALLINALES Silva & Johansen (1986), p.250.

Plants calcified and hard, of variable form, either semi-endophytic in other corallines, or growing on various substrates, or unattached; thalli multiaxial, pseudoparenchymatous, or rarely of partially or largely unconsolidated filaments; meristematic cells (initials) either terminal, or intercalary and then producing one to several epithallial cells with partly calcified walls outwardly and other vegetative cells inwardly; walls of most other vegetative cells completely impregnated with calcite; primary pit plugs (where studied) containing an outer dome-shaped cap layer; cells of contiguous vegetative filaments often united by fusions or direct secondary pit connections; trichocytes often present.

Gametangial and tetra/bisporangial plants of similar organization; gametangia and sporangia borne within conceptacles; gametangial plants monoecious or dioecious, conceptacles uniporate; spermatangial systems simple to dendroid, borne on conceptacle floor and sometimes on walls, carpogonial conceptacles with supporting (auxiliary) cells aggregated on conceptacle floor, each bearing 1–2 two-celled carpogonial branches; carposporophytes forming after presumed karyogamy and transfer of zygotic nucleus or its derivative(s) to supporting cell; supporting cells and sometimes other adjoining cells fusing to form a single fusion cell (possibly several in some Melobesioideae), producing short unbranched gonimoblast filaments each with a terminal carposporangium; tetra/ bisporangial conceptacles uniporate or multiporate; tetraspores formed as a result of presumed meiosis followed by simultaneous zonate (very rarely cruciate) divisions of tetrasporocytes; sometimes replaced by binucleate or uninucleate bispores, the former presumably formed following meiosis and the latter following mitosis, populations reproducing apparently exclusively by uninucleate bispores sometimes common.

Most plants in the Corallinales are heavily calcified, producing calcite within the walls of vegetative cells. Until recently it has been a well circumscribed order in the red algae (Silva & Johansen, 1986; Cabioch, 1988; Woelkerling, 1988), containing one extant family, the Corallinaceae. In 1993 Verheij proposed the family Sporolithaceae to accommodate the extra-European genus *Sporolithon*. The most significant feature setting this family apart is simultaneously cleaved but cruciately arranged tetrasporangia borne in sori rather than conceptacles. Although Verheij included this family in the Corallinales, it does not fall within the circumscription given above and further evaluation is needed. The fossil family Solenoporaceae may represent ancestral stock from which taxa of the Corallinaceae evolved (Woelkerling, 1988).

CORALLINACEAE Lamouroux

CORALLINACEAE Lamouroux (1812), p.185 [as Corallineae].
With the characters of the order.

Representatives of five of the seven subfamilies accepted by Woelkerling (1988, table 5.2) occur in the British Isles: Choreonematoideae, Corallinoideae, Lithophylloideae, Mastophoroideae, Melobesioideae. As shown in his table, each subfamily possesses a unique combination of diagnostic features. The features concerned are the presence or absence of genicula, secondary pit connections, cell fusions and tetra/bisporangial plugs, combined with either uni- or multiporate tetra/bisporangial conceptacle roofs. The Corallinoideae contains the only geniculate genera present in the British Isles. Cabioch's (1971b, 1972, 1988) views on relationships within the family caused her to include both geniculate and nongeniculate genera in some subfamilies. For example, mastophoroid genera such as *Fosliella* (incl. *Pneophyllum*) were placed with *Corallina* in the subfamily Corallinoideae.

CHOREONEMATOIDEAE Woelkerling

by Yvonne M.Chamberlain and Linda M.Irvine

CHOREONEMATOIDEAE Woelkerling (1987a), p.125.

Type genus: *Choreonema* Schmitz (1889), p.455.

Thallus nongeniculate; cells of contiguous vegetative filaments apparently not joined either by secondary pit connections or by cell fusions; tetrasporangia possessing apical plugs and borne within uniporate conceptacles.

This subfamily was created by Woelkerling (1987a) for the endophytic genus *Choreonema* which he recognized had a grouping of characters not found in the previously described subfamilies; in particular, the tetrasporangia bear apical plugs similar to those seen in the Melobesioideae but the tetra/bisporangial conceptacles are uniporate. Both Minder (1910) and Suneson (1937) had recognized the presence of tetrasporangial plugs within a uniporate conceptacle but did not appreciate the taxonomic significance of this so far unique combination. *Choreonema* is one of only two known genera of Corallinaceae outside the Melobesioideae where spermatangia occur on the floor, walls and roof of the conceptacles; the other is *Neogoniolithon* (Penrose, 1992b) in which they are also in simple systems. In all other genera they are on the floor only.

Cabioch (1972) considered *Choreonema* to be closely related to *Cheilosporum, Haliptilon* and *Jania*, genera of the Corallinoideae which it is known to infect, and is therefore an adelphoparasite. Afonso-Carrillo *et al.* (1986) regarded the presence of a *Jania*-type external surface to the conceptacle of *Choreonema* to be an indication of a taxonomic relationship with its host. Johansen (1981), however, drew attention to Bressan's (1974) record of *Choreonema* on *Corallina,* which belongs to a different Tribe of the Corallinoideae. Woelkerling (1987b) showed that the enigmatic *Chaetolithon* is congeneric with *Choreonema.*

The single genus, *Choreonema*, has been recorded for the British Isles.

CHOREONEMA Schmitz

CHOREONEMA Schmitz (1889), p.455 emend. Woelkerling (1987a), p.122.

Type species: *C. thuretii* (Bornet) Schmitz (1889), p.455.

Chaetolithon Foslie (1898b), p.7.

Endosiphonia Ardissone (1883), p.450, nom. illeg., non *Endosiphonia* Zanardini (1878), p.35 (ICBN, Art. 64.1).

Vegetative thallus endophytic, filamentous, filaments simple or branched, possibly uncalcified, lacking cell fusions or secondary pit connections, with occasional lenticular cells, possibly representing epithallial cells.

Conceptacles (all types) calcified, globular to domed, produced on host surface, developing from pad of appressed filaments just below host surface, conceptacle roofs uniporate, walls composed of a layer of cells with long axes parallel to roof surface; epithallial cells lacking.

Gametangial plants dioecious; spermatangial systems simple, spermatangia in chain-like rows; carpogonial conceptacles containing 2-celled carpogonial branches, with supporting cell functioning as auxiliary cell; carposporangial conceptacle with central fusion cell from which gonimoblast filaments arise peripherally, bearing terminal carposporangia; tetrasporangial conceptacles containing zonately divided tetrasporangia across conceptacle floor, each bearing a distinct apical plug, plugs collectively blocking single pore prior to spore release; bisporangial plants unknown.

For an account of this monotypic genus see Woelkerling (1987a, 1988).

Choreonema thuretii (Bornet) Schmitz (1889), p.455.
Figs 10; 11

Lectotype: PC(!). E. Bornet, Pointe de Querqueville, France, on *Haliptilon squamatum*, 31 December 1855.

Melobesia thuretii Bornet in Thuret & Bornet (1878), p.96.
Melobesia deformans Solms-Laubach (1881), p.57.
Chaetolithon deformans (Solms-Laubach) Foslie (1898b), p.7.

Key characters: Semi-endophytic in *Haliptilon* and *Jania,* globular conceptacles protruding from surface of host.

Illustrations: Bornet in Bornet & Thuret (1878), pl.50, figs 1–8; Solms-Laubach (1881), pl. III, figs 1,4–9; Suneson (1937), figs 33–35; Hamel & Lemoine (1953), pl.XXII, fig.3; Woelkerling (1987b), figs 1–25.

Plants visible externally only as sessile conceptacles on intergenicular surface of *Haliptilon* and *Jania*; *spermatangial, carposporangial and tetrasporangial conceptacles* globular to conical, to *c.*125μm diameter, 150μm high, with surface composed of a reticulum of ridges enclosing flat plates.

Endophytic filaments branched, possibly uncalcified; cells 8–39μm long × 6–20μm diameter, lacking cell fusions or secondary pit connections; occasional lenticular cells present, possibly being equivalent to epithallial cells.

Gametangial plants dioecious; *spermatangial conceptacle* chambers flask-shaped, *c.*60μm diameter, *c.*45μm high, with the roof *c.*20μm thick, spermatangia borne on all surfaces, spermatangial systems simple, spermatangia spherical to ovoid in chain-like rows; *carpogonial conceptacle* chambers high-domed; *carposporangial conceptacle* chambers globular, *c.*56μm diameter, *c.*58μm high, with walls *c.*30μm thick, walls composed of layer of elongate cells, epithallial cells lacking, fusion cell central, *gonimoblast filaments* borne from periphery, to *c.*5 cells long, gradually enlarging towards terminal carposporangium; *tetrasporangial conceptacle* chambers globular, *c.*70μm diameter, *c.*74μm high, with walls *c.*13µm thick, wall structure as for carposporangial conceptacles, *tetrasporangia* borne across conceptacle floor, *c.*55µm long × *c.*20µm diameter, each tetrasporangium terminating in a mucilaginous plug, collectively enclosed in uniporate roof; *bisporangial conceptacles* unknown.

Semi-endophytic in *Haliptilon* and *Jania*, mainly in younger branches; recorded from lower littoral and upper sublittoral.

Cornwall to Dorset; Channel Isles; in Ireland northwards to Mayo, eastwards to Wexford. Widespread in temperate and tropical waters.

Possibly occurring mainly in summer, but Woelkerling (1987a) found no seasonality in vegetative or reproductive development in southern Australia.

This inconspicuous plant has been subjected to several of the most detailed studies and illustrations of any coralline species (see references above and Minder, 1910).

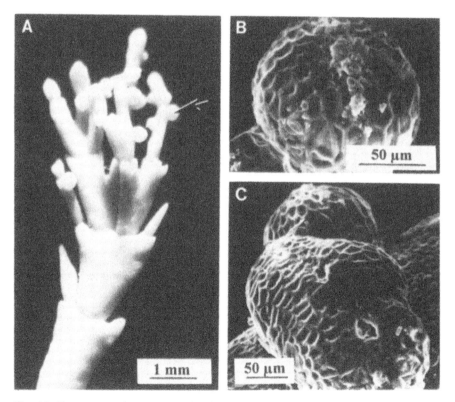

Fig. 10 *Choreonema thuretii* on *Haliptilon squamatum* (BM, *YMC*, Sark, 5.ix.1960). (A) Conceptacles (arrow) protruding from host. (B) Conceptacle showing characteristic surface. (C) Fused conceptacles.

Some authors (Johansen, 1981 and refs.) interpret the small lenticular cells sometimes found on vegetative cells as vestigial epithallial cells.

Spore germination is reported to be of the *Naccaria* type (Cabioch, 1972), so far found elsewhere in the Corallinales only in two species of *Amphiroa*. She described spore germination on a glass surface, noting (Cabioch, 1972, pl.I, figs 1–4) that a large germination tube develops, and Woelkerling (1987a, figs 9–11) also observed early stages in thallus and conceptacle growth.

Bisporangia have been recorded only in California (Abbott & Hollenberg, 1976).

The parasitic nature of this species needs further investigation; severe distortion and reduction of host occurs (LMI obs.; Cabioch, 1980), but no haustoria or other connections with host cells were seen by Woelkerling (1987a) or reported elsewhere. Host plants appear to be confined to the tribe Janieae of the Corallinoideae and it is easy to spot this species on *Haliptilon* and *Jania* because its conceptacles are unlike those of the hosts. Pseudolateral conceptacles of *Corallina officinalis*, however, look remarkably similar to those of *C. thuretii* and it is advisable to check records of the latter to make sure that the identification was correct.

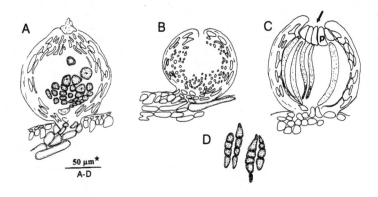

Fig. 11 Vertical sections of conceptacles of *Choreonema thuretii* growing on *Haliptilon squamatum*. (A) Carposporangial conceptacle (BM, *YMC*, Sark, 5.ix.1960). (B) Spermatangial conceptacle (BM, *YMC*, Sark, 5.ix.1960). (C) Immature tetrasporangial conceptacle; note pit plugs (p) within the single pore (arrow) (BM, *YMC*, Sark, 5.ix.1960). (D) A group of mature tetrasporangia extruded from a damaged conceptacle (*M.D.Guiry* no.146).

CORALLINOIDEAE

by Linda M. Irvine and H. William Johansen*

CORALLINOIDEAE

Type genus: *Corallina* Linnaeus (1758) p.805.

Thallus geniculate, fronds attached by crustose bases which in some taxa are ephemeral and replaced by stolons, bases monomerous and resembling *Lithothamnion*; each frond consisting of one to hundreds of calcified intergenicula alternating with uncalcified genicula, branching dichotomous and/or pinnate, sometimes irregular, adventitious branches sometimes present; intergenicula terete to compressed to flat, sometimes with wings or lobes; structure multiaxial, intergenicular medulla consisting of one to many arching tiers of cells of uniform length, central medullary filaments unbranched, peripheral medullary filaments branching pseudodichotomously to produce short cortical filaments with intercalary initials;

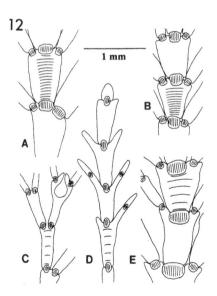

Fig. 12 Intergenicula and genicula in the Corallinoideae. (A) *Corallina officinalis.* (B) *Corallina elongata.* (C) *Jania rubens* var. *rubens.* (D) *Jania rubens* var. *corniculata.* (E) *Haliptilon squamatum.* Arching lines indicate tiers of intergenicular medullary cells.

*Department of Biology, Clark University, Worcester, Mass., USA.

trichocytes present in many taxa; contiguous cells united by fusions, secondary pit-connections lacking; epithallial cells covering all calcified surfaces except branch apices and crustose base margins; each geniculum a single tier of unbranched medullary cells, genicular cell walls thick but uncalcified except where ends adjoin intergenicular cells; each geniculum partly covered by flanges of intergenicular cortex.

Conceptacles either axial, marginal, lateral, or pseudolateral, branches growing from roofs of certain types of conceptacles in some genera; each conceptacle uniporate with a central or excentric pore, roof formed by centripetal growth of filaments surrounding reproductive cells on conceptacle floor; spermatangia lining floor and walls of spermatangial conceptacles, spermatangial systems simple; gonimoblast filaments arising from periphery, upper surface, or both, of the single fusion cell in each carposporangial conceptacle; germination of *Corallina*-type, producing crustose bases from which fronds arise.

Three of the twelve genera in this subfamily occur in the British Isles. *Corallina* is the type genus of the subfamily and is the oldest name in coralline literature, being mentioned by Pliny in the First Century AD. It belongs to the tribe Corallineae; the other two genera in the British Isles, *Jania* and *Haliptilon*, belong to the tribe Janieae (Johansen & Silva, 1978). Although *Haliptilon* superficially resembles *Corallina*, significant differences in the organization of the conceptacles and their contents (Table 1, Johansen, 1970, 1981), together with less obvious details of medullary filaments, intergenicular surface features (Fig.18, see p. 54), and genicula show the closer relationship of the former to *Jania*. In addition, *Choreonema q.v.* occurs only in *Haliptilon* and other members of the Janieae (Johansen, 1981; but see Bressan, 1974).

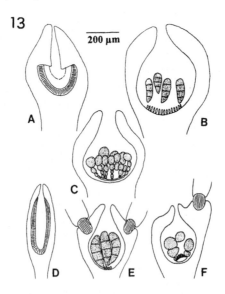

Fig. 13 Conceptacles in the Corallinoideae. (A) *Corallina officinalis*, spermatangial. (B) *Corallina officinalis*, tetrasporangial. (C) *Corallina elongata*, carposporangial. (D) *Jania rubens* var. *rubens*, spermatangial. (E) *Haliptilon squamatum*, tetrasporangial. (F) *Haliptilon squamatum*, carposporangial.

The pinnate branching invariably present in *Haliptilon* distinguishes the genus from *Jania*, in which most species have exclusively dichotomous branching. In the NE. Atlantic, however, *J. rubens* var. *corniculata* appears intermediate, having some secondary pinnate branching; the similarity of this variety to *Haliptilon* spp. suggests the need for a reassessment of these genera.

Crustose bases (Fig.14) are similar in structure to *Lithothamnion* (Rosenvinge, 1917) and appear to be uniform throughout the subfamily (Johansen, 1969), although their size and rate of development vary between genera (Cabioch, 1966b; Jones & Moorjani, 1973).

KEY TO SPECIES OF CORALLINOIDEAE

The taxa keyed below are separated by groups of characters since individual characters do not necessarily hold for every specimen. It should be noted that spermatangial conceptacles are easy to recognize; in all genera they lack surmounting branches, in *Corallina* they have pronounced beaks (see Fig.13A), and in *Jania* and *Haliptilon* they are elongate (see Fig.13D). When identifying geniculate corallines it is better to rely on characteristics of vegetative rather than fertile intergenicula; far more vegetative intergenicula are typically present in a frond and they are usually more uniform in structure. Additionally, intergenicula near branch apices (except partially developed terminal intergenicula) are least affected by age and erosion.

1   Branching pinnate throughout; epilithic, attached by extensive crustose bases >10mm in diameter; tetra/ bisporangial conceptacles each containing >30 sporangia; carposporophytic fusion cells broad and shallow, with concave upper surfaces   2
    Branching entirely or partly dichotomous; epiphytic and/or epilithic, attached by small crustose bases <5mm in diameter, with or without stolons; tetrasporangial conceptacles each containing <10 sporangia; carposporophytic fusion cells narrow and thick, with convex upper surfaces   3

2   Carpo- and tetra/ bisporangial conceptacles infrequently branched (<20%); average length of intergenicula in main axes >1mm; gaps between successive lateral branches conspicuous .... *C. officinalis*
    Carpo- and tetra/ bisporangial conceptacles frequently branched (>20%); average length of intergenicula in main axes <1mm; gaps between successive lateral branches absent or inconspicuous... *C. elongata*

3   Fronds up to 15(20)cm long; main axes dichotomous and densely pinnate; gaps between successive lateral branches absent or inconspicuous .... *H. squamatum*
    Fronds up to 5cm long; main axes dichotomous, sometimes also pinnate; gaps between successive lateral branches conspicuous   4

4   Branching strictly dichotomous; intergenicula rarely with distal lobes which, when present, are short (<150μm) and present only near base of plant ....
    .... *J. rubens* var. *rubens*
    Branching tending to become pinnate, especially in vegetative parts; intergenicula mostly with distal lobes up to 250μm long, sometimes bearing branches ....
    .... *J. rubens* var. *corniculata*

Tribe CORALLINEAE

Bases crustose, stolons rare; branching pinnate, or sometimes dichotomous or irregular; intergenicular medullary cells 50–90μm long, in >7 tiers per intergeniculum; genicular cells mostly of uniform length; conceptacles axial, marginal, lateral, or sometimes pseudolateral; spermatangial conceptacles broad (350-450μm), canals long (200–500μm or more); carpogonial conceptacles with extensive fertile areas each containing >200 supporting cells; carposporophytic fusion cells broad (90–300μm) and shallow (<12μm) with concave upper surfaces, gonimoblast filaments peripheral or arising from upper surface of fusion cell; tetrasporangial conceptacles containing >200 initials and, after maturation, >30 tetrasporangia, bisporangia occasionally present, sterile filaments present; never infected by *Choreonema*.

CORALLINA Linnaeus

CORALLINA Linnaeus (1758), p.805.

Type species: *C. officinalis* Linnaeus (1758), p.805 (see Schmitz, 1889).

Titanephlium Nardo (1834a), p.674.

Thallus consisting of a crustose base bearing erect, pinnately branched, geniculate fronds, stolons rare or absent; intergenicula terete to compressed, usually wedge-shaped; intergenicular medulla consisting of arching tiers of straight cells, genicula consisting of single tiers of long medullary cells; trichocytes round in surface view, pores central; most conceptacles axial, at maturity a single swollen conceptacle terminating each fertile intergeniculum, pores central, conceptacles sometimes also pseudolateral; gametangial plants dioecious, spermatangial conceptacle roofs beaked, lacking branches, carposporangial and tetra/ bisporangial conceptacles sometimes bearing branches.

Spermatangial conceptacle roofs more than 100μm thick, spermatangia simple, lining floor and sides of broad chambers with low ceilings; gonimoblast filaments mostly peripheral; spores germinating to form spreading crusts from which several fronds arise.

Most plants in the British Isles are clearly assignable to either *C. officinalis* or *C. elongata* although apparent intergrades occur in southern areas where the two species are sympatric. Successful discrimination between these two species often depends, however, on using several characters in combination (see key). We agree with Børgesen (1929, p.69), who stated that "...to point out any fixed character on which to rely does not seem possible." For example, one to several small branches consisting of one or more occasionally fertile intergenicula may be produced distally on a conceptacle. These branches have been called horns, antennae, or cornicles. Their occurrence is sometimes considered to be of taxonomic significance, and *C. elongata* was distinguished from *C. officinalis* (Areschoug, 1852, as *C. mediterranea*) by having such horns, but later workers have had difficulty in applying this character strictly (Ardré, 1970, and refs.). In *C. officinalis* the fronds are relatively stiff and tetrasporangial conceptacles usually lack branches, whereas in *C. elongata* they are limp and 20–80% of the tetrasporangial conceptacles bear branches. There is also a greater tendency for fused intergenicula to occur in branch tips in *C. elongata*. Hamel & Lemoine (1953) found that carposporangial conceptacles are more likely to bear branches, especially in *C. elongata*, and these branches are sometimes fertile.

Baba *et al.* (1988) and Johansen & Colthart (1975) studied species of *Corallina* in Japan and NW Altantic respectively, and found that whilst intergenicular shape (cylindrical to flat) and branching (pinnate to whorled) depended on environmental factors, the number of medullary tiers per intergeniculum provided a stable taxonomic character. Andrake, however (pers. comm.), found that the number of tiers was greater in longer intergenicula, with medullary cell length remaining constant. Bressan & Benes (1978), using plants from the Channel coast of France and the Mediterranean, attempted to use biometric characters including intergenicular size, shape and anatomy to differentiate between *C. elongata* and *C. officinalis*. *Corallina elongata* grows in warmer waters and is less widely distributed in the British Isles than *C. officinalis*.

Chromosome counts in both species of *Corallina* have determined that n=24 (Suneson, 1937; Magne, 1964; Yamanouchi, 1921).

Corallina elongata Ellis & Solander (1786), p.119.
　Figs 12B; 13C; 14A,B; Table 1

Lectotype: Ellis (1755), pl.24, fig. 3, in the absence of a specimen (see Greuter *et al.*, 1988, Art.7.4). Cornwall.

?*Corallina granifera* Ellis & Solander (1786), p.120.
Corallina mediterranea Areschoug (1852), p.568.

Illustrations (all as *Corallina mediterranea*): Ardré (1970), pl.7, figs 4–8; Bressan (1974), figs 4a–4b; Bressan & Benes (1978), fig.D; Cabioch (1966a), figs 1A-B; Cabioch (1966b), fig.A1; Cabioch (1968), pl.2, fig.9; Cabioch (1988), figs 1C,1D, 2C; Cabioch (1972), fig.6D, pl.2, fig.7, pl.12, fig.5; Gayral (1958), pl.71; Cabioch *et al.* (1992), fig.149; Gayral (1966), pl.92; Hamel & Lemoine (1953), pl.1, fig.3, pl.2, figs 1,2,6; Stegenga & Mol (1983), pl.72, fig.1.

Thallus with a firmly attached crustose base up to 15 or more cm in diameter bearing tufts of branched, limp fronds, up to 20cm long, *branching* usually dense, simply to compoundly pinnate, occasionally irregular, gaps between successive lateral branches absent or inconspicuous, resulting from narrow branch-angles combined with short intergenicula in the main axes; *fronds* consisting of compressed intergenicula which in the main branches are 0.6–1.0mm long and 0.4–0.8mm wide, tending to be nearly as wide as long; *conceptacles* axial, pseudolateral conceptacles unknown in the British Isles; carpo- and tetrasporangial conceptacles frequently bearing branches; *spermatangial conceptacles* (300)375–570μm outside diameter, *carposporangial conceptacles* 400–650μm outside diameter, *tetrasporangial conceptacles* 465–600μm outside diameter.

Intergenicular *medullary cells* 50–90μm long, in 7–12 tiers in each axial intergeniculum; *genicula* in main branches 140–210μm long and 100–250μm wide.

Spermatangial conceptacle chambers 185–390μm in diameter, 165–280μm high, with the roofs 110–415µm thick; *carposporangial conceptacle* chambers 250–360µm in diameter, fusion cell up to 180µm in diameter, 6–11µm thick, *carposporangia* 40–80µm diameter; *tetrasporangial conceptacle* chambers 150–350(400)µm diameter, 200–320µm high, with the roofs 80–150µm thick, *tetrasporangia* 95–225µm long.

Epilithic in pools and hanging from rock faces in both shady and well-illuminated damp places; lower littoral to upper sublittoral; recorded to 3m deep in the Canary Isles (L!).

Northwards at least to Anglesey. A report of *Haliptilon squamatum* from the Isle of Man by Knight & Parke (1931, as *Corallina*), was probably based on *C. elongata*. However, a report of *C. elongata* from the Outer Hebrides (Norton, 1972) was based on specimens of *C. officinalis* with a few branched conceptacles (BM!); similar plants are known from the Faroes and Iceland (BM!, C!). Eastwards to Kent; Channel Isles. In Ireland northwards to Mayo, eastwards to Waterford; Antrim. British Isles to Sénégal; Mediterranean; Canary Isles; Argentina.

Little is known about growth and reproductive periodicity. Haas *et al.* (1935, misidentified as *C. squamata*) described and illustrated the growth of Dorset fronds (BM!) from 3cm long in winter to 6cm in July, an average rate of 5mm/month, followed by much shedding and new growth in autumn; they recorded conceptacles throughout the growing season but did not illustrate them. Ardré (1970, as *C. mediterranea*) recorded tetrasporangia from February–July in Portugal.

There is considerable variation in frond length; this may be due to factors such as light intensity and exposure to wave action (Ardré, 1970). *C. elongata* has a tendency to produce expanded winglike intergenicular lobes near branch apices, as described by Mendoza (1976) for Argentina and Hamel & Lemoine (1953) for the Mediterranean.

C. elongata was specifically distinguished from *C. officinalis* by Ellis & Solander (1786) and has priority over *C. mediterranea*, a name which is widespread in the literature. Specimens of this species have often been recorded as *C. squamata* (e.g. Haas *et al.* 1935) (BM!) and, conversely, specimens described as *C. elongata* by Johnston (1842) (BM!) belong to *Haliptilon squamatum*. Sometimes a single name has been given to mixed gatherings of *C. elongata, H. squamatum* (e.g. Cotton, 1912, as *C. squamata*) (BM!) and even *C. officinalis* (BM!). All records and information in the literature must be treated with caution.

Although the original illustrations of *C. granifera* Ellis & Solander (1786, p.120, tab.21, fig. c,C) from the Mediterranean African coast show a plant closely similar to *C. elongata*, the name has often been treated as a synonym of *Haliptilon virgatum* (Zanardini) Garbary & Johansen (e.g. Ardré, 1970).

Reports of pseudolateral conceptacles in the Mediterranean (Bressan & Benes, 1978) require confirmation.

Cabioch (1966a) described spore germination and later (1966b) the crustose bases, which she found to be less extensive than in *C. officinalis*.

Yamanouchi (1921) described the life history of this species in the Mediterranean.

Titanoderma corallinae and *T. pustulatum* var. *confine* (*q.v.*) occur as epiphytes on *C. elongata*.

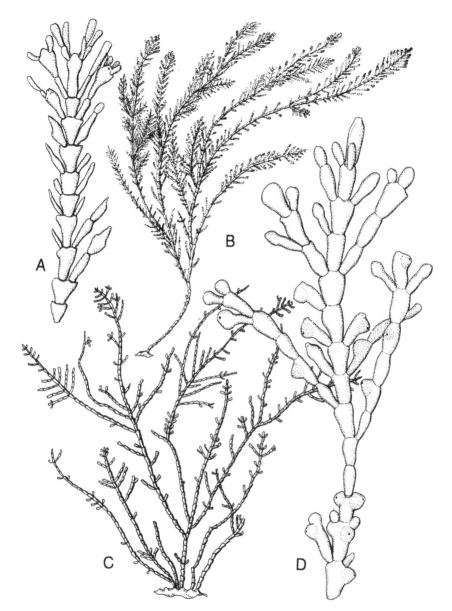

Fig. 14 *Corallina elongata* and *C. officinalis*. (A) Habit of *C. elongata* (March) ×2.
(B) Branch of *C. elongata* with axial conceptacles bearing one or more surmounting branches
(March). ×2. (C) Habit of *C. officinalis* (March) ×2. (D) Branch of *C. officinalis* with axial
and lateral conceptacles (March) ×8.

TABLE 1. Characteristics distinguishing *Corallina elongata* from *Haliptilon squamatum*

CHARACTER	*C. elongata*	*H. squamatum*
Primary branching	Pinnate	Dichotomous, 1–6 dichotomies per plant
Medullary tiers/axial intergeniculum	8–12	5–9
Medullary tier height	50–90μm	90–120μm
Holdfast	Crustose	Stoloniferous
Spermatangial conceptacles, chamber diam.	185–390μm	90–130μm
Carposporophytic fusion cell, depth	Shallow (6–11μm)	Deep (20–35μm)
Tetrasporangial conceptacles	Sometimes branched	Always branched
Number of tetrasporangia per conceptacle	>30	<10
Tetrasporangial length	95–225μm	200–260μm

Corallina officinalis Linnaeus (1758), p.805.
Figs Front cover; 12A; 13A,B; 14C,D; 15A,B; 18A

Lectotype: LINN 1293.9. (!). Manza's (1940) citation of BM is incorrect. 'In O[ceano] Europaeo'.

Illustrations: Andrake & Johansen (1980), figs 1–2; Baba *et al.* (1988), figs 2,3,5; Borowitzka & Vesk (1978), figs 1–16; Bressan (1974), fig.1; Bressan & Benes (1978), fig.E; Cabioch (1971a), pl.2, fig.4, fig.3D, fig.8C; Cabioch (1968), pl.1, figs 7–8; Cabioch (1966b), figs A2–4; Cabioch (1972), fig.7A, figs 35,36C, pl.12, figs 6–8; Cabioch (1966a), fig.2; Colthart & Johansen (1973), figs 1–2; Garbary (1978), fig.1; Garbary & Johansen (1982), figs 2,3; Gayral (1958), pl.70; Gayral (1966), pl.91; Giraud & Cabioch (1976), figs 9–11; Hamel & Lemoine (1953), pl.1, fig.1; Harvey (1849b), pl.222; Johansen (1969), fig.28A, pl.17A; Johansen (1970), figs 1,3,5,6,9,11,13; Johansen & Colthart (1975), figs 3–10; Johnstone & Croall (1859), pl.37; Kornmann & Sahling (1977), fig.114; Kylin (1944), pl.9, figs 29–32; Levring (1937), pl.3, fig.5; Matty & Johansen (1981), figs 1–14; Rosenvinge (1917), figs 192–197, pl.4, figs 5–8; Stegenga & Mol (1983), pl. 72, fig.2; Suneson (1937), figs 18–22, pl.1, figs 1–4; Suneson (1943), figs 40–43.

Thallus with a firmly attached crustose base up to 70mm in diameter (LMI. obs.) bearing tufts of branched, stiff, usually erect fronds up to 120mm long, *branching* dense to sparse, simply to compoundly pinnate but often irregular, successive lateral branches separated by conspicuous gaps resulting from wide branch-angles combined with long intergenicula in the main axes, *fronds* consisting of unlobed intergenicula which in the main branches are 1–2(4)mm long and 0.5–1(1.5)mm wide, tending to be longer than broad, cylindrical to compressed, especially near genicula. *Conceptacles* axial, 1–3 pseudolateral conceptacles per intergeniculum sometimes also present, protruding markedly; axial conceptacles occasionally bearing branches in carposporangial and tetra/ bisporangial plants; *spermatangial conceptacles* beaked, 390–600µm outside diameter; *carposporangial conceptacles* 390–540µm outside diameter, *tetra/ bisporangial conceptacles* 450–600µm outside diameter.

Intergenicular *medullary cells* 50–90µm long, in (9)12–20(25) tiers in each intergeniculum in the main axes; *genicula* in main branches 180–350µm long and 200–350µm diameter.

Spermatangial conceptacle chambers 250–480µm in diameter, 100–300µm high, with the roofs 200–600µm thick; *carposporangial conceptacle* chambers 250–400µm diameter, 300–320µm high, with the roofs 150–240µm thick; fusion cells 90–200µm diameter, 6–11µm thick, *carposporangia* 30–80µm diameter; *tetra/ bisporangial conceptacle* chambers 300–550µm diameter, 300–500 (600)µm high, roofs (50) 80–240 (280)µm thick, *tetrasporangia* 110–225µm long, 40–60µm in diameter, spores after release (30) 40–80µm in diameter.

Epilithic, or occasionally on mollusc shells or macroalgae, such as non-geniculate corallines, *Furcellaria*, and *Fucus* (Rosenvinge, 1917; Suneson, 1937); in pools and damp places, often as a turf, throughout the littoral, carpeting channels issuing from pools or hidden by overlying vegetation (Goss-Custard *et al.*, 1979; Knight & Parke, 1931); sublittoral in scattered clumps to at least 18m (29m in Europe, see Rosenvinge, 1917 and 33m off eastern North America, see Vadas & Steneck, 1988); especially prevalent in areas exposed to moderate to heavy wave action. At a depth of 12m in Port Erin Bay 5% of the total flora was attached *Corallina* (Burrows, 1958).

Generally distributed throughout the British Isles.

Norway (Finnmark) to Morocco, limit in W Africa uncertain (see Price *et al.*, 1986); Iceland; Faroes; Helgoland; W. Baltic; Mediterranean. Greenland; Arctic Canada (Adlavik Is. 55°N) to USA (South Carolina); in Caribbean upwelling areas of Venezuela and Colombia (Schnetter & Richter, 1979, as *C. pannizoi*); Argentina. Widely reported elsewhere, especially in temperate areas. An entity questionably assigned to *C. officinalis* as var. *chilensis* (Decaisne) Kützing occurs on the west coast of North America.

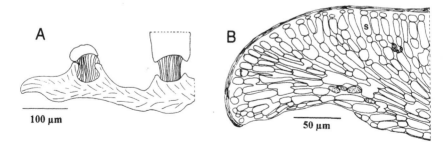

Fig. 15 *Corallina officinalis.* (A) Diagrammatic drawing of VS of thallus base showing origin of two erect branches (YMC 93/45). (B) Crustose thallus margin; note terminal initials (t) and subepithallial initials (s) (YMC 93/45).

Plants with perennial vegetative bases producing caducous reproductive fronds (Berner, 1979). There is some evidence (LMI obs.; Rosenvinge, 1917) for a pattern of frond initiation in autumn followed by growth and reproduction, with senescence and frond loss the following autumn, although many fronds overwinter under suitable conditions. Gametangial conceptacles are rare for the British Isles (Buffham, 1888, BM!; Knight & Parke, 1931; Blackler, 1956) and surrounding areas (Rosenvinge, 1917; Suneson, 1943; Hamel & Lemoine, 1953). Tetrasporangia have been recorded throughout the year; Blackler (1956) reported bisporangia for June, November-December. Jaasund (1965) reported cystocarps and tetrasporangia June-August in Norway.

Matty & Johansen (1981) and Colthart & Johansen (1973) outlined the development of intergenicula and genicula in an axis and showed that, at 12h light:12h dark daylength, optimum growth (about 2mm/month, 1 intergeniculum/12 days, 1.5 tiers/day) occurred between 12–18°C, most plants dying at 25°C. This correlates with rapid growth reported for NW Atlantic (New England) in spring and autumn (Conover, 1958); such growth presumably continues in the lower summer temperatures around the British Isles (see Hoek, 1982a, fig.5, 1982b). The reduced growth rate occurring at 6°C was less marked at lower light intensities (cf. winter conditions in the British Isles); Rosenvinge (1917) reported no winter growth in Denmark, i.e., below 5°C. In N. Japan Baba *et al.* (1988) reported perennial fronds which increased in length and dry weight when they became fertile in April to November; senescence occurred in October, fronds becoming rough and discoloured with remaining conceptacles white or broken.

Corallina officinalis is a very common species which adapts to a wide range of habitats. Descriptions of rocky shore ecology (e.g. Evans, 1957, in Atlantic France) have even depicted '*Corallina* zones' (*C. officinalis*). The fronds vary in morphology with contrasting forms side by side or intermingled; in general, however, stunted specimens occur in high pools (Knight & Parke, 1931), much-branched vigorous forms (f. *vulgaris* Kützing, 1858) in lower littoral and thick elongate fronds with few short branches in the sublittoral, especially in flowing water (f. *profunda* Farlow, 1881; f. *robusta* Kjellman, 1883). A cushion-like turf, typical of areas exposed to wave action, was described as f. *compacta* (P.& H. Crouan) Batters (1902) (CO!) and the frequent occurrence of curved branches with secondary attachments was noted by Rosenvinge (1917). Maggs *et al.* (1983) recorded a sublittoral turf of *C. officinalis* in County Cork. In Norway a low littoral turf of fronds 1–2cm long contrasts with those in pools 10–17cm long (Jaasund, 1965). This morphological variability has resulted in the use of many names that are probably heterotypic synonyms of *C. officinalis*; examples are *C. laxa* and *C. longicaulis*, both described by Lamarck (1815) from the north and south coasts of France, respectively, and *C. hemisphaerica* Foslie (1887) from Norway. Verification of these awaits location and study of type material within the context of a monograph of the genus *Corallina*.

In New England Andrake (pers.comm.) found moderate water movement necessary for optimum growth; in pools and sublittoral areas plants have long intergenicula, but they become shorter and more branched with greater exposure to wave action. Water motion may promote multiple geniculum formation at apices, producing up to seven coplanar or verticillate branches (Rosenvinge, 1917); intergenicular flattening, possibly due to wave action, would favour a coplanar organization, but hydrodynamic experimentation is required.

Gametangial plants are usually smaller and more irregular in shape than tetrasporangial plants, and bear densely crowded conceptacles (Rosenvinge, 1917); many irregularities also appear to be induced by animal occupants (molluscs, crustacea; pers. obs.; Dommasnes, 1968). Fist-sized ball forms have been found in Loch Sween and Ballynakill Harbour (Hiscock and Maggs, pers. comm.). Von Stosch (1962) and Johansen & Colthart (1975) have noted that plants in laboratory culture were strikingly different from their counterparts in the field. The latter, working in New England, USA, found that specimens in culture had shorter and narrower intergenicula, shorter medullary cells, thinner cortices, and shorter genicula.

Corallina officinalis has been the subject of many studies on anatomy (Rosenvinge, 1917; Suneson, 1937), cytology, growth, and development (Cabioch, 1971a, 1972), ultrastructure (Giraud & Cabioch, 1976, 1977; Borowitzka & Vesk, 1978), and calcification (e.g. Digby, 1977a, 1977b, 1979; Pentecost, 1978). As early as 1876, Nelson & Duncan carefully removed a layer of epithallial cells from an intergeniculum; these shallow polygonal cells resembled a 'beautiful pavement epithelium.' More recently SEM has revealed the nature of the epithallial layer (Garbary & Johansen, 1982; Fig.18A).

Peel *et al.* (1973) and Peel & Duckett (1975) studied fertile material in Anglesey and found elongate spermatangial initials (30–40µm long), the spermatangia shortening after release with the excess coat material forming a 'tail.' During tetrasporogenesis each post-meiotic nucleus becomes surrounded by a unique nuclear endoplasmic reticulum which may be involved with ribosome assembly in cellular reorganization and enlargement as the tetrasporangia mature (Peel *et al.*, 1973).

Jones & Moorjani (1973) found that an average of 25 tetraspores per conceptacle were released in 12h and became attached within 48h; they were calcified by the 4-celled stage and produced bases which grew most quickly (3.6µm/day) at 17–20°C; frond initiation occurred after 3 weeks, much more slowly than in *Jania rubens*, *q.v.*, and the first intergeniculum appeared after 13 weeks (see also Cabioch, 1966b).

Munda (1977) contrasted the occurrence of *C. officinalis* in N. Adriatic as a belt of compact cushions on steep slopes with strong wave action, with its discontinuous occurrence in Iceland, where it was restricted to pools and gentle slopes with relatively high winter temperatures. Colthart & Johansen (1973) noted that this species has been recorded for many places where temperatures exceed 20°C (e.g., W. Atlantic, Taylor, 1960); these reports either represent another species, or races with different temperature requirements. Novaczek & McLachlan (1986), studying recolonization by algae following the demise of sea urchins off Nova Scotia, found that *C. officinalis* became a dominant understorey species down to a depth of 8m.

Littler & Kauker (1984) demonstrated the adaptive significance of heterotrichy (in S. California, probably *C. officinalis* var. *chilensis*). They suggested that the crustose bases evolved for increased persistence in the face of herbivory and the fronds for greater reproductive capacity. The frondless bases of *C. officinalis* can be a significant component of crustose coralline cover in shallow water, as noted for Rhode Island (Harlin & Lindbergh, 1977); they resemble nongeniculate corallines, but scars of dehisced fronds are often visible (pers. obs.; Rosenvinge, 1917, fig.192). Harlin & Lindbergh (1977) also studied substrate preferences of *C. officinalis* in New England, using variously textured artificial surfaces; optimum settlement and growth occurred on finely roughened surfaces (0.1–0.5mm particle diam.).

Numerous non-coralline epiphytes as well as the coralline epiphytes *Mesophyllum lichenoides q.v.* and *Titanoderma pustulatum* var. *confine* and *T. corallinae q.v.*, the last causing tissue destruction (Cabioch, 1979, as *Dermatolithon corallinae*), grow on *C. officinalis*. The report of infection by *Choreonema thuretii* (Bressan, 1974) in the Mediterranean requires confirmation.

Coull & Wells (1983) found that *Corallina* provided a better refuge for meiofauna (mainly copepods) than other algae tested, probably because its complexity and rigidity provided more niches (see also Dommasnes, 1968, 1969 and Goss-Custard *et al.*, 1979).

Although traditionally used as a vermifuge since the time of Theophrastus (Berner, 1979), Blunden (pers. comm.) has found no effective ingredients.

Tribe JANIEAE Johansen & Silva (1978), p.414.

Bases crustose, often ephemeral and augmented by stolons; primary branching dichotomous; intergenicular medullary cells 90–130μm long, in 1–12(–19) tiers per intergeniculum; genicular cells longer in the centre than at the periphery; conceptacles exclusively axial; spermatangial conceptacles narrow (90–250μm), canals short (30–120μm); carpogonial conceptacles with limited fertile areas each containing <200 supporting cells; carposporophytic fusion cells narrow (40–130μm) and deep (up to 35μm) with convex upper surfaces, gonimoblast filaments strictly peripheral; tetrasporangial conceptacles containing <200 initials and, after maturation, <10 mature tetrasporangia, bisporangia occasional in Australia (HWJ obs.), sterile filaments absent; sometimes infected by *Choreonema*.

HALIPTILON (J.Decaisne) J.Lindley

HALIPTILON (J.Decaisne) J.Lindley (1846), p.25.

Type species: *H. cuvieri* (Lamouroux) Johansen & Silva (1978), p.417.

Jania sect. *Haliptilon* J.Decaisne (1842), p.123 (reprint p.111), (as *Haliptylon*).

Thallus initially consisting of a crustose base which later becomes obscured by stolons or lost, bearing branched fronds, primary branching dichotomous, secondary branching pinnate, secondary branches often pinnately and/or dichotomously branched, adventitious branches sometimes present, intergenicula terete to compressed and wedge-shaped; intergenicular medulla consisting of arching tiers of straight cells of equal length, genicula consisting of single tiers of long unbranched medullary cells; trichocytes elongate in surface view, pores excentric; all conceptacles axial, at maturity a single swollen conceptacle terminating each fertile intergeniculum, spermatangial conceptacle elongate, lacking branches, carpogonial, carposporangial, and tetrasporangial conceptacles bearing branches.

Gametangial plants monoecious, spermatangial conceptacle roofs less than 100μm thick, spermatangia lining a long narrow chamber; carpogonial conceptacles each with fewer than 100 supporting cells; tetrasporangial conceptacles with fertile areas limited, each with fewer than 100 sporocytes, fewer than 10 mature sporangia.

Haliptilon squamatum, the most robust species in the genus, is essentially subtropical and tropical and the British Isles appears to be its northern limit. *H. virgatum* (Zanardini) Garbary

& Johansen, common in Portugal (Ardré, 1970) and the Mediterranean (Hamel & Lemoine, 1953), has been erroneously reported for the British Isles (Batters, 1897; Lyle, 1937; Blackler, 1951, as *C. granifera*, BM!: misidentifications of *Jania rubens* and *Corallina* spp.). There are no chromosome counts for *Haliptilon*.

Haliptilon squamatum (Linnaeus) Johansen, L.Irvine & Webster (1973), p.212.
Figs 12E; 13E,F; 15A,B; Table 1.

Lectotype: Ellis (1755), pl.24, fig.C. 'in O [ceano] Europaeo'.

Corallina squamata Linnaeus (1758), p.806.

Illustrations: Cabioch (1980), fig.A; Cabioch (1966b), fig.B (as *Corallina squamata*); Cabioch (1972), figs 5D,7B,36A-B, pl.12, figs 1–4 (as *Corallina squamata*); Gayral (1958), pl.72 (as *Corallina squamata*); Gayral (1966), pl.93; Hamel & Lemoine (1953), pl.1, fig.2, pl.2, fig.5 (as *Corallina squamata*); Harvey (1849b), pl.201 (as *Corallina squamata*); Johnstone & Croall (1859), pl.38 (as *Corallina squamata*); Cabioch *et al.* (1992), fig.148.

Thallus attached by a small, inconspicuous crustose base and a mass of entangled stolons producing secondary adhesion discs 500–750μm diameter, bearing tufts of branched, relatively stiff, *fronds* up to 20cm long, main axes sparsely dichotomously branched, lateral branches arising from most intergenicula except near base of frond, simple to pinnate; main axis intergenicula compressed, becoming progressively broader in upper parts of branches, often appearing sagittate because of small lobes below insertions of lateral branches, occasionally producing adventitious branchlets, intergenicula 0.6–1.0mm long and 0.6–1.0mm broad across the bases of the lobes; lateral intergenicula often as long as main axis intergenicula but only 230–330 um broad. Gametangial plants monoecious; *spermatangial conceptacles* rare, outside diameter not recorded, *carposporangial conceptacles* 300–600(750)μm outside diameter, *tetrasporangial conceptacles* 420–750μm outside diameter.

Intergenicular *medullary cells* 90–120μm long, in (4)5–8(9) tiers in each main axis intergeniculum; *genicula* 110–240μm long, 240–300μm diameter in main axes, 120μm diameter or less in lateral branches.

Spermatangial conceptacle chambers 90–130μm diameter, 300–400μm high, with the roof less than 80μm thick; *carpogonial conceptacle* chambers 150μm diameter, 180μm high, *carposporangial conceptacle* chambers 200–450μm in diameter, 300–360μm high, with the roofs 60μm thick; fusion cells (45)60–80μm diameter, 20–35μm thick, *carposporangia* 70–90μm diameter; *tetrasporangial conceptacle* chambers 250–420μm diameter, 360–450μm high, with the roofs 90–120μm thick, *tetrasporangia* 200–260μm long, 100–130μm diameter, spores after release 60–90μm diameter.

Epiphytic, usually on *Cystoseira*, and occasionally epilithic; in protected niches such as deep clear pools in areas exposed to strong wave action, upper sublittoral to 3m (Webster, pers. obs.).

Recorded northwards to Isle of Man, eastwards to Sussex, Channel Isles; in Ireland northwards to Mayo, eastwards to Waterford.
British Isles to Sénégal; Mediterranean; Canary Isles.

Little information is available on growth, frond longevity and reproduction; spermatangia and carposporangia recorded for November, tetrasporangia for September and March (BM!). Hamel & Lemoine (1953) reported gametangial plants in December in Cherbourg and Ardré (1970) found conceptacles in March and June in Portugal.

This species is less variable in external appearance than the British species of *Corallina*. When infected by *Choreonema thuretii*, *q.v.*, branches can be reduced in size and resemble *Jania rubens* (see Cabioch, 1980).

There are no specimens of *H. squamatum* in the Linnaean Herbarium; Ellis's (1755) illustration, based on a plant from Cornwall, can be chosen as the lectotype (Greuter *et al.*, 1988, Art. 7.4) since it was cited by Linnaeus (1758); it shows an axis with characteristic intergenicula having about seven medullary tiers per intergeniculum.

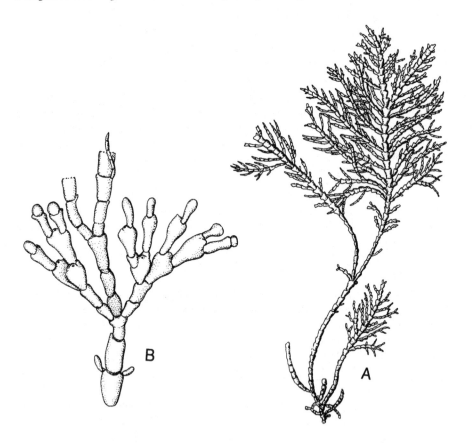

Fig. 16 *Haliptilon squamatum*. (A) Habit showing branched base and dichotomous primary branching (June) ×2. (B) Branch with concatenate tetrasporangial conceptacles (February) ×8.

Cabioch (1966b, 1971b, 1972, as *Corallina squamata*) included comparative cytological and developmental information for this and other coralline species. Spores germinate to form small pulvinate crustose bases from which 1–2 fronds arise.

Haliptilon squamatum superficially resembles *Corallina elongata*, but differs in reproductive characteristics. Spermatangial conceptacles are narrower, longer, and have short canals, fusion cells are narrower and deeper, tetrasporangia are longer, and fewer than ten tetrasporangia are present in each conceptacle. Also, fronds of *H. squamatum* have occasional dichotomies and crustose bases are small and augmented or replaced by stolons (Cabioch, 1972, p.254, fig.36B).

Distribution within the British Isles appears to be more restricted than previous records suggest; this is because of a series of misidentifications resulting from the misapplication of the name *Corallina squamata* to specimens of *C. elongata* (BM!) and *vice versa* (Johnston, 1842); for differences between the two species see Table 1 under *C. elongata*.

Epiphytes of this species include *Titanoderma corallinae*, *q.v.*; *Choreonema thuretii* (Bornet) Schmitz *q.v.* is a semiendophyte producing conceptacles which are visible on the intergenicular surfaces.

JANIA Lamouroux

JANIA Lamouroux (1812), p.186 [as *Iania*].

Lectotype species: *Jania rubens* (Linnaeus) Lamouroux (1816), p.272 (see Manza, 1937, p.47).

Thallus consisting of a crustose base with or without stolons and erect, dichotomously branched fronds, lateral and/or adventitious branches sometimes also present, intergenicula terete to subterete, or wedge-shaped, or with distal lobes; intergenicular medulla consisting of arching tiers of straight cells, genicula consisting of single tiers of long unbranched medullary cells; trichocytes elongate in surface view, pores excentric; all conceptacles axial, at maturity a single swollen conceptacle terminating each fertile intergeniculum, but often appearing in a dichotomy because of surmounting branches; gametangial plants monoecious or dioecious, spermatangial conceptacles elongate, lacking branches, carposporangial and tetrasporangial conceptacles bearing branches.

Spermatangial conceptacles with roofs less than 100μm thick, spermatangia lining long narrow chambers; carpogonial conceptacles each with fewer than 100 supporting cells, tetrasporangial conceptacles each with fewer than 100 sporocytes, fewer than 10 mature sporangia.

One species, *J. rubens*, with two varieties (var. *rubens* and var. *corniculata*) has been recorded from the British Isles. Intergenicula of *J. rubens* var. *rubens* are cylindrical to subcylindrical with distal lobes rare and slight (see Harvey, 1849b, pl.252). In contrast, large collections from the British Isles and surrounding areas (BM!) contain many plants with pronounced distal lobes regularly present on wedge-shaped vegetative intergenicula, the lobes sometimes being surmounted by secondary branches so that the fronds appear pinnate; such plants are typical of *J. rubens* var. *corniculata* (see Harvey, 1849b, pl.234, as *J. corniculata*). Harvey pointed out, however, that the 'articulations of lesser branches and

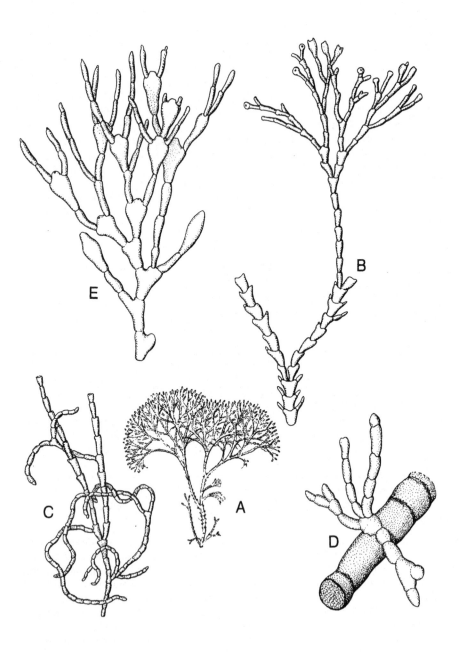

ramuli [remained] cylindrical' as did those associated with conceptacles in our material. Because of the differences between parts of the same plant, and also because of occasional plants with intergenicula of both kinds in the main branches, we have refrained from considering *J. rubens* var. *corniculata* as a distinct species. The influence of environment on intergenicular shape and branching pattern is not known and further studies are needed.

Plants with lobed intergenicula and a tendency towards pinnate branching appear to be confined to the NE Atlantic. A Pacific species described and illustrated by Dawson (1953, p.115, pl.9, fig.2) as *J. subpinnata* appears to be a *Haliptilon, q.v.*, as the pinnate branches are not subtended by distal intergenicular lobes.

Chromosome counts for *J. rubens* revealed n=24 (Suneson, 1937).

Jania longifurca Zanardini has been reported as far north as Brittany (Dizerbo, 1969) and might possibly be encountered in SW Britain; it is a robust species with intergenicula up to 500μm diameter at the base (Hamel & Lemoine, 1953).

Choreonema thuretii is a semi-endophyte, sometimes producing conceptacles on intergenicular surfaces of both varieties of *J. rubens*.

Jania rubens (Linnaeus) Lamouroux (1812), p.186.
Jania rubens var. **rubens**
Figs 12C; 13D; 17C,D,E

Lectotype: UPS (Herb. Burser vol. XX, p.72).

Type locality: 'in Oceano Europaeo'.

Corallina rubens Linnaeus (1758), p.805.

Illustrations: Bressan (1974), fig.9; Cabioch (1971a), fig.13A; Cabioch (1980), fig.B; Cabioch (1966b), fig.C; Cabioch (1972), figs 6C,7C; Garbary (1978), fig.4; Gayral (1958), pls 74,75; Gayral (1966), pls 94,95, fig.47C; Hamel & Lemoine (1953), pl.3, figs 1–2; Harvey (1849b), pl.252; Johnstone & Croall (1859), pl.39; Jones & Moorjani (1973), fig.1; Kylin (1944), pl.9, fig.33; Newton (1931), fig.191 (as *Corallina rubens*); Rosenvinge (1917), figs 198–199 (as *Corallina rubens*); Suneson (1937), figs 23–27, pl.2, figs 5–9 (as *Corallina rubens*); Cabioch et al. (1992), fig.150.

Thallus with a small, firmly attached crustose base up to 600μm in diameter, sometimes becoming obscured by stolons, bearing tufts of *fronds* up to 25mm long, *branching* dichotomous at intervals of 1–2 intergenicula, intergenicula terete to subterete, unlobed or with occasional small lobes on lowermost intergenicula, intergenicula 350–1000(1300)μm long and (60)90–240μm diameter, up to 350μm diameter in basal parts; stolons as little as 50μm diameter, adventitious, straight or contorted. Gametangial plants monoecious;

Fig. 17 *Jania rubens*. (A) Habit of plant of *J. rubens* var. *corniculata* (April) ×2. (B) Branch of *J. rubens* var. *corniculata* with young conceptacles (April); note lower genicula with distal lobes, upper ones unlobed ×8. (C) Mature branch of *Jania rubens* var. *rubens* with stolon-like adventitious branchlets (April) ×8. (D) Young plant of *J. rubens* var. *rubens* on *Cladostephus* (November) ×35. (E) Branch of *J. rubens* var. *rubens* with mature spermatangial conceptacles (lacking surmounting branches) and carposporangial conceptacles (with surmounting branches) in concatenate series (July) ×20.

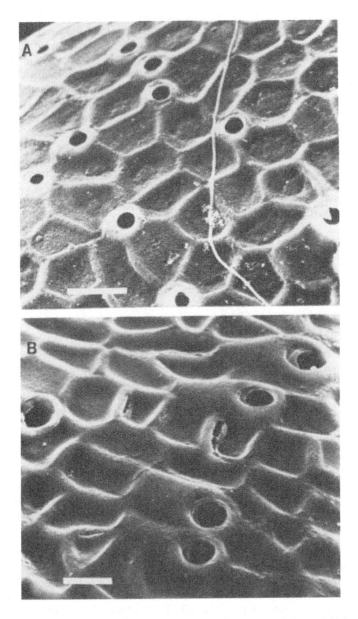

Fig. 18 SEMs of intergenicular surfaces showing epithallial concavities and trichocytes with pores from which hairs dehisced. (A) *Corallina officinalis*. Scale bar = 15μm. (B) *Haliptilon* sp. Scale bar = 10μm. [From Garbary & Johansen, 1982].

spermatangial conceptacles 180–225 (325)µm outside diameter, *carpogonial conceptacles* 200–225µm outside diameter, *carposporangial conceptacles* 315–410(600)µm outside diameter, *tetrasporangial conceptacles* 375–500(600)µm outside diameter.

Intergenicular *medullary cells* (80)100–170µm long, in 4–5(6) tiers in each intergeniculum; *genicula* (80)90–140(160)µm long and (70)80–120µm diameter, ultimate genicula 45–75µm diameter.

Spermatangial conceptacle chambers 110–150(175)µm diameter, 170–225(300)µm high, with the roofs 30–35(50)µm thick; *carpogonial conceptacle* chambers 120–150µm in diameter, 110–115µm high, with the roofs 35–115µm thick; *carposporangial conceptacle* chambers 150–200µm in diameter, 185–260µm high, with the roofs 35–75µm thick; fusion cells (30)40–100µm in diameter, 8–35µm thick, *carposporangia* (20)40–60(90)µm diameter; *tetrasporangial conceptacle* chambers (150)225–340µm diameter, 260–340µm high, with the roofs 35–75µm thick, *tetrasporangia* 140–225µm long, 35–55(70)µm diameter, spores after release (50)60–80(100)µm diameter.

Epiphytic, mainly on *Cladostephus, Cystoseira,* and *Halopithys* (Duerden & Jones, 1981) in British Isles, occurring on other hosts and frequently epilithic further south (Børgesen, 1929; André, 1970); upper sublittoral to 8m (Maggs *et al.,* 1983; 20m in Norway, Åsen, 1976); known from a wide range of habitats from extreme shelter to pools exposed to strong wave action.

In the British Isles northwards to W. Inverness, Aberdeen (report for Orkney not confirmed); in Ireland from Cork to Mayo; Channel Isles.

Norway (Hordaland) to Cameroun; W. Baltic; Mediterranean; Azores; Canary Isles. In the W. Atlantic from USA (North Carolina) to Argentina.

Field and herbarium (BM!) studies suggest an annual cycle of development, the young fronds arising in autumn, overwintering and becoming fertile in April-May; more conceptacles may be produced successively until late autumn when most fronds are shed. Some plants regenerate from lateral or adventitious branches, apparently especially in warm water areas, and produce large tangled masses. Jones & Moorjani (1973) reported that, in 12h light: 12h dark daylength, plants could grow between 10–30°C, most spores failing at 10°C; between 17–20°C 98% spores attached in 4.5h and crustose bases were complete in 6 days, with the first intergeniculum appearing about 48h later. Plants became fertile between 11–17°C and overwintered as adults.

In Anglesey, Jones & Moorjani (1973) found tetrasporangial plants predominant, but in general gametangial plants appear equally common; they are monoecious with spermatangial conceptacles unbranched and longer than carposporangial. According to Jayasinghe (pers. comm., 1982), fertilisation occurs at the cuplike stage, the conceptacle walls then growing up to surround the developing carpospores. Sixty percent of carpo- and tetrasporangial conceptacles bear 2 branches and 40% bear 1 or 3 or more; spores are discharged 30–35 days after conceptacle initiation. Surmounting branches produce further conceptacles in a repeated sympodial series of up to about 5. Branching varies from fastigiate to divaricate, producing in summer the familiar corymbose tufts (Fig.17C) held together by entangled branches, secondary holdfasts, or stolons; these sometimes detach with or without host fragments and persist as balls in the drift. Slender, elaborately curled branches, or even complete fronds, are sometimes produced adventitiously from the bases or fronds; they may function as propagules or entanglement structures (Cabioch, 1972).

The lectotype was chosen by one of us (LMI) after a study of specimens available to Linnaeus at the time of publication of the binomial *Corallina rubens* (Linnaeus, 1758). Material in the Burser Herbarium (UPS) supporting Bauhin's polynomial synonyms quoted by Linnaeus seemed the most appropriate.

Jania rubens has been reported and/or studied by many researchers. Anatomical details have been elucidated by Suneson (1937), Cabioch (1971a, 1972), Giraud & Cabioch (1976, 1977), and others.

Von Stosch (1969) reported that although the species is monoecious, fertilization experiments with cultures failed, although carpospores were produced apparently apomictically in one conceptacle. He also found (von Stosch, 1969) that *J. rubens* grew best with only 25% as much phosphate as required by other marine algae.

Thuret & Bornet (1878), Chemin (1937), and Cabioch (1966b), described and/or illustrated sporelings; Cabioch (1966b) noted that young sporelings often became detached from the substrate only to become entangled in nearby host branches where they became firmly anchored by stolons.

Jania rubens var. *corniculata* (Linnaeus) Yendo (1905), p.38.

Figs 12D; 17A,B

Lectotype: LINN 1293.19. 'in Oceano Europaeo'.

Corallina corniculata Linnaeus (1758), p.806.
Jania corniculata (Linnaeus) Lamouroux (1812), p.186.
Jania plumula Zanardini (1844) p.1025 (reprint p.21). (VEN!)
Corallina elegans Lenormand in Areschoug (1852), p.570. (CN!) non Kützing (1849), p.707, nom. illeg.
Jania nitidula Meslin (1976), p.418.

Illustrations: Bressan (1974), fig.10 (as *Jania corniculata*); Harvey (1849b), pl.234 (as *Jania corniculata*); Garbary (1978), figs 2,3 (as *Jania corniculata*); Hamel & Lemoine (1953), pl.3, figs 4–5 (as *Jania corniculata*), pl.4, fig.1 (as *Corallina elegans*); Johnstone & Croall (1859), pl. 40 (as *Jania corniculata*).

Thallus with a small crustose base up to 600μm diameter (sometimes becoming obscured by stolons), bearing tufts of *fronds* up to 25mm long, *branching* dichotomous at intervals of 1–6 intergenicula, intergenicula terete to subterete, or wedge-shaped, shoulders of intergenicula usually with distally projecting lobes up to 250μm long, sometimes bearing branches, intergenicula 450–900μm long and 120–440μm wide below lobes, breadth greater in basal parts; gametangial plants monoecious; *spermatangial conceptacles* 170–240(290)μm outside diameter, *carpogonial conceptacles* 330–375μm outside diameter, *carposporangial conceptacles* 250–400μm outside diameter, *tetrasporangial conceptacles* 260–375μm outside diameter.

Intergenicular *medullary cells* 75–150μm long, in 4–5(6) tiers in each intergeniculum; *genicula* 110–180μm long and 80–160μm diameter, ultimate genicula narrower.

Spermatangial conceptacle chambers 75–140μm diameter, 200–260μm high, with the roofs less than 80μm thick; *carpogonial conceptacle* chambers 80–110μm diameter; *carposporangial conceptacle* chambers 150–285μm diameter, 170–300μm high, with the roofs 35–70μm thick; fusion cells 60–80μm diameter, 10–20μm thick, *carposporangia* 35–75 × 65–100μm diameter; *tetrasporangial conceptacle* chambers 185–225 (250)μm diameter, 225–300µm high, with the roofs 150µm thick, *tetrasporangia* (140)180–220µm long, 50–60µm diameter, spores after release 50–100, mostly 60–80µm diameter.

The habitat appears to be the same as for *J. rubens* var. *rubens*.

Northwards to Anglesey and Norfolk; Channel Isles; in Ireland from Cork to Mayo; Dublin.
Sweden to Portugal; Mediterranean; Canary Isles. Suneson (1943) and Børgesen (1929) found very few specimens tending towards J. rubens var. corniculata in Sweden and the Canary Isles respectively.

Field and herbarium (BM!) studies showed that young plants appear in early autumn and overwinter; spermatangial conceptacles develop in spring, followed by carposporangial and tetrasporangial conceptacles, and reproduction reaches a climax in summer with the conceptacles often becoming concatenate.

Variation in overall appearance of plants included in this variety is similar to that described for *J. rubens* var. *rubens*, *q.v.* The lobed intergenicula characteristic of this variety are restricted to sterile parts of the fronds; they are easily seen in autumn and winter but are obscured by the fertile distal parts of fronds in spring and summer. Intergenicula in fertile parts are usually more cylindrical and unlobed, resembling those of *J. rubens* var. *rubens*. Ellis (1755, pl.24, fig.7) illustrated this variation: fig.E shows unlobed intergenicula forming the distal part of a sterile frond which could belong to either variety. Fig.D is typical of var. *corniculata* with lobed intergenicula below and unlobed intergenicula in the upper (younger) parts which are not yet fertile. Fig.F (quoted under *Corallina cristata* Linnaeus, 1758) is the distal part of a frond of either variety with one set of conceptacles subtending a second immature set, whilst fig.G (quoted under *Corallina spermophoros* Linnaeus, 1758) shows a concatenate later stage. In older parts, and occasionally also distally on sterile fronds, lobes may bear branches giving a pinnate appearance; similar plants are apparently widespread in Europe (Rosenvinge 1917, p.275; *J. plumula* Zanardini, 1844; Dizerbo, 1969; Furnari & Scammacca, 1971; Giaccone, 1969); all such fronds are sterile, however, and probably result from environmental influences. *J. nitidula* Meslin (1976) (=*Corallina elegans* Lenormand in Areschoug, 1852, p.570, CN! see Kützing, 1858, pl.87, fig.I), which has pinnate branches that are di- or tri-chotomously branched, was reported by Meslin from Brittany growing on *Zostera* rhizomes; he suggested that this habitat fostered the unusual branching. Plants resembling *J. nitidula* occur also in Jersey and Guernsey (BM!); we include this entity in *J. rubens* var. *corniculata*.

Few studies have been carried out specifically on *J. rubens* var. *corniculata*; Giraud & Cabioch (1977) included this entity (as *J. corniculata*) in an ultrastructural study.

Ardré (1970, p.96, as *Corallina granifera*) discussed the difficulty in distinguishing this variety from *Haliptilon virgatum* (Zanardini) Garbary & Johansen. However, the intergenicular lobes typical of *J. rubens* var. *corniculata* and upon which the pinnate branches are borne are absent in *H. virgatum*.

Choreonema thuretii occurs also on *J. rubens* var. *corniculata*.

LITHOPHYLLOIDEAE Setchell

by Yvonne M.Chamberlain and Linda M.Irvine

LITHOPHYLLOIDEAE Setchell (1943), p.134 (as Lithophylleae).

Type genus: *Lithophyllum* Philippi (1837), p.387.

Thallus nongeniculate; some but not all cells of contiguous filaments normally joined by secondary pit connections, cell fusions apparently lacking or comparatively rare; tetra/ bisporangia lacking apical plugs and borne within uniporate conceptacles.

Although the subfamily is very distinct, united by the presence of secondary pit connections and the almost complete absence of cell fusions, generic concepts remain uncertain. *Titanoderma* (= *Dermatolithon*) is reasonably distinct (Chamberlain, 1991b), differing from *Lithophyllum* by the universal presence of a unistratose layer of basal filaments composed of usually oblique, sinuate, palisade cells (VS) in dimerous portions, and a bistratose thallus margin. *Lithophyllum* rarely shows palisade cells and they occur only sporadically. Bistratose thallus margins, composed of basal and epithallial cells only, do not occur. While both genera are primarily dimerous many species, especially in *Lithophyllum*, show secondary monomerous development. In the British Isles the two genera are clearly demarcated except that *T. corallinae* sometimes lacks palisade cells and older thalli may have thickened margins. New evidence of intermediate forms in Australia led Campbell & Woelkerling (1990) and Woelkerling & Campbell (1992) to combine the two genera.

The distinction between *Lithophyllum* and *Pseudolithophyllum* requires more study. *Pseudolithophyllum* Lemoine (1913b) was originally based on a combination of characters:- 1) hypothallium unistratose, 2) perithallial cells non-aligned, 3) secondary pit connections present in all perithallial cells. Subsequently, the second of these assumed particular importance (see Lemoine, 1929a; Hamel & Lemoine, 1953, pp.27, 69; Lemoine, 1978) and Lemoine's generic concepts came to revolve around the presence or absence of a tiered appearance. It is important to understand this as the criterion is still used by palaeontologists. *Pseudolithophyllum* was not at that time characterised solely by the presence of a unistratose, nonpalisade hypothallium; Lemoine continued to place some such species in *Lithophyllum*. Adey (1970a) considered that *Pseudolithophyllum* should be typified by a species in the Mastophoroideae.

Cabioch (1969a, 1972) suggested that *Pseudolithophyllum orbiculatum* represented a neotenic stage of *Lithophyllum incrustans*. The relationship between the species in both the British Isles and France (YMC & LMI, obs.) is close and this proposal merits further, perhaps experimental, investigation.

Although Mendoza & Cabioch (1984, p.142) quoted Lemoine's original definition, *Pseudolithophyllum* was redefined by them (p.144) listing only the first of Lemoine's

(1913b) characters in their diagnosis; elsewhere (p.152) they explained that they restricted *Lithophyllum* itself to those species in which the adult thallus shows extensive, monomerous, secondary development (='faux hypothalle', Cabioch, 1972; see also Chamberlain *et al.*, 1988). With respect to cell alignment, they considered (p.144) that this character is significant at species rather than genus level. They did not comment on the consistency or otherwise of the presence of secondary pit connections in all perithallial cells.

Of the three genera recognized by Woelkerling (1988), *Lithophyllum* and *Titanoderma* are recorded for the British Isles; the third genus, *Ezo*, is semi-endophytic and known only from Japan. Species attributed by Cabioch (e.g. 1972, 1992) to *Pseudolithophyllum* are placed in *Lithophyllum*.

LITHOPHYLLUM Philippi

LITHOPHYLLUM Philippi (1837), p.387.

Type species: *L. incrustans* Philippi (1837), p.388, (see Woelkerling 1983b).

Pseudolithophyllum Lemoine (1913b), p.45.

Plants nongeniculate and attached or unattached; varying from flat, crustose thalli to bushy plants composed entirely of protuberances; all conceptacle types uniporate.

Thallus pseudoparenchymatous, dorsiventral, primarily dimerous and composed of a single layer of mainly non-palisade basal filaments, erect filaments that may become very long, and non-flared epithallial cells; erect filaments in some species giving rise to secondary monomerous dorsiventral growth, or monomerous protuberances with radial construction; margin of dimerous thallus thickening immediately behind terminal initial, not bistratose. Cells of basal filaments squarish to taller than wide (VS) but not consistently palisade, cells of erect filaments short, to squarish, to elongate with secondary pit connections usually halfway or more down cell, cell fusions unknown; trichocytes rare; spore germination characteristics poorly known.

Gametangial plants normally dioecious; spermatangial systems simple, borne on conceptacle floor only; carposporangial conceptacle with central fusion cell bearing gonimoblast filaments on periphery or on lower surface at periphery only; tetrasporangial conceptacles each arising from discoid subepithallial primordium which produces zonate tetrasporangia either across conceptacle floor or from periphery only leaving a central columella; bisporangial conceptacles as tetrasporangial ones but producing bisporangia with two uninucleate bispores.

For an account of the genus as recognized here see Woelkerling (1988, p.97). Eight species are included in the present treatment; three of these, *L. crouanii*, *L. incrustans*, *L. orbiculatum*, are very common, widespread and between them contribute significantly to the extent and appearance of coralline cover throughout the littoral zone, but only the first penetrates appreciably into the sublittoral. *L. nitorum* is rarely encountered whilst *L. duckeri* and *L. hibernicum*, each known only from old collections from a single locality, await rediscovery. The names *L. dentatum* and *L. fasciculatum* have been used for certain maerl-rhodolith specimens (see Introduction 4.1); the correct application of these names is uncertain but the material appears to be closely related to *L. incrustans*. *L. lichenoides* Philippi has not yet been recorded for the British Isles, although Adey & Adey (1973, as *L. tortuosum*) suggested that it might occur in the south-west. Further comments are given in the notes for each species. The taxonomic importance accorded to branch morphology varies from one author to another. Steneck & Adey (1976) noted that its dependence on macro- and microenvironment is obvious even to a casual observer; their experiments in the Caribbean with *L. congestum* (Foslie) Foslie showed that it is most abundant on algal reefs subject to medium wave action, the greatest cover (89%) being on the highest elevations (up to 50cm above mean LW) continuously wetted by heavy wave action, but the species cannot withstand really high energy wave action without some protection and is also restricted by desiccation and intense sunlight. Under moderate to highly turbulent conditions with reduced light (e.g. under overhangs) no branches form and accretion rate is low. In turbulent areas with moderate light intensities branches develop and increase in length to breaking point. In turbulent areas with high light intensity complex anastomoses occur to form distinctive heads. As primary branch apices grow and broaden, a central depression may develop; such apical depressions are commonly seen in plants exposed to high light intensity. Broadening to form anastomosing heads provides internal support to withstand breaking waves and an enlarged surface for spore production. A similar range of form occurs in the British Isles but so far there is no experimental evidence relating it to particular environmental conditions except that plants of *L. incrustans* growing under fucoids (i.e. with reduced light) do not produce excrescences (LMI obs.).

KEY TO SPECIES

1   Plants known only unattached in British Isles 2
    Plants rarely unattached in British Isles 5
2   Thallus a compact ball.. 3
    Thallus a more open branch system 4
3   Composed of a solid ball with knob-like branches.. *L. duckeri*
    Composed of densely interlocked, lamellate branches................................... *L.dentatum*
4   Branches up to 3(5)mm thick, LS as in Fig.25B.................................... *L. fasciculatum*
    Branches up to 1mm thick, excluding expanded apices, LS as in Fig.26B
    .... *L. hibernicum*
5   VS showing erect filament cells laterally aligned throughout 6
    VS showing erect filament cells not laterally aligned, or only so in discrete areas ...   7
6   Colour orange-pink; conceptacles markedly raised, conical, tetra/ bisporangial chambers
    not globular .... *L. nitorum*
    Colour reddish-violet to grey; conceptacles scarcely raised, always small, tetra/
    bisporangial chambers globular, gametangial ones flask-shaped.............. *L. crouanii*
7   Bisporangia unknown; tetrasporangial conceptacles with columella little developed, non-
    calcified or absent; old conceptacles not buried; cells not laterally aligned
    .... *L. orbiculatum*
    Bi- and tetrasporangial conceptacles known, columella pronounced, calcified; old
    conceptacles buried; thallus with areas of laterally aligned, sometimes coaxially
    arranged cells .... *L.incrustans*

See Table 2 for distinguishing characters of epilithic species.

Lithophyllum crouanii Foslie (1898c), p.17 (as *Crouani*).
Figs 4A; 19A; 20A-C; 21; Table 2; outgrowth: Figs 20D-E; 22.

Lectotype: TRH! (see Adey & Lebednik, 1967, p.67; Adey, 1970c, p.5; Chamberlain *et al.*, 1988, p.178). E. *Batters*, Berwick-upon-Tweed, March 1896, on *Laminaria*.

Lithophyllum orbiculatum auct., Foslie (1900c), p.19.
Pseudolithophyllum orbiculatum (Foslie) Lemoine (1929a), p.3, pro parte excl. typ.; Adey (1966a), p.481.
Dermatolithon crouanii (Foslie) Hamel & Lemoine (1953), p.62, pro parte, type only - see below.

Key characters: Thallus with strongly aligned cells, without secondary growth, tetrasporangial conceptacles small and flush with surface, appearing in darkish thalli as pale discs with central pore, usually in densely crowded groups, rounded in VS, pore surrounded by rosette of enlarged cells, old conceptacles buried, bisporangial conceptacles unknown. Overgrowing *Titanoderma laminarieae*, *Phymatolithon lenormandii*, *P.laevigatum*; overgrown by *P. purpureum*, *Lithothamnion glaciale*, *Lithophyllum incrustans*.

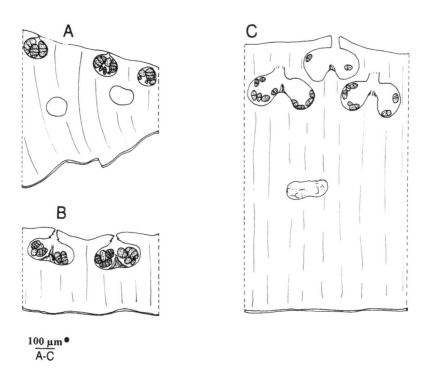

100 μm •
A-C

Fig. 19 *Lithophyllum* spp.: diagrammatic drawings to compare VS of three species. (A) Tetrasporangial thallus of *L. crouanii* (*R.Walker*). (B) Tetrasporangial thallus of *L. orbiculatum* (*R.Walker*). (C) Bisporangial thallus of *L. incrustans* (YMC 88/12).

TABLE 2. Characters distinguishing British *Lithophyllum* spp.

CHARACTER	crouanii	orbiculatum	incrustans	nitorum
Erect filament cells horizontally aligned	+	−	−	+
Tetra/bisporangia occurring	T	T	TB	B
Approximate number of sporangia per conceptacle	20	40+	32+	?
Tetra/bisporangial conceptacle shape 100 μm	(drawing)	(drawing)	(drawing)	(drawing)
(Conceptacle chamber diameter) × (height) (μm)	(120–160) × (81–130)	(180–260) × (60–115)	(244–342) × (91–140)	(289–380) × (91–156)
(Tetra/bisporangial length) × (diam.) (μm)	(60–143) × (23–75)	(73–91) × (37–60)	(55–146) × (29–78)	(46–78) × (32–39)
Typical thallus colour	Ruddy purple	Beige to dark grey-purple	Lavender to greyish	Bright red-pink

Illustrations: Rosenvinge (1917), p.258; Printz (1929), pl.LVII, figs 1,2,5; Suneson (1943), p.35, pl.VI, fig.29; (all as *Lithophyllum orbiculatum*); Adey (1966a), pp.494–496 (as *Pseudolithophyllum orbiculatum*); Chamberlain *et al.* (1988), figs 2–28; outgrowth: Chamberlain (1988), figs 1–16.

Thallus encrusting, adherent, to at least 15cm diameter, to 2 (4)mm thick, typically flat or somewhat lumpy but without branches or protuberances, sometimes bearing pale outgrowths described below, orbicular when young, becoming confluent, confluences flat or slightly crested; *margin* entire, thick, without orbital ridges; *surface* smooth, cell surface *Phymatolithon*-type (SEM), colour typically ruddy purple often with a milky bloom, bleaching to yellowish or greyish pink in well-lit situations (Methuen greyish red, violet brown, dark ruby), texture matt; *conceptacles* appearing as pale discs with sometimes protruding central pores, usually in densely crowded groups; *spermatangial conceptacles*

Fig. 20 Features of *Lithophyllum crouanii* and its outgrowth. (A) Population of young *L. crouanii* thalli overgrowing *Phymatolithon lenormandii* (YMC 84/401). (B) Tetrasporangial thallus with crowded, non-raised conceptacles (YMC 84/401). (C) Vertical fracture of thallus with tetrasporangial conceptacle (C, *L.K.Rosenvinge*, Denmark). (D) Thallus surface with emergent outgrowth (YMC 86/136). (E) Vertical fracture of carposporangial (c) 'host' thallus with outgrowth (arrow) on surface (YMC 86/164).

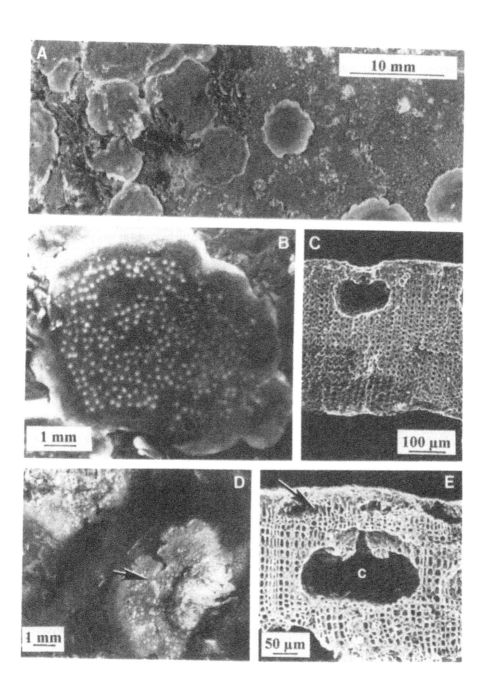

to 75μm diameter, *carposporangial and tetrasporangial* ones to 110μm, tetrasporangial pore surrounded by 6-celled rosette (SEM).

Basal filaments unistratose with cells varying in shape and larger than adjacent erect filament cells, (4)8–13(26)μm long × 7–33μm diameter; *erect filaments* to *c*.50 cells long, individual filaments distinct, contiguous cells usually laterally aligned with resulting cell rows accentuated by concentration of cell contents at top of cells, cells of erect filaments varying erratically in length between but not within lateral rows, 6–28μm long × (4)7–16μm diameter with younger cells heavily calcified in corners, causing lumens to appear diamond-shaped; *epithallial cells* flattened, up to 4(6) cells deep; old, thick thalli often degenerate below and invaded by epiphytic algae and worms, new basal filaments able to regenerate from old or damaged erect filament cells; *trichocytes* not seen but recorded for North American plants (Adey, 1966a).

Gametangial plants dioecious; *spermatangial conceptacle* chambers shallowly elliptical, 90–156μm diameter, 19–52μm high, with the roof 8–31(45)μm thick, pore prolonged into a spout with ring of cells at base, sunken, flush or slightly raised, spermatangial systems simple, borne on floor, old conceptacles buried; *carpogonial conceptacle* chambers broadly triangular with pore canal tapering upwards, protruding trichogynes giving slightly raised effect; *carposporangial conceptacle* chambers elliptical, (100)182–208(225)μm diameter, 57–91(100)μm high, with the roof 21–39(52)μm thick, the roof composed of 2–3 layers of cells with diamond-shaped lumens, papillae round pore, conceptacle floor composed of layers of squashed cells, central fusion cell wide and fairly thin, *gonimoblast filaments* borne peripherally, composed of 1–3 gradually enlarging cells and a terminal carposporangium, old conceptacles usually becoming buried; *tetrasporangial conceptacle* chambers globose, (85) 120–160(180)μm diameter, (55)81–130μm high, with the roof (18)21–39μm thick, the roof composed of 2–3 layers of cells similar to vegetative cells and a ring of 6 enlarged cells (2 visible in VS) surrounding pore (SEM), no columella seen, old conceptacles either buried up to 12 layers deep, or breaking out to leave an open pit; *tetrasporangia* plump, 60–143μm long × 23–75μm diameter, *c*.20 per conceptacle (3–5 seen per VS); *bisporangial conceptacles* unknown.

Common. On rocks, stones and occasionally on shells, *Laminaria* stipes and holdfasts, on shores both exposed to and sheltered from wave action; lower littoral and sublittoral to 50m, most abundant at 0–10m in the south, common to 40m in Scotland (Adey & Adey, 1973, fig.49).

Throughout the British Isles, increasing in abundance northwards.

Arctic Russia to Spain (Galicia), Iceland, Faroes, Baltic; Arctic eastern Canada to USA (Long Island).

Perennial; no information on vegetative growth available but plants with up to 12 layers of buried conceptacles are known. Reproducing principally between October and March, some carposporangial conceptacles observed between April and September, gametangial plants considerably rarer than tetrasporangial ones. Adey (1966a) and Adey & Adey (1973) found that spermatangial plants occurred about twice as often as carpogonial/ carposporangial ones but we have found the latter even more rarely.

Plants usually discrete but confluent and forming extensive rock cover in lower littoral in Scotland.

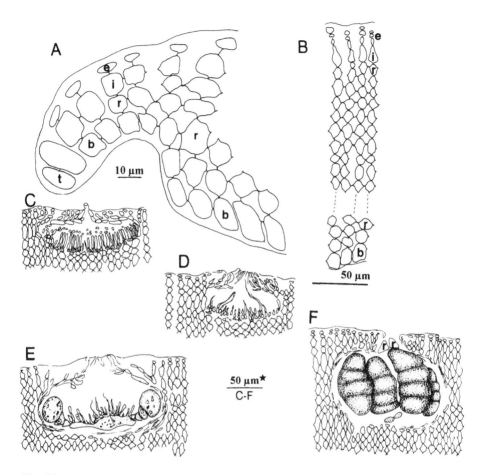

Fig. 21 Vertical sections of *Lithophyllum crouanii*. (A) Thallus margin showing terminal initial (t), cells of basal filaments (b), cells of erect filaments (r), subepithallial initials (i) and epithallial cells (e) (YMC 83/502). (B) Older thallus with cells indicated as in Fig. A (YMC 83/502). (C) Spermatangial conceptacle (YMC 83/502). (D) Carpogonial conceptacle (YMC 84/162). (E) Carposporangial conceptacle (*R.Walker*). (F) Tetrasporangial conceptacle; note enlarged pore cells (r) (YMC 83/2).

In Dorset and Devon thalli of *L. crouanii* have been found bearing a pale outgrowth (Fig.20D) to 80μm diameter, covered with minute, flat conceptacles (Chamberlain, 1988). The cells of the lower part of the outgrowth are similar to those of *L. crouanii*, towards the surface tall cells, probably initials, develop measuring to 30μm long × 5–20μm diameter, and above these filaments of up to 5 small cells are produced which seem to be the equivalent of epithallial cells. All conceptacle types, except bisporangial which have not been seen, may be borne on a single outgrowth; conceptacle roofs are flush with the thallus surface and the conceptacles are similar in shape and structure to the equivalent ones of *L. crouanii* but much smaller. On these outgrowths spermatangial conceptacles measure 79–111μm diameter, 39–52μm high, with the roof 6–33μm thick, a carposporangial one 104μm diameter, 59μm high, with the roof 20μm thick, and tetrasporangial ones 70–89μm diameter, 58–78μm high, with the roof 13–26μm thick and the pore surrounded by a rosette of enlarged

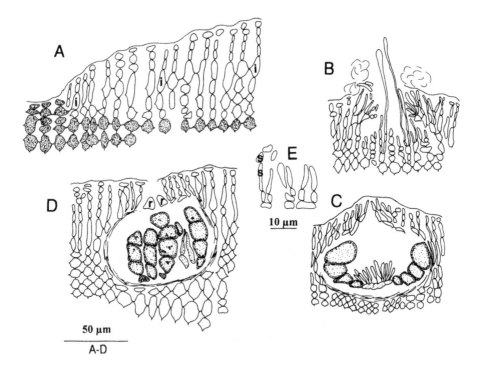

Fig. 22 Vertical sections of *Lithophyllum crouanii* outgrowth. (A) Outgrowth emerging from 'host' thallus which is indicated by shaded cells, i = subepithallial outgrowth initials (YMC 86/152). (B) Carpogonial conceptacle with trichogynes protruding through pore (YMC 86/152). (C) Carposporangial conceptacle with gonimoblast filaments (YMC 86/136). (D) Tetrasporangial conceptacle; note irregularly divided sporangia and pore surrounded by enlarged rosette cells (r) (YMC 86/152). (E) Spermatangial systems with attached and released spermatangia (s) (YMC 86/152).

cells. Tetrasporangia 41–46μm long × 20–24μm diameter, mostly divided irregularly into three rather than the usual four spores. The outgrowth clearly erupts from the *L. crouanii* thallus with which it is in full continuity (Fig.22A) and no sign of spore settlement on the host thallus was seen. This is in contrast to adelphoparasites such as *Kvaleya* Adey & Sperapani (1971) and *Ezo* Adey *et al.* (1974) which grow on *Leptophytum laeve* and *Lithophyllum yessoense* respectively and make contact with their hosts by means of haustoria. It is possible, however, that another parasite, *Masakia bossiellae* Kloczcova (1987), described from Pacific Russia, is in continuity with host thalli, *Bossiella* spp. Chamberlain (1988) suggested that the self-generated outgrowth of *L. crouanii* represents the first stage in the evolution of an adelphoparasite.

The misidentification by Foslie (1905a) of a Danish specimen of *L. crouanii* as *L. orbiculatum* caused longstanding confusion (Rosenvinge, 1917; Suneson, 1943; Adey, 1966a; Adey & Adey, 1973), although Lemoine (1929a, 1931), Hamel & Lemoine (1953) and Cabioch (1972) continued to recognize *L. orbiculatum* correctly in France. Chamberlain *et al.* (1988) examined the type specimens of the two taxa and resolved the confusion. Hamel & Lemoine (1953) made the combination *Dermatolithon crouanii* but their description refers to *Titanoderma laminariae, q.v.*

Although Adey & Adey (1973) considered the species to be most abundant in areas showing least variation between winter and summer temperatures, plants up to 15cm diameter and with up to 12 layers of buried conceptacles have been found (LMI) on flint in Kent which probably experiences the most extreme sea surface temperature range in the British Isles.

Small (to 2mm diameter), more or less orbicular thalli which occur on *Laminaria* stipes and holdfasts in deep pools in the south can be mistaken for *Titanoderma pustulatum* var. *pustulatum* or *T. laminariae q.v.* The smooth thallus of *L. crouanii* may resemble that of *Phymatolithon lamii* and broken out conceptacles strongly resemble the deeply sunken young conceptacles of the latter. A thallus fracture or section should show the cell alignment and single basal layer characteristic of *L. crouanii*.

Lithophyllum crouanii is closely similar to *L. yessoense* Foslie which plays a similar ecological role in Japan. The latter species is the host of the adelphoparasite *Ezo* (see Adey *et al.*, 1974; Chamberlain *et al.*, 1988).

Lithophyllum dentatum (Kützing) Foslie (1900a), p.10.
Fig.23.

Holotype: L (943.7.69; see Woelkerling, 1985a, figs 1–8). Italy (Naples).

Spongites dentata Kützing (1841), p.33.
Lithothamnion dentatum (Kützing) Areschoug (1852), p.525.

Key characters: Ball-formations referred to this species are composed of dense, interlocked, lamellate branches in contrast to the more open, terete branching systems of *L. fasciculatum* and the knobbly balls of *L. duckeri*.

Illustrations: Printz (1929), pl.LXII, figs.1–13; Hamel & Lemoine (1953), pl.VII; Woelkerling (1988), figs 1–8.

Thallus encrusting or unattached, to 20mm thick, irregularly shaped, coarsely tuberculate with various excrescences; *margins* unknown; *surface* smooth, colour of living material unknown but originally described as greenish red; *conceptacle* appearance unknown.

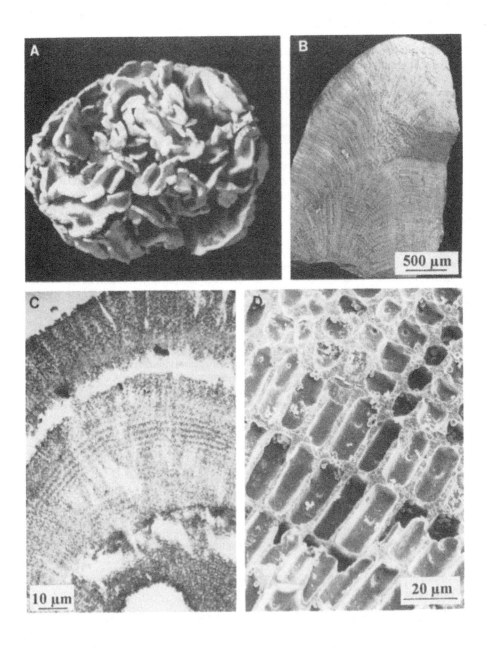

Basal filaments unistratose, cells 10–22µm long × 6–11µm diameter, *erect filaments* loosely aggregated with some large gaps, sometimes conspicuously tiered with cells laterally aligned, transitions between tiers abrupt with cells varying in length between but not within tiers from elongate and 12–41µm long × 8–11µm diameter to squat and 6–16µm long × 6–11µm diameter, secondary pit connections much more common in squat cells, size of initials not recorded, *epithallial cells* single, rounded-elliptical, 3–6µm long × 7–11µm diameter.

Gametangial conceptacles unknown, a single possibly *tetra/bisporangial conceptacle* was immersed, with a scarcely protruding roof of 7–10 cell layers, a chamber 330µm in diameter and columella remnants.

Bosence (pers. comm., as *L. fasciculatum,* BM!) reported occasional balls of this species on mud surfaces at 2m depth in quiet areas, e.g. protected by a bar across a creek (Mannin Bay, Galway, core samples indicating a possible buried bank of the species); Blunden *et al.* (1981) found similar material (BM!) in a ridge and furrow maerl bed system at 4–18m near Kilkieran pier, Galway.

Cork, Galway, Mayo.

Reported from British Isles to Spain and Mediterranean but species limits uncertain.

Seasonal behaviour unknown.

Foslie (1900a, p.31) and Printz (1929) listed a number of forms of which f. *aemulans*, f. *dilatata*, f. *gyrosa* and f. *macallana* were recorded for the west of Ireland; see also Lemoine (1911, 1913a).

The description given here is based on that of the type (Woelkerling, 1985a). Foslie (1900a) used the name for material showing a range of forms which were more or less branched, some attached but others unattached, sometimes as balls; the specimens came from W. Ireland and E. Mediterranean. His comment that the plants nearly always had a well-developed hypothallium suggests their identity with Kützing's type is questionable. From her knowledge of the species in the Mediterranean, Lemoine (1911) applied the name to attached material of characteristic appearance, composed of a closely adherent encrusting thallus giving rise to thin erect plates scarcely 1mm thi
ck and 10–30mm in height, disposed in all directions; she did not believe it could detach and form rounded balls as described by Foslie. Later (1913a) she transferred the Irish balls identified by Foslie (1900a) as f. *macallana* to *L. incrustans* f. *subdichotomum, q.v.,* but they were again referred to *L. dentatum* by Hamel & Lemoine (1953).

The relationship between Foslie's and Lemoine's concepts and Kützing's type needs to reassessed. Lemoine's concept has been adopted by both Cabioch (1968) and Feldmann (1937) but material of this kind has not been found in the British Isles. Material currently attributed to *L. dentatum* in the British Isles was growing on or itself forming mobile substrata

---

Fig. 23 *Lithophyllum dentatum.* (A) Unattached thallus (BM Box 1693. *A.Lees*, Roundstone, Ireland) ×1. (B) Longitudinal fracture of a protuberance (BM. Roaringwater Bay, Ireland, *B.Picton*)(photo W.J.Woelkerling). (C) TS protuberance of specimen in A. (D) Longitudinal fracture of thallus cells; note mainly rectangular cell lumens (specimen in B)(photo W.J.Woelkerling).

and could well be a form of *L. incrustans*, as suggested by Lemoine (1913a); further study awaits the discovery of fertile specimens.

Lithophyllum duckeri Woelkerling (1983a), p.184.
Fig.24

Holotype: L(943.10...34). Sicily (see Woelkerling, 1983a, figs 17–22).

Lithothamnion crassum Philippi (1837), p.388.
Lithophyllum crassum (Philippi) Heydrich (1897b), p.411, nom. illeg., non Rosanoff (1866), p.93 (ICBN, Art.64.1).
Lithothamnion calcareum f. *crassum* (Philippi) Lemoine (1909), p.52.
Lithophyllum racemus f. *crassum* (Philippi) Foslie (1900a), p.9.

Key characters: Ball-formations referred to this species have a knobbly surface unlike the lamellate or terete branching of *L. dentatum* and *L. fasciculatum* respectively.
Illustrations: Printz (1929), pl.LXIII, figs 14–21; Hamel & Lemoine (1953) pl.VII, fig.6; Woelkerling (1983a), figs 17–22.

Thallus unattached, forming a branched subglobular ball to 43mm in diameter, branches arising in all planes from a central core, irregularly forked, 3–8mm in diameter, to 15mm long with rounded apices; *gametangial and bisporangial conceptacles* unknown, *tetrasporangial conceptacles* each with slightly raised roof and depressed pore.

Basal filaments not examined, *erect filaments* loose with some large gaps, not conspicuously tiered, with cells 8–27μm long × 6–14μm diameter, initials readily distinguishable but size not recorded, *epithallial cells* single, 3–6μm long × 6–16μm diameter.

Gametangial and bisporangial conceptacles unknown, *tetrasporangial conceptacles* with large elliptical chambers to 450μm in diameter, 225μm high, with columella remnants, old conceptacles becoming buried.

Cornwall (Falmouth). The single record of this species (BM Box 598!, as *L. racemus* f.*crassa*) was collected at Falmouth 'on a muddy sandbank uncovered at low tide in the middle of the docks' (Traill, letter to Batters in BM, 15 Aug. 1892).
Sicily (holotype).

Specimens from the Falmouth area identified as *L. racemus* and available for study are knobbly balls of dead material consisting only of inorganic remains. SEM study has shown that although the majority (e.g. Holmes: *Algae Britannicae Rariores Exsiccatae* no. 294, BM!) appear referable to *Phymatolithon*, probably mainly *P. calcareum, q.v.*, one specimen shows old conceptacles with a columella and closely resembles the description and illustrations of the type of *L. duckeri* (Woelkerling, 1983a) upon which the account given here is based. The conceptacles recall those of *L. incrustans, q.v.*, of which the specimen may be an unattached form; see also *L. dentatum*. The name given to the specimen (the first knobbly ball found by Traill in 1892) was *Lithothamnion crassum* Phil. (BM Box 598!, Batters's label). The type of *Lithothamnion crassum* belongs to the genus *Lithophyllum* but (Woelkerling (1983a) noted that the combination *Lithophyllum crassum* (Philippi) Heydrich (1897b) is a later homonym (Rosanoff, 1866) and so renamed the species *L. duckeri*. Following Foslie (1900a), Batters revised the Box label to *L. racemus* f. *crassa* and published it as the first record of *L. racemus* for the British Isles (Batters, 1902). Similar nodules found

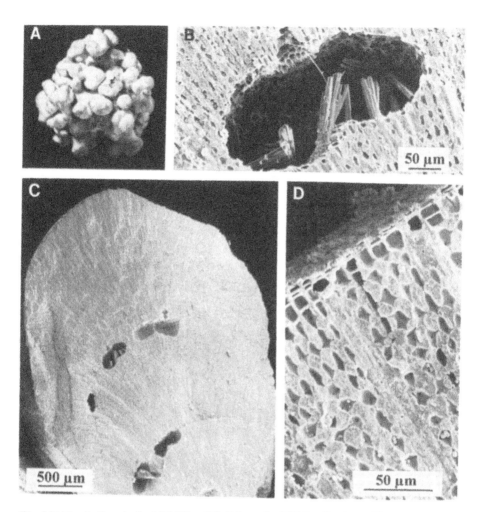

Fig. 24 *Lithophyllum duckeri* (BM Box 598. Falmouth, *G.W.Traill*) (photos W.J.Woelkerling). (A) Unattached thallus ×1. (B) Vertical fracture of tetra/ bisporangial conceptacle. (C) Longitudinal fracture of a protuberance with tetra/ bisporangial conceptacles. (D) Vertical fracture of upper thallus; note flattened epithallial cells at surface and cortical cells with diamond-shaped lumens.

on top of the Falmouth maerl beds (Farnham, pers. comm.) have also been recorded as *L. racemus* (Newton, 1931; Farnham & Bishop, 1985); this identification is unlikely to be correct as the type was probably from French Guiana. Hamel & Lemoine (1953, p.57) considered it to be a Mediterranean species recorded in error by P.& H.Crouan (1867, p.151) from France (Brest). Lemoine (1913a) treated Adams's (1908) W. Ireland record as *L.fasciculatum, q.v.* The combination *Lithothamnion calcareum* f. *crassum* (Philippi) Lemoine (1909) was made without examining the type; Lemoine assigned the name to sublittoral material from Brittany on the basis of external features only and its actual identity is unknown.

Lithophyllum fasciculatum (Lamarck) Foslie (1900a), p.10.
   Fig.25

Lectotype: not selected; no original material known. Locality: 'habite différentes mers'.

Nullipora fasciculata Lamarck (1816), p.203.
Lithothamnion fasciculatum (Lamarck) Areschoug (1852), p.522.
Melobesia fasciculata (Lamarck) Harvey (1847), pl.74.

Key characters: Ball-formations referred to this species have more or less terete branches unlike the knobbly or lamellate balls seen in *L. duckeri* and *L. dentatum.*

Illustrations: Johnston (1842), pl.24, fig.6; Harvey (1847), pl.74; Printz (1929), pl.LXIII, figs 5–10.

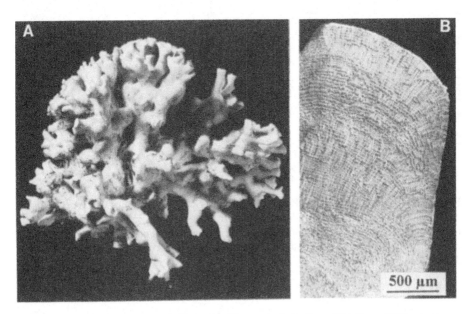

Fig. 25 *Lithophyllum fasciculatum.* (A) Unattached thallus (BM Box 1689, *A. Lees*, Roundstone Bay, Ireland) ×1. (B) Longitudinal fracture through protuberance (BM Box 944, Roundstone Bay, Ireland, *A.D.Cotton*)(photo W.J.Woelkerling).

Thallus crustose, 300–500μm thick; *margin* thick and little contorted; *surface* smooth; later producing erect branches that break away and persist as unattached plants.

Medulla multistratose, to 300μm thick, cells in RVS in conspicuous tiers, 20–25μm long × 5–8μm diameter, *cortex* loose, conspicuously tiered in RVS, with cells laterally aligned, 10–15μm long × 5–8μm diameter and *epithallial cells* up to 3 cells deep; branches arising by localised proliferation of erect filaments.

Unattached plants consisting of subglobular to flattened masses to 80mm in diameter, sparsely to densely branched, branches to 5mm in diameter below tapering to 2–3mm above, sometimes compressed, sometimes expanded at the apices which may be concave, dimpled or flat and tabular, to 5mm in diameter; colour violet when fresh, lilac-grey when dry.

Structure radial, monomerous, *medullary filaments* with rectangular cells 20–30μm long × 6–10μm diameter, *cortical filaments* loose, with cells 10–15μm long × 7–9μm diameter. *Spermatangial conceptacles* unknown, *carpogonial conceptacles* developing centrifugally on both crusts and branches in groups of about 50, forming white patches *c.*10mm in diameter, *carposporangial conceptacles* slightly protruding, with chambers to 200μm in diameter; superficial dead cells sloughing away after spore release, leaving no trace of conceptacles. *Tetra/ bisporangial conceptacles* unknown.

In maerl beds and gravel mixed with shells, recorded between 2 and 9m sublittorally.

W. Ireland (Cork, Galway, Mayo). Material supporting a number of old records for Scotland, Wales and Co. Waterford has not been available for study.

British Isles to France (Brittany).

Only unattached plants without conceptacles recorded for British Isles and little is known about their growth. Cabioch (1968) did not specify when gametangial conceptacles were found in Brittany.

The form range (described by Foslie, 1900a, as f. *incrassata*, f. *divaricata*, f. *compressa* and f. *eunana*) is similar to that found in *Phymatolithon calcareum* and *Lithothamnion corallioides, q.v.*, and is probably likewise influenced by environmental factors.

Johnston (1842) appears to have been the first to apply Lamarck's name *Nullipora fasciculata* to a species of *Lithophyllum* found in maerl beds in W. Ireland; similar material was illustrated by Harvey (1847) and referred to by Areschoug (1852). Cotton (1912) recorded this species, locally called Wild Coral, as common in Clew Bay and at Roundstone; he pointed out that Foslie (1900a) did not attempt to interpret Lamarck's concept, which probably included fasciculate forms of several taxa, but regarded Harvey's description and illustration as the first satisfactory account. Since later workers such as Lemoine (1913a) and Cabioch (1968) have also accepted Harvey's concept, we have used Cabioch's description of material from France (Brittany: Baie de Morlaix [dead] and Brest Harbour) as the basis of our account. No material of Lamarck's nongeniculate corallines has been found and the Irish material appears to be sterile; a neotype might be selected from Cabioch's gametangial specimens, of which she gave a good description. She also discussed features which distinguish it from the closely similar species *L. dentatum* and *L. incrustans, q.v.*, and commented on anatomy, endogenous branching and the phenomenon of delamination.

Traill's (1890) record for Orkney is based on a misidentification (BM Box 1456!). Lemoine (1913a) treated Adams's (1908) record of *L. racemus* (= *L. duckeri q.v.*) from W. Ireland as this species. Kleen's (1875) record for Norway was considered by Foslie (1895) to be a misidentification of *Lithothamnion tophiforme* Unger. For comments on the balls treated as this species by Bosence (pers. comm.) see *L. dentatum*.

This species has a smooth matt surface with a powdery appearance and few encrusting epiphytes; other branched twiggy species which contribute to maerl beds (see Introduction 4.1) have surface irregularities of various kinds, frequently bear epiphytes, and lack secondary pit connections between the cells.

Lithophyllum hibernicum Foslie (1906), p.24.
Fig.26

Holotype: TRH (see Printz, 1929, pl.63, figs 11–13; Adey & Lebednik, 1967, p.44). Co. Galway (Ballynakill Harbour).

Lithophyllum fasciculatum f. *subtilis* Foslie (1897), p.8.

Key characters: Thalli branched, branches terete, slender, with cuplike apices.

Illustrations: Printz (1929), pl.LXIII, figs 11–13.

Thallus only known unattached, forming subglobular branch systems to 35mm in diameter, branches arising in all planes from a central core, terete, 0.5–1mm in diameter, often with broad, flattened or cup-shaped apices to 6mm in diameter with irregular rims.

Structure monomerous, radial; *filaments* loose but large gaps rare, conspicuously tiered, cells very regularly laterally aligned in LS, up to twice as long as broad, seldom quadrate, 9–18µm long × 7–9(11)µm diameter, *epithallial cells* up to 4 cells deep with cells 2µm long, to 5µm in diameter.

Reproduction unknown.

On mobile substrata in harbours, sublittoral at 3–4m.

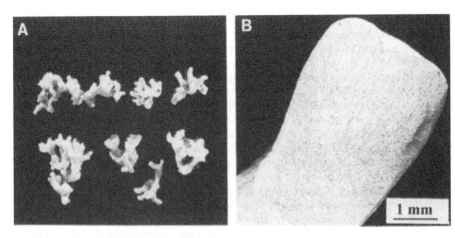

Fig. 26 *Lithophyllum hibernicum* (BM Box 578, Roundstone Bay, Ireland, *H.Hanna*)(photos W.J.Woelkerling). (A) Unattached thalli ×1. (B) Longitudinal fracture of protuberance.

Galway (Ballynakill Harbour and Roundstone Bay).
No information available on seasonal behaviour or form variation.

The internal structure differs noticeably from that of *L. fasciculatum*, of which it was originally thought to be a form (f. *subtilis* Foslie), and recalls that of *Titanoderma byssoides* (Lamarck) Chamberlain & Woelkerling (see Woelkerling, 1988, figs 40,90,91) in some respects.

Known only from three nineteenth century collections, TRH, BM Boxes 535! and 578!; the material in Box 578 closely resembles the type externally to the extent that it looks like part of the same gathering from Ballynakill, even though it was labelled 'Roundstone Bay, H. Hanna' by E.M. Holmes.

Lithophyllum incrustans Philippi (1837), p.388.
Figs Frontispiece; 6; 7C; 19C; 27–29; Tables 2; 3

Lectotype: L! (943.10.34) (see Woelkerling, 1983b, figs 15–23). Sicily, on a shell.

Corallium cretaceum lichenoides Ellis (1755), p.27, figs 2d,D.
Crodelia incrustans (Philippi) Heydrich (1911a), p.12.
Lithothamnion depressum P.& H.Crouan (1867), p.151 (CO!).

Key characters: Thallus cells non-aligned, secondary growth extensive, often coaxial, margin thick, tetra/ bisporangial conceptacles with conspicuous, calcified columella, pore canal of equal width throughout and not tapering, old conceptacles buried. Overgrowing most other epilithic crustose coralline algae; about equal in competition with *Phymatolithon purpureum* and *Lithothamnion glaciale*.

Illustrations: Printz (1929), pl.LVIII; Hamel & Lemoine (1953), pl.VI; Edyvean & Ford (1986b), pl.I.; Cabioch *et al.* (1992), fig.146.

Thallus encrusting, weakly adherent, to at least 10cm diameter, to 800µm thick, old parts degenerating and upper parts regenerating, sometimes becoming several mm thick, flat or becoming undulating, knobbly, or much elaborated and labyrinthic, or with overlapping lobes, young plants orbicular, becoming confluent, confluences slightly to magnificently crested; *margin* entire, thick, without orbital ridges; *surface* smooth, cell surface *Phymatolithon*-type but with relatively small concavities (SEM), colour typically chalky pink to lavender, or greyish when bleached, or violet brown to reddish in shady places (Methuen from brown to dark magenta but predominantly greyish brown), texture matt, rather soft and often scratched by grazers; *conceptacles* of all types usually appearing as slightly sunken, pin-point pores only, occasionally slightly raised or with the roof circumference visible, patches of spermatangial and bisporangial conceptacles sometimes on same thallus; *spermatangial* pores less than 250µm apart, *carposporangial and tetra/ bisporangial* pores c. 280–500µm apart, roofs of damaged conceptacles sometimes falling out leaving a circular crater with calcified columella visible in tetra/ bisporangial conceptacles.

Basal filaments unistratose with cells tall-rectangular to trapezoid, 7–20µm long × 13–26µm diameter; *erect filaments* becoming very long, loose, contiguous cells not usually conspicuously laterally aligned, cells mainly elongate, 5–33µm long × 5–12µm diameter; *epithallial cells* rectangular to flattened, to 4 cells deep; bulk of thallus often composed of filaments

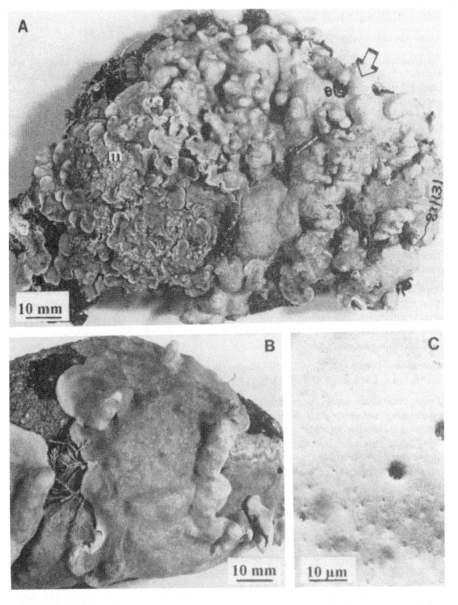

Fig. 27 *Lithophyllum incrustans*. (A) Protuberant thallus (arrow) adjacent to *Mesophyllum lichenoides* (m) (YMC 83/131). (B) Thallus with flat surface and free margins (YMC 84/305). (C) Thallus surface with pinpoint bisporangial conceptacle pores (type specimen [L], Sicily).

secondarily derived from erect filaments growing monomerously, with medullary cells often coaxial and with sporadic areas of short and very elongate cells; *trichocytes* not seen.

Gametangial plants usually dioecious; *spermatangial conceptacle* chambers shallowly elliptical, (90)117–169μm diameter, (22)31–52μm high, with the roof (22)52–57μm thick, composed of filaments up to *c*.9 cells long, without a spout; *carpogonial conceptacle* chambers flask-shaped, usually well below thallus surface; *carposporangial conceptacle* chambers broadly elliptical, (250)265–338μm diameter, 117–169μm high, with the roof (58)78–130μm thick, composed of upwardly orientated filaments to *c*.15 cells long, pore canal notably wide and more or less straight, lined with papillae, fusion cell wide and fairly shallow, borne on layers of squashed thallus cells, *gonimoblast filaments* borne from periphery or underside at periphery, each composed of up to 6 gradually enlarging cells plus carposporangium, old conceptacles buried; *tetra/bisporangial conceptacle* chambers dumb-bell-shaped, (230)244–342(360)μm diameter, 91–140(210)μm high, with the roof (36)52–120(125)μm thick and similar to carposporangial one, pronounced calcified columella present, sometimes with tuft of uncalcified filaments on top; *tetra/bisporangia* borne peripherally, usually on different plants, but sometimes both in same conceptacle, *tetrasporangia* 55–146μm long × 29–78μm diameter, *bisporangia* 72–104μm long × 46–72μm diameter, up to 32 sporangia per conceptacle.

Very common; midlittoral to at least 8m, in shallow pools or margins of deep pools and extensively under fucoid cover; occurring as independent plants or coalescing to a more continuous cover. Tolerating well-illuminated situations where it becomes bleached; best developed on wave-exposed sites. In higher pools often settling on, and eventually ousting, *L. orbiculatum, q.v.*

Throughout the British Isles, rarer between Yorkshire and east Kent but records include East Runton, Norfolk and Shakespeare Cliff, Kent.

Faroes, Norway (Trondheimsfjord) to Spain (Cabo Blanco), Mediterranean. Records for Morocco and Mauritania (Hamel & Lemoine, 1953) unconfirmed.

All reproductive types occur predominantly from October to April, but tail off into summer and gradually increase again towards October, with some conceptacles apparent throughout year but not always containing viable spores. Edyvean and collaborators (see references below) have provided much information on pool-inhabiting individuals: conceptacles containing bisporangia with uninucleate bispores usually predominant, gametangial and tetrasporangial plants common on some shores in south Cornwall and Devon but rare in the north of the country. Assuming that one layer of conceptacles is produced each year, plants up to 30 years old were reported. Edyvean (pers. comm.) suggests that thalli grow *c*.3mm diameter and thicken about 260μm per year. In pools, local habitat differences such as surface/volume ratio influence populations, an unstable age structure being associated with shallow pools. Reproductive effort varies clinally with latitude and is highest in the south. It is calculated that 1mm^2 of reproductive thallus produces 549,920 conceptacles with 17,597,434 individual bispores per year, with an average settlement of only 54.89 sporelings. Further information is available in Edyvean, Ford & Hardy (1983); Ford, Hardy & Edyvean (1983); Edyvean & Ford (1984a,b, 1986a,b); and Edyvean & Moss (1984); a summary of the population studies is presented in Edyvean & Ford (1984b).

Young plants are flat and orbicular, with a thick margin from a very early stage; older

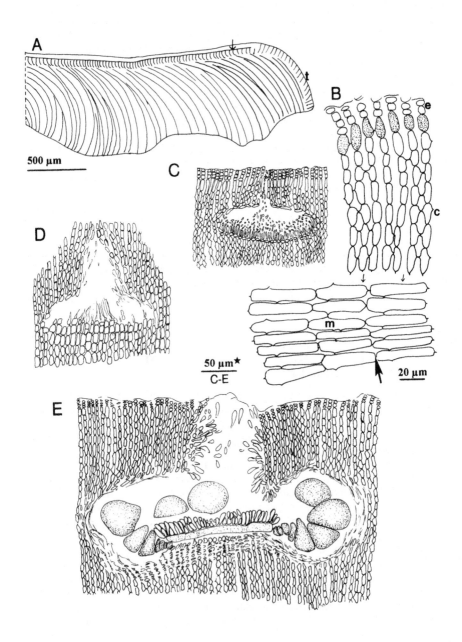

plants gradually develop into crested and convoluted plants that are highly distinctive. Lami (1932) gave a graphic description of the formation of walnut- to cauliflower-sized micro-reefs and atolls in the Mediterranean in non-draining pools; similar forms are seen in the British Isles, particularly in south and west Ireland (e.g. Cotton, 1912) and the Channel Isles. Rock-encrusting plants in shady places are much less elaborate. Plants frequently grow on shells and thick plants may form nodules round small pebbles in maerl populations.

Flat forms of *L. incrustans* have been attributed to f. *depressum* (P.& H.Crouan) Foslie (e.g. Heydrich, 1911a; Hamel & Lemoine, 1953; Cabioch, 1969a). The type collection (CO!, basionym *Lithothamnion depressum* P.& H.Crouan) comprises three different crustose coralline algae on flat stones. Two are species of *Phymatolithon*; the third, a *Lithophyllum*, is here selected as lectotype. The thallus measures *c*.20mm² × 1.5mm thick and is hard and entirely flat, with groups of spermatangial and bisporangial pores on the surface. The thallus is composed mainly of basal and erect filaments, but some secondary monomerous development is present. Cell dimensions are close to those recorded here for *L. incrustans* and swollen cells that may be trichocytes replace some subepithallial initials. Spermatangial and bisporangial conceptacles are similar in size and shape to those of *L. incrustans* but bisporangial conceptacles are somewhat larger, measuring *c*.390μm diameter, 190μm high. Buried conceptacles occur in abundance. Although the Crouans' (1867) description mentions the presence of bispores surrounding a conical axis, the columella is rather small. Similar plants lacking a well-developed, calcified columella occur in the south of England; they have flat, hard thalli with rather thin margins and are overgrown by typical *L. incrustans*. Because of the great variation in external and anatomical features in *L.incrustans*, we regard the differences described for f. *depressum* as insufficent to justify its acceptance as a distinct taxon.

Form names such as f. *harveyi* Foslie (1895; Harvey, 1850, pl.345, isotype BM Box 591!) and f. *subdichotomum* Heydrich (1899) have been given to British Isles specimens on the basis of their external appearance. Plants of f. *subdichotomum* have furrowed margins, an early stage of thallus elaboration under high light intensity; crests are also furrowed because they are formed by the confluence of two thalli or convoluted parts of the same thallus.

It was thought that the coaxial filaments produced by secondary growth represented a multistratose hypothallium until Cabioch's studies (1969a, 1972, 1988) showed that unistratose basal filaments develop from the germinating spore, the multistratose thallus (which she terms a false hypothallium) arising from erect filaments later . Cabioch (1969a) has suggested that *L. orbiculatum* (*q.v.*), which is morphologically and ecologically similar to *L. incrustans,* may represent a neotenous juvenile stage.

Well-grown, crested, greyish or lavender plants of *L. incrustans* are usually easy to identify

---

Fig. 28 Vertical sections of *Lithophyllum incrustans*. (A) Diagrammatic drawing of secondary, monomerous thallus lobe with terminal initials (t), the cortex develops from subepithallial initials (arrow) and medullary filaments have cells arranged coaxially (YMC 83/121). (B) Representative filaments from thallus in Fig. A: cells of medullary filaments (m) divide more or less synchronously and the lines of end walls (arrow) form the coaxial arches; at the upper surface subepithallial initials (shaded) produce cortical cells (c) inwardly and epithallial cells (e) outwardly. (C) Spermatangial conceptacle (YMC 84/95). (D) Carpogonial conceptacle (YMC 84/204). (E) Carposporangial conceptacle (YMC 83/488).

but otherwise the species is not always easy to distinguish from well-grown plants of *L. orbiculatum* in higher pools and the differences between them are tabulated (Table 3). The most difficult plants to recognize are flat ones growing in shade. These may be reddish or purplish and are easily confused with, for example, *Phymatolithon purpureum, q.v.* Fortunately uniporate bisporangial conceptacles are present for much of the year in *L. incrustans*: a broken thallus showing dumb-bell-shaped, buried conceptacles is a very useful guide to identification.

Reproductive abnormalities occur quite often: as mentioned thalli may bear both spermatangial and bisporangial conceptacles, and tetrasporangia and bisporangia occur in

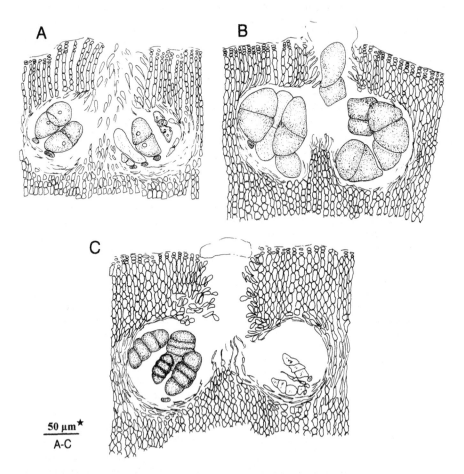

Fig. 29 Vertical sections of *Lithophyllum incrustans*. (A) Young bisporangial conceptacle (YMC 85/198). (B) Mature bisporangial conceptacle (YMC 86/152). (C) Tetrasporangial conceptacle (YMC 87/73).

TABLE 3. Differences between *Lithophyllum orbiculatum* and *L. incrustans*.

CHARACTER	*L. orbiculatum*	*L. incrustans*
Thallus margin	Thin	Thick
Epithallial filaments	Usually 5–6 cells long	Not usually >4 cells long
Secondary coaxial filaments developing	No	Yes
Becoming fertile	On very small thalli	On thalli >10mm diam.
Conceptacle periphery normally evident on thallus surface	Yes	No
Carposporangial conceptacles becoming buried	No	Yes
Bisporangial conceptacles	Absent	Present
Calcified columella in tetra/bisporangial conceptacles	No	Yes
Pore canal	Tapering	Parallel-sided
Tetra/bisporangial conceptacles becoming buried	No	Yes
Thallus overgrowth	Always overgrown by *L. incrustans*	Always overgrowing *L. orbiculatum*

the same conceptacle, one such conceptacle was found to contain carpogonia as well. Conceptacles do not usually develop until the young thallus is several mm in diameter thus contrasting with *L. orbiculatum* in which they develop on very small thalli. The labyrinthic thallus of well-grown *L. incrustans* acts as a nursery for quantities of young shellfish, marine worms such as *Dodecaceria* and *Polydora* and other animals and senescent plants become invaded by blue-green algae and other epiphytes. At Fanore, W. Ireland, cups of the sea urchin *Paracentrotus lividus* (Lamarck) (Frontispiece) are lined with *Phymatolithon purpureum*, with knobbly plants of the softer *L. incrustans* developing on the rock surface between the cups (Chamberlain & Cooke, 1984).

Lithophyllum nitorum W.& P.Adey (1973), p.386.
   Figs 4B; 7D; 30; 31; Table 2

Holotype: USNC! (70–10B, 10–30) (Adey & Adey, 1973, tab.XII), Port Erin Harbour, Isle of Man, 3–9m.

Pseudolithophyllum nitorum (W.& P.Adey) Mendoza & Cabioch (1984), p.145.

Key characters: Secondary growth lacking, bisporangial conceptacles raised. Overgrowing *Lithothamnion sonderi*.

Illustrations: Adey & Adey (1973), tab XII (diagrammatic).

Thallus encrusting, adherent, to at least 20mm diameter, to *c.*500μm (2mm) thick, more or less flat, orbicular to confluent, confluences slightly crested; *margin* entire, thin, without orbital ridges; *surface* smooth, cell surface *Leptophytum*-type (SEM) (Fig.7D), colour bright red-pink, texture somewhat glossy; *conceptacles* (only bisporangial ones seen) raised, truncated cone-shaped, to *c.*420μm diameter, raised to *c.*120μm above surface.

Basal filaments unistratose, cells nearly square in VS to somewhat higher than wide, 10–20μm long × 9–27μm diameter; *erect filaments* to *c.*30 cells long, individual filaments conspicuous with contiguous cells laterally aligned and varying erratically in length between but not within the rows, cells (3)5–41μm long × (4)7–20μm diameter, mature cells heavily calcified in corners, often with diamond-shaped lumen; *epithallial cells* triangular at thallus margin, flattened in mature parts, 1–2 cells deep; *trichocytes* unknown.

Gametangial and tetrasporangial plants unknown in British Isles; *bisporangial conceptacle* chambers elliptical, situated below thallus surface, (245)289–380(385)μm diameter, 94–156μm high, with the roof (22)52–120μm thick, and raised above thallus surface, composed of upwardly orientated filaments to 5 cells long, cells long and thin with diamond-shaped lumens, middle three cells often laterally aligned giving three-layered appearance, pore canal broad at bottom tapering to thallus surface, lined by small papillae and downward growing papillae surrounding canal base, columella small, old conceptacles not becoming buried; *bisporangia* plump, borne peripherally, 46–78μm long × 32–39μm diameter.

Rare. On rocks and shells, sublittoral to 40m, most abundant at 3–9m (Adey & Adey, 1973, fig.52).

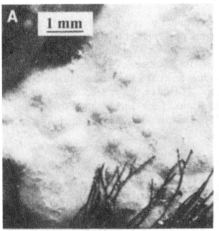
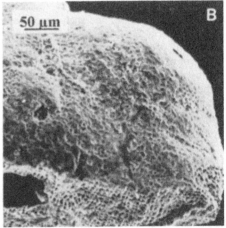

Fig. 30 *Lithophyllum nitorum*. (A) Thallus with bisporangial conceptacles (YMC 82/83). (B) Bisporangial conceptacle (YMC 84/166).

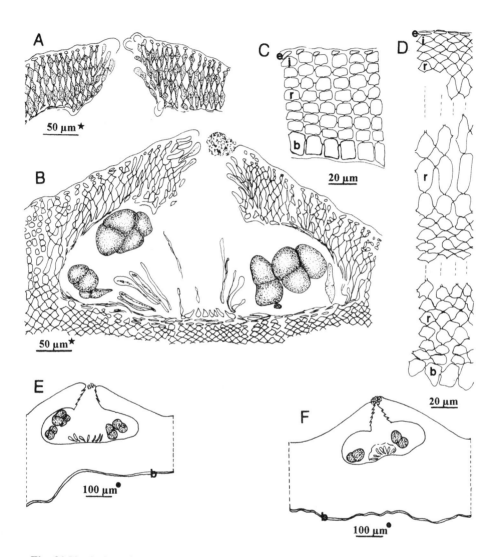

Fig. 31 Vertical sections of *Lithophyllum nitorum* (YMC 82/83). (A) Roof of bisporangial conceptacle showing rows of aligned cells with elongate, diamond-shaped lumens (shaded). (B) Bisporangial conceptacle. (C) Young thallus, cells indicated as in Fig. D. (D) Representative thallus cells showing unistratose basal filament cells (b), variable erect filament cells (r), short subepithallial initials (i) and flattened epithallial cells (e). (E & F) Diagrammatic drawings of bisporangial conceptacles; note single layer of basal filaments (b).

Cornwall, Devon, Pembroke, Isle of Man; Channel Isles. Recorded by Adey & Adey (1973) for W. Scotland, eastern Ireland; particularly abundant off the Lizard, Cornwall. British Isles to northern Spain.

Bisporangia observed in April, October and December.

Adey & Adey (1973) described only tetra/ bisporangial conceptacles, but illustrated spermatangial conceptacles (tab.XII) with flask-shaped buried chambers measuring 9–118µm diameter, 18µm high and a pore protruding to 22µm above the thallus surface. We have not seen spermatangial conceptacles and it is not known whether the conceptacles measured by Adey & Adey (1973) came from the British Isles, France or Spain.

Although the type description (Adey & Adey, 1973) is illustrated only by means of diagrams, Norris & Townsend (1984) concluded that this could nevertheless be construed as a valid species publication.

Adey et al. (1982) described *L. ganeopsis* from Hawai'i as very similar to *L. nitorum*. They did not record bisporangial conceptacles for comparison, and their illustration of a tetrasporangial conceptacle does not show whether it has the shape characteristic of *L. nitorum*.

The raised bisporangial conceptacles of *L. nitorum* are most likely to be confused with those of *Titanoderma pustulatum*, particularly var. *macrocarpum, q.v.*, which may also be reddish; the presence of tall, obliquely orientated basal cells in *Titanoderma* distinguishes the taxa. They may also be confused with carposporangial conceptacles of *Phymatolithon lenormandii, q.v.*, from which they can be distinguished by spore type, the presence of multistratose medullary filaments in *P. lenormandii* being a further distinguishing character.

Lithophyllum orbiculatum (Foslie) Foslie (1900c), p.19.
   Figs 19B; 32; 33; Tables 2; 3

Lectotype: TRH! (see Foslie, 1895, pl.22, fig.103; Adey & Lebednik, 1967, p.21; Chamberlain *et al.*, 1988). *Ekman*, ex Herb. Areschoug, Kristiansund, Norway.

Lithothamnion orbiculatum Foslie (1895), p.171 (reprint p.143).
Goniolithon subtenellum Foslie (1898c), p.8 (see Chamberlain, 1993, p.112).
Pseudolithophyllum orbiculatum (Foslie) Lemoine (1913b), p.3.

Key characters: Thallus without secondary growth, cells non-aligned, margin thin, tetrasporangial conceptacles lacking calcified columella and not becoming buried, bisporangial conceptacles unknown. Overgrowing *Phymatolithon lenormandii, P. laevigatum;* overgrown by *L. incrustans, Corallina officinalis* basal crust.

Illustrations: Printz (1929), pl.LVII, fig.5; Lemoine (1929a), pl.1; Hamel & Lemoine (1953), pl.X; Chamberlain *et al.* (1991), figs 1–11,13–38.

Thallus encrusting, strongly adherent, to at least 25mm diameter, to 1mm thick, flat to extravagantly knobbly with excrescences to 5mm high, plants orbicular when young, becoming irregularly lobed and confluent, confluences flat or sometimes crested; *margin* crimped, thin, without orbital ridges; *surface* smooth, cell surface *Phymatolithon*-type (SEM), colour dark grey-purple in shady places but usually greyish, yellowish to beige (Methuen greyish brown to purplish grey), texture grainy and usually scratched by grazers;

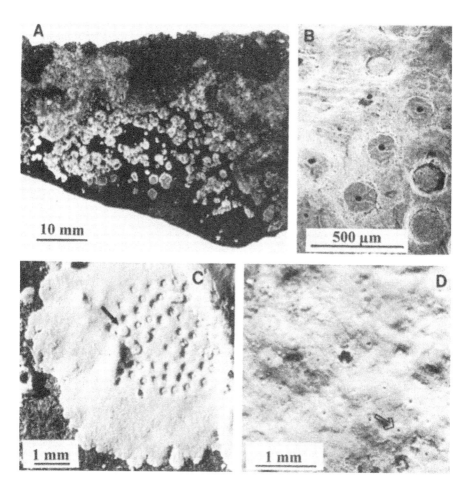

Fig. 32 *Lithophyllum orbiculatum.* (A) Stone with knobbly thalli showing spore-settlement producing young orbicular thalli (YMC 85/72). (B) Surface of tetrasporangial thallus showing uniporate conceptacle roofs and craters of shed conceptacles (YMC 85/90). (C) Young thallus that is already fertile, with old conceptacle roofs (arrow) becoming detached (YMC 84/349). (D) Mature thallus surface with rough appearance and conceptacles (arrow) (YMC 84/108).

conceptacles of all types flat with central pore surrounded by a disc that is slightly paler than thallus, *spermatangial conceptacles* to *c*.80μm diameter, *carposporangial and tetrasporangial* ones to *c*.200μm; senescent conceptacle roofs becominge chalky and falling out to leave circular depressions that are later infilled by thallus growth.

Basal filaments unistratose, cells larger than adjacent erect filament cells, tall and wide in RVS, narrow in TVS, 8–20μm long × 11–21μm diameter; *erect filaments* becoming long with individual filaments distinct but contiguous cells not laterally aligned, cells more or less square to elongate in VS, 10–21μm long × 7–14μm diameter; *epithallial cells* rectangular, to 6 cells deep; large thick-walled cells that may be *trichocytes* sometimes occurring in erect filaments, endophytic algae often present.

Gametangial plants dioecious; *spermatangial conceptacle* chambers wide and shallow, 95–150μm diameter, 20–40μm high, with the roof 30–80μm thick and bearing a spout, conceptacles buried when old but often containing apparently normal spermatangia;

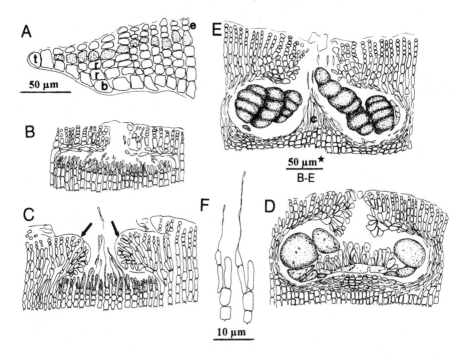

Fig. 33 Vertical sections of *Lithophyllum orbiculatum*. (A) Thallus margin showing terminal initial (t), cells of basal filaments (b), cells of erect filaments (r), subepithallial initials (shaded) and epithallial cells (e) (YMC 84/196). (B) Spermatangial conceptacle (*R.Walker*). (C) Carpogonial conceptacle with surrounding roof filaments (arrow) (YMC 86/76). (D) Carposporangial conceptacle with gonimoblast filaments (*R.Walker*). (E) Tetrasporangial conceptacle; note filamentous columella (c) (YMC 84/196). (F) Carpogonial branches (YMC 86/76).

carpogonial conceptacle chambers very wide and shallow; *carposporangial conceptacle* chambers elliptical, 175–215μm diameter, 60–95μm high, with the roof 40–62μm thick and composed of upwardly orientated filaments up to 7 cells long, pore canal quite broad at base and tapering to surface, with downward-growing papillae surrounding canal base, central fusion cell quite wide, fairly shallow, bearing *gonimoblast filaments* composed of up to 4 gradually enlarging cells plus large carposporangium from underside at periphery, old conceptacles not buried; *tetrasporangial conceptacle* chambers shallowly elliptical, 180–260μm diameter, 60–115μm high, with the roof 30–70μm thick and similar in structure to carposporangial conceptacles but filaments up to 9 cells long with papillae round pore canal base less pronounced, columella filamentous, giving chamber a dumb-bell shape; *tetrasporangia* borne peripherally, 73–91μm long × 37–60μm diameter, c.40+ per conceptacle, old conceptacles not buried; *bisporangial conceptacles* unknown; thalli of all reproductive types becoming fertile when only a few mm in diameter.

Common. Occurring principally on hard rocks in mid to upper littoral pools and gullies on fairly- to very-exposed shores, occupying the same niche as *Phymatolithon lenormandii* which it overgrows and dominates. Often overgrown and sometimes ousted by *L. incrustans*. Not able to grow on soft or slatey rocks. Not yet known from the sublittoral.

Throughout the British Isles, abundant in Scotland and dominating many Shetland shores. Norway (Trondelag) to Morocco; W. Mediterranean, Adriatic.

All reproductive types present throughout the year but occurring predominantly from October to February; seasonal periodicity to be assessed. In southern England tetrasporangial plants are common, spermatangial ones quite common and carpogonial/carposporangial ones rather infrequent (rare in northern British Isles).

There is considerable variation in external form from flat, smooth plants to convoluted, knobbly ones. In France and Spain (YMC obs.) this species is abundant; in Normandy it forms extensive, convoluted populations on midlittoral rocks, while on the Basque coast smaller, flat, orbicular plants are a conspicuous feature of littoral rocks and stones.
  A population typically comprises an area of coalescent thalli surrounded by orbicular young plants (Fig.32A), many of which are white and presumably dead. This indicates that although recruitment occurs continually many sporelings on shallow surfaces emersed at low tide are killed, probably by frost in winter or desiccation and insolation in summer (Chamberlain *et al.*, 1991).

For many years the presence of *L. orbiculatum* in the British Isles was overlooked because of nomenclatural confusion (see Chamberlain *et al.*, 1991) with *L. crouanii* (*q.v.*) and its misidentification as a thin form of *L. incrustans* in upper littoral pools (Cotton, 1912, pp.41–42). Cabioch (1969a) discussed the relationship between *L. incrustans* and *L. orbiculatum*.
  Conceptacle initiation and development is particularly easy to observe in this species and Chamberlain *et al.* (1991) showed that all types of conceptacle are produced from subepithallial initials.
  The lectotype (TRH!) of *Goniolithon subtenellum* Foslie (1898c) from Guéthary, France, was found to be conspecific with *L. orbiculatum* (Chamberlain, 1993, p.112) and with recently observed populations from Guéthary (YMC obs).

Comments on the genus *Pseuodlithophyllum* Lemoine (1913b; 1978), to which this species was referred, are given under Lithophylloideae, *q.v.*
This species may be confused with *L. incrustans* (see Tables 2; 3) and Cabioch (1969a) discusses the relationship between the two species. It may also be confused with *Phymatolithon lenormandii* in high pools; although the latter has a multistratose thallus base and multiporate tetra/ bisporangial conceptacles, the two can be difficult to distinguish in the field and the species shown in a pool in the Isle of Man (Knight & Parke, 1931, pl.VI) is almost certainly *L. orbiculatum*, not *P. lenormandii* as labelled.

EXCLUDED SPECIES

Lithophyllum expansum Philippi (1837), p.389.

L. expansum was recorded from Guernsey by Van Heurck (1908) on the basis of material collected by Marquand. Similar material collected by Marquand from Jersey (BM Box 577, slides 10710, 10711!) was examined and illustrated by Lemoine (1965) who identified it as a form of *L. incrustans*, *q.v.*, with overlapping lobes. The type of *L. expansum* has been shown to belong to the genus *Mesophyllum* (Woelkerling, 1983b).

TITANODERMA Nägeli

TITANODERMA Nägeli (1858), p.352.

Type species: *T. pustulatum* (Lamouroux) Nägeli (1858), p.352.

Melobesia Heydrich (1897b), p.408, non *Melobesia* Lamouroux (1812) nec *Melobesia* Foslie (1898b).
Dermatolithon Foslie (1898b), p.11.
Lithophyllum Philippi subgenus *Dermatolithon* (Foslie) Foslie (1904b), p.3.
Litholepis Foslie (1905b), p.5 (see Woelkerling, 1986).
Tenarea Bory (1832), p.207 pro parte.

Plants nongeniculate, attached or unattached, crustose and flat, or composed of regenerating imbricating layers, or mainly of branches; all conceptacle types uniporate; structure pseudoparenchymatous, dorsiventral, dimerous or both dimerous and monomerous, with branches, when present, more or less radial and monomerous; dimerous thallus composed of unistratose basal filaments, with or without erect filaments that may become very long; epithallial cells borne on basal cells or terminally on erect filaments, non-flared. Margin of dimerous thallus composed of palisade basal cells plus epithallial cells only (Fig.40B). Basal cells palisade and often sinuate and oblique (VS), cells of erect filaments short to very tall, contiguous cells joined by secondary pit connections usually in upper third of cell, cell fusions very rare; trichocytes occasional; spore germination pattern showing two rows of 4 cells.
Gametangial plants dioecious; spermatangial systems simple, borne on conceptacle floor only; carposporangial conceptacle with central fusion cell producing gonimoblast filaments from periphery or lower surface at periphery only; tetrasporangial conceptacles uniporate, arising from subepithallial disc of initials, producing zonate tetrasporangia peripherally and with central columella; bisporangial conceptacles as tetrasporangial ones but normally

producing bisporangia with two uninucleate bispores; occasional conceptacles containing a mixture of tetrasporangia with two or four nuclei and bisporangia with from 2–8 nuclei (Suneson, 1950).

For an account of the genus, see Woelkerling *et al.* (1985), Woelkerling (1988, p.109) and Chamberlain (1991b). Campbell & Woelkering (1990) subsumed the genus in *Lithophyllum* but we consider that the presence of a bistratose dimerous thallus margin and predominantly palisade basal filament cells in *Titanoderma* but not in *Lithophyllum* justifies maintaining *Titanoderma* as a distinct genus. In the British Isles plants are crustose or with imbricating lobes; branching species do not occur here.

The name *Titanoderma* was almost completely overlooked after its establishment by Nägeli (1858) until resurrected and more fully characterised by Woelkerling *et al.* (1985). Three species occur in the British Isles. One of these, *T. pustulatum*, shows much variation. It is regarded in the present account as having four varieties: *pustulatum, canellatum, confine* and *macrocarpum* which are mainly clearly demarcated in the British Isles. Nevertheless intermediate forms occur and the varieties are best interpreted as focal points within a continuum. Woelkerling & Campbell (1992) found similar variability in southern Australian populations of *T. pustulatum* but were not able to identify stable varieties.

The degree of immersion of tetra/ bisporangial conceptacles in the thallus has been used as a diagnostic character for *Titanoderma* species (e.g. Chamberlain 1991b). Woelkerling & Campbell (1992, p.11, as *Lithophyllum*) expressed this as the number of cell layers below the thallus surface that the conceptacle floor was situated. This has proved a precise character in British Isles species where *T. corallinae* and *T. laminariae* have conceptacle floors six or more cells below the surface while in *T. pustulatum* agg. they are two to three cell layers below the surface. The same character is often expressed as degree of prominence of the conceptacle roof but that is susceptible to modification due to the age and condition of the plant and, while useful, is less precise.

Athanasiadis (1989) described a distinctive, centripetal growth pattern in Greek plants of *T. cystoseirae* (Hauck) Woelkerling, Y. Chamberlain & P.Silva. This species does not occur in the British Isles and this growth pattern has not been seen in British Isles plants.

KEY TO SPECIES AND VARIETIES

1 Conceptacles immersed in the thallus, tetra/ bisporangial conceptacle floor at least 6 cell layers below surface.. 2
  Conceptacles raised, tetra/ bisporangial conceptacle floor 2–3 cell layers below surface (*T. pustulatum* agg.).. 3
2 Conceptacle chambers tetrasporangial, over 300μm internal diameter..... *T. laminariae*
  Conceptacle chambers bisporangial, under 250μm internal diameter .... *T. corallinae*
3 Conceptacle roofs three cells thick, upper and lower cells small, central cell tall 4
  Conceptacle roofs more than three cells thick 5
4 Conceptacles without distinct periphery, usually over 500μm external diameter
  .... *T. pustulatum* var. *pustulatum*
  Conceptacles with distinct periphery, prominent, under 400μm external diameter
  .... *T. pustulatum* var. *canellatum*

(Key continued p. 90)

5   Erect filaments absent, plants composed of imbricating lobes ....
....growth form of *T. pustulatum* var. *macrocrocarpum*
     Erect filaments present, plants of another form 6
6   Plants bluish, conceptacles mainly abruptly prominent, rounded-conical with roofs to 4
     cells thick and cells irregularly-sized; basal cells tall, thin and oblique, erect cells
     mainly 9–13μm diameter.. *T. pustulatum* var. *confine*
     Plants orange-pink to mauvish, conceptacles raised to broadly prominent with
     roofs usually more than 4 cells thick and cells more or less equally sized; areas of
     short, wide, vertical basal cell often present, erect cells mainly 12–23μm diameter
     .... *T. pustulatum* var. *macrocarpum*

Chamberlain (1991) gave the following identification aids to *T. pustulatum* varieties: list of
substrata (p.2); varietal characteristics (table 1); comparative drawings (fig.52).

Titanoderma corallinae (P.& H.Crouan) Woelkering *et al.* (1985), p.333.
     Figs 34; 35

Lectotype: CO! (see Chamberlain, 1991b, p.66). Herb. P.-C. & H.-M. Crouan, Banc du
     Chateau et Baie de la Ninon, rade de Brest, France.

Melobesia corallinae P.& H.Crouan (1867), p.150.
Lithophyllum corallinae (P.& H.Crouan) Heydrich (1897a), p.47.
Dermatolithon corallinae (P.& H.Crouan) Foslie in Børgesen (1903), p.402.
Lithophyllum pustulatum f. *corallinae* (P.& H.Crouan) Foslie (1905a), pp.118 & 127.

Key characters: Epiphytic, mainly on *Corallina officinalis*, thallus margin often thickened,
conceptacles immersed to at least 6 cell layers below thallus surface, to 235μm internal
diameter. Overgrowing *Melobesia membranacea*; about equal in competition with
Titanoderma pustulatum var. *confine*.

Illustrations: P.& H.Crouan (1867), p.23, 133 bis, figs 6–11 (as *Melobesia*; Suneson (1943),
figs 24–26, pl.6, fig.38 (as *Lithophyllum*); Masaki & Tokida (1960a), pl.1, figs 1–4, pls 2, 4,
5 (as *Dermatolithon*); Cabioch (1979), figs a-d, pl.1 (as *Dermatolithon*); Chamberlain
(1991b), figs 208–224; Woelkerling & Campbell (1992), figs 22–32 (as *Lithophyllum*).

Thallus encrusting, flat and closely adherent or a centrally attached disc with free margins,
to at least 6mm diameter, to 370μm thick, orbicular or encircling the host; *margin* sometimes
bistratose but often thickened; *surface* smooth, colour mauvish (Methuen reddish lilac to
violet brown), texture somewhat chalky; *conceptacles* of all types either immersed and
visible only as a pore, or slightly raised and visible as a pinkish roof surrounded by a white
circle, 150–200μm diameter, with roof surface somewhat ridged (SEM).
     Basal filaments unistratose, cells varying from palisade, oblique and sinuate, often tapering
at the base, to short and vertically orientated, cells 7–16μm long × 7–98μm diameter; *erect
filaments* up to 20 cells long with cells 10–63μm long × 5–20μm diameter, usually laterally
aligned; *epithallial cells* single, domed; thallus cells usually with sparse contents, lower cells
sometimes with starch grains; *trichocytes* not seen.
     Gametangial plants monoecious or spermatangial only; *spermatangial conceptacles* (from
two plants only) immersed, conceptacle chamber broadly triangular, 78–130μm diameter,
39–52μm high, with simple spermatangial systems on the floor only; *carpogonial*

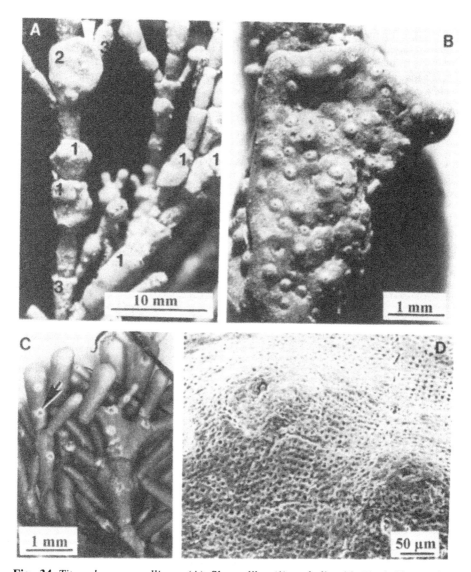

Fig. 34 *Titanoderma corallinae.* (A) Sleeve-like (1) and discoid (2) thalli growing on *Corallina.* Damaged surfaces (3) remain after epiphyte has fallen off (YMC 78/305). (B) Sleeve-like bisporangial thallus, the conceptacles appearing somewhat raised in dried material (YMC 79/208). (C) Sporelings (arrow) settling on adaxial surface of *Corallina* causing thallus bleaching as they develop (YMC 78/305). (D) Surface view of slightly raised bisporangial conceptacles (YMC 78/305).

conceptacles (from one population only) immersed, conceptacle chambers flask-shaped, 90–104μm diameter, 39–59μm high; *carposporangial conceptacle* (only one seen) slightly raised, conceptacle chamber rounded, 182μm diameter, 130μm high, with the roof 47μm thick, fusion cell small, thick, with peripheral *gonimoblast filaments* up to 4 cells long plus the carposporangium; *tetrasporangial conceptacles* not seen; *bisporangial conceptacles* completely immersed to somewhat raised with chamber floors to at least six cell layers below thallus surface, chambers rounded to elliptical, 169–235μm diameter, 75–117μm high, with the roof 26–52μm thick, roof composed of varying number of smallish cells, tapering to the pore, columella slightly- to well-developed; *bisporangia* 50–87μm long × 39–59μm diameter, *c.*20 per conceptacle; old conceptacles of all types becoming infilled, not buried in the thallus. Epiphytic, mainly on *Corallina officinalis*, occasionally on *C. elongata* and *Haliptilon*, quite frequently on *Furcellaria*, occasionally on other red and brown algae, often overgrowing *Melobesia membranacea* thalli on *Furcellaria*; common in rock pools in lower littoral and sublittoral to at least 15m (Suneson, 1943).

Common. Widely distributed throughout the British Isles.

British Isles to Canary Isles; Faroes; W. Baltic; Mediterranean; Pacific USA; Japan; Australia; New Zealand; West Africa; South Africa; probably cosmopolitan.

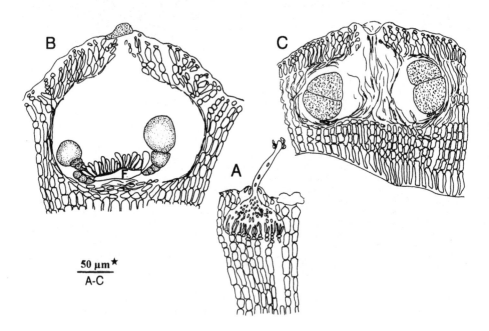

Fig. 35 Vertical sections of *Titanoderma corallinae*. (A) Spermatangial conceptacle (YMC 76/167). (B) Carposporangial conceptacle; note fusion cell (f) (YMC 76/167). (C) Bisporangial conceptacle (YMC 83/99).

Monoecious spermatangia and carpogonia/ carposporangia recorded for August, dioecious spermatangia recorded for January; tetrasporangia not seen; bisporangia recorded throughout the year. Young sporelings were seen in abundance in November on one occasion.

Varying in form from applanate to somewhat leafy. There is anatomical variation in the basal filament cells that may be palisade and oblique as is typical for *Titanoderma*, or squarish and vertically orientated as is usual for *Lithophyllum*. In bistratose margins, however, basal cells are always palisade and oblique.

Gametangial/ carposporangial plants were reported by Suneson (1943) from Sweden in June and July, and by Masaki & Tokida (1960a) from Japan between March and July; both monoecious and spermatangial gametophytes were seen in both areas. In most areas bisporangia predominate, but in Japan only tetrasporangial conceptacles were observed. Suneson (1950, as *Lithophyllum*) found occasional tetrasporangia with two or four nuclei and bisporangia with from four to eight nuclei among binucleate bispores in Swedish plants. He established that the haploid chromosome number was n=16 and that production of tetraspores and binucleate bispores involved meiosis whereas uninucleate bispores were mitotically derived and diploid.

Cabioch (1979) found that the area surrounding sporelings on *Corallina officinalis* became permanently colourless and regarded this as a form of destructive epiphytism; old intergenicula, which had shed the epiphyte, still retained a damaged and irregular surface.

Goss-Custard *et al.* (1979) found fluctuating quantities of *T. corallinae* (as *Lithophyllum*) on *Corallina* in littoral pools at Lough Ine, southern Ireland, but no apparent seasonal correlation.

Woelkerling & Campbell (1992, as *Lithophyllum*) reported that *T. corallinae* occurs in the littoral and sublittoral to 9m depth in southern Australia. It shows morphological variation from flat to warty thalli and grows eplithically and epizoically. It was not observed as an epiphyte and its habit in southern Australia is therefore markedly different from the British Isles where it is exclusively epiphytic.

For field observations on *T. corallinae* growing together with *T. pustulatum* var. *confine* see entry for the latter.

Titanoderma laminariae (P.& H.Crouan) Y.Chamberlain (1991b), p.69.
Figs 36; 37; 38

Lectotype: CO! Herb. P.-L. & H.-M. Crouan, Cherbourg, February 1859, on *Laminaria digitata* (see Chamberlain, 1991b).

Melobesia laminariae P.& H.Crouan (1867), p.150.
Dermatolithon laminariae (P.& H.Crouan) Foslie (1900b), p.13.
Lithophyllum macrocarpum f. *laminariae* (P.& H.Crouan) Foslie (1900c), p.22.
Lithophyllum pustulatum f. *laminariae* (P.& H.Crouan) Foslie (1905a), pp.118, 128.
Dermatolithon crouanii (Foslie) Hamel & Lemoine (1953), p.62, pro parte, excl. type.

Key characters: Epiphytic, almost exclusively on *Laminaria*, tetrasporangial conceptacles immersed, floor to at least 6 cell layers below thallus surface, to 364μm internal diameter. Overgrowing *Melobesia membranacea*; overgrown by *Lithophyllum crouanii*.

Illustrations: Hamel & Lemoine (1953), fig.24 (as *Dermatolithon crouanii*); Chamberlain (1991b), figs 225–247.

Thallus encrusting, adherent, fragile, to at least 20mm diameter, to 500μm thick, flat, orbicular becoming confluent; *margin* entire, sometimes bistratose, often thickened; *surface* very smooth, cell surface honeycomb-like (SEM), colour mauvish (Methuen colour not recorded), texture somewhat chalky; *conceptacles* of all types visible only as pores.

Basal filaments unistratose, cells palisade, oblique, often sinuate, 6–13μm long × 13–52μm diameter; *erect filaments* to 20 cells long, cells 6–39μm long × 7–15μm diameter, often laterally aligned; *epithallial cells* single, flattened to somewhat domed; thallus cell contents mainly sparse, small starch grains (to 3.5μm diameter) frequent in lower thallus; *trichocytes* not seen.

Gametangial plants mainly dioecious, occasionally monoecious; *spermatangial conceptacle* chambers broadly triangular, 97–127μm diameter, 22–33μm high, with or without a spout, with simple spermatangial systems on conceptacle floor only; *carpogonial conceptacle* chambers flask-shaped; *carposporangial conceptacle* chambers elliptical, 208–290μm diameter, 52–117μm high, with the roof 39–52μm thick, pore canal wide at base and tapering towards thallus surface, lined by papillae, roof near pore mainly three cells thick comprising epithallial cell and two equal sized cells, fusion cell wide and shallow with *gonimoblast filaments* up to four cells long plus carposporangium emanating from lower surface at the periphery; *tetrasporangial conceptacle* chamber floors at least six cell layers below thallus surface, chambers elliptical, 312–364μm diameter, 104–156μm high, with the roof 42–94μm thick, structure as for carposporangial conceptacles, columella usually well-developed and pore canal tapering towards surface but not lined by papillae; *tetrasporangia* 75–112μm long × 39–68μm diameter; *bisporangial* plants unknown; old spermatangial conceptacles becoming buried in the thallus, other types becoming infilled.

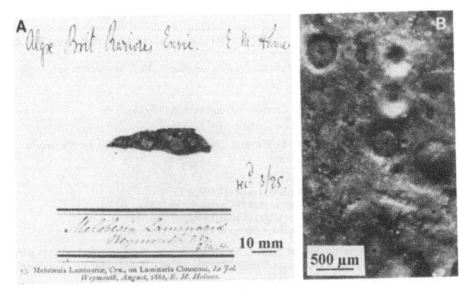

Fig. 36 *Titanoderma laminariae*. (A) Specimen collected by E.M.Holmes (BM). (B) Thallus surface with conceptacles (YMC 86/134).

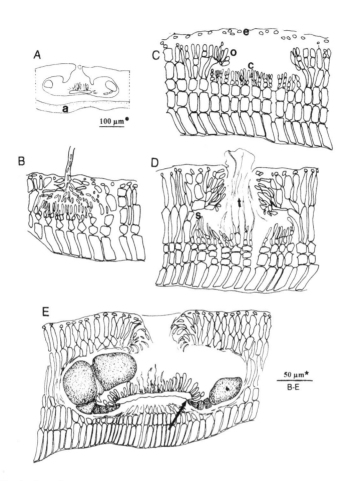

Fig. 37 Vertical sections of *Titanoderma laminariae*. (A) Diagrammatic drawing to show characteristic shape of carposporangial conceptacle, b = basal filament (YMC 86/134/14). (B) Spermatangial conceptacle (YMC 86/134/14). (C) Early carpogonial conceptacle showing carpogonial branches (c), surrounding roof filaments (o) and shedding epithallial layer (e) (YMC 78/80). (D) Mature carpogonial conceptacle; note sterile peripheral branches (s) and trichogynes (t) (YMC 78/80). (E) Mature carposporangial conceptacle; note position at which gonimoblast filaments are attached (arrow).

Rare. Epiphytic on stipes and holdfasts of *Laminaria digitata* and *L. hyperborea*, rarely on *Palmaria palmata* and *Laurencia pinnatifida*. Mainly sublittoral, to at least 10m, sometimes in low littoral pools. Often found as drift.

Shetland Isles, Dorset, Devon (north and south); inconspicuous and possibly more widespread. British Isles to northern France.

Fertile plants collected throughout the year; spermatangia, carposporangia and tetrasporangia found in about equal numbers.

Thalli of *T. laminariae* are superficially similar to, and have been confused with, those of *Lithophyllum crouanii q.v.* which may also grow on *Laminaria* (see Chamberlain, 1991b). *Lithophyllum crouanii* may be distinguished by its non-palisade basal cells and much smaller (to 130μm internal diameter), rounded tetrasporangial conceptacles.

The description of *Lithophyllum pustulatum* f. *laminariae* in Newton (1931, p.306) refers to *Titanoderma pustulatum* var. *pustulatum*, whilst the description of *Dermatolithon crouanii* in Hamel & Lemoine (1953, p.62) refers to *T. laminariae* (see also under *Lithophyllum crouanii*). For further details see Chamberlain (1991b).

Titanoderma pustulatum (Lamouroux) Nägeli (1858), p.532 var. *pustulatum*
Figs 39; 40

Lectotype: CN! (see Woelkerling *et al.*, 1985). Hb. Lamouroux, on *Chondrus crispus*, France.

Melobesia pustulata Lamouroux (1816), p.315.
Dermatolithon pustulatum (Lamouroux) Foslie (1898b), p.ll.
Lithophyllum pustulatum (Lamouroux) Foslie (1904b), p.8.
[An extended synonymy is given in Woelkerling & Campbell (1992, tables 12, 13)].

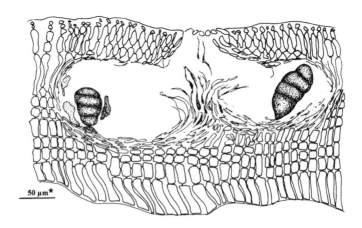

Fig. 38 Vertical section of mature tetrasporangial conceptacle of *Titanoderma laminariae* (YMC 85/191).

Key characters: Plants robust, bright mauve-pink, conceptacles convex, without distinct periphery, bisporangial conceptacle floor 2–3 cell layers below thallus surface, roof 3 cells thick. Overgrowing *Melobesia membranacea* and *Pneophyllum* spp.

Illustrations: Lamouroux (1816), pl.XII, fig.2 (as *Melobesia*); Kornmann & Sahling (1977), fig.115 (as *Dermatolithon*); Woelkerling *et al.* (1985), figs 29–39; Chamberlain, (1991) figs 55–81; Woelkerling & Campbell, (1992, as *Lithophyllum*) figs 50–60.

Thallus encrusting, adherent, to at least 30mm diameter, to 500µm thick, flat and non-imbricating, orbicular, becoming confluent, confluences flat; *margin* entire, bistratose, usually areally extensive; *surface* of thallus smooth, colour bright mauve-pink to brown-pink (Methuen greyish-ruby to dark magenta), chalky; *bisporangial conceptacles* widely spaced to densely crowded, convex, without a distinct periphery, 400–800µm diameter, surface honeycomb-like (SEM).

Basal filaments unistratose, cells palisade, oblique and usually sinuate, often very high, 8–21µm long × 13–190µm diameter; erect filaments absent or to 12 cells long, cells 8–118µm long × 6–22µm diameter, laterally aligned; *epithallial cells* triangular or domed with cell contents mainly very sparse and starch grains uncommon; *trichocytes* seen very rarely.

Gametangial and tetrasporangial plants not seen; *bisporangial conceptacle* floors two to three cell layers below thallus surface, chambers low-elliptical, 156–468µm diameter, 78–234µm high, with the roof 23–73µm thick, three cell layers thick throughout, epithallial and inner cells small, middle cell tall and thin, roof tapering to pore, columella slightly- to well-developed; *bisporangia* often pear-shaped, 72–156µm long × 33–104µm diameter, up

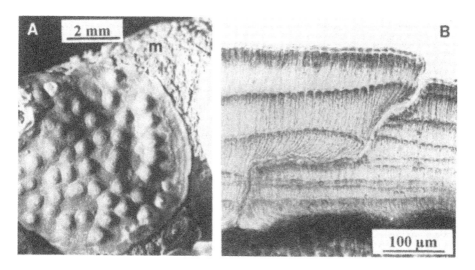

Fig. 39 *Titanoderma pustulatum* var. *pustulatum*. (A) Bisporangial thallus overgrowing *Melobesia membranacea* (m) on *Chondrus* (YMC 88/126). (B) VS of adjacent thick thalli that lack bistratose margins (YMC 85/27).

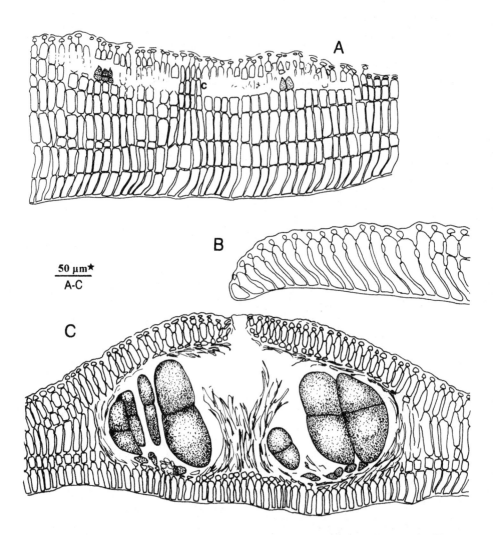

Fig. 40 Vertical sections of *Titanoderma pustulatum* var. *pustulatum*. (A) Young bisporangial conceptacle; note peripheral bisporangial initials (shaded) and central columella (c) (YMC 75/143). (B) Thallus margin (YMC 75/143). (C) Mature bisporangial conceptacle (YMC 85/30).

to *c*.60 per conceptacle; old conceptacles usually becoming infilled, rarely becoming buried in the thallus.

Common. Epiphytic, mainly on *Chondrus*, *Mastocarpus* and stipes and holdfasts of *Laminaria* spp, frequently overgrowing *Melobesia membranacea* thalli; also on other red and brown algae and occasionally on rocks and shells in lower littoral pools; predominantly littoral, recorded rarely from sublittoral, to at least 2m.

Throughout the British Isles.

Southern Norway to northern France, Faroes, Helgoland; southern Australia; South Africa; Canary Isles, Sénégal (records from southern Australia to Sénégal refer to *T. pustulatum* agg.).

Bisporangia recorded throughout the year; prolific spore settlement seen June-August.

The type specimen of *T. pustulatum* var. *pustulatum*, the generitype of *Titanoderma*, was described in detail by Woelkerling *et al.* (1985). The type locality is France (probably the Cherbourg area); bisporangial plants occur commonly as flat, robust, mauve-pink, epiphytes in northern France and the British Isles and are unusually consistent both morphologically and anatomically. In the Mediterranean area (Bressan, 1974, as *Dermatolithon*) gametangial, tetrasporangial and bisporangial plants are recorded. Woelkerling & Campbell (1992, as *Lithophyllum*) have given a detailed description *T. pustulatum* from southern Australia where gametangial, tetrasporangial and bisporangial plants all occur. They have found that while individual plants of the varieties described here from the British Isles can be recognized, variation is so great and continuous that it was impossible to recognize varieties. As Woelkerling & Campbell (1992) also suggest, it is possible that the greater stability of varieties in the British Isles is due to the fact that only apomictic, bisporangial populations are present.

Titanoderma pustulatum var. **canellatum** (Kützing) Y.Chamberlain (1991b), p.61.
Figs 41; 42

Holotype: CN! (see Chamberlain, 1986; 1991b). Lamouroux, Mediterrranean, on *Rytiphlea*.

Melobesia verrucata Lamouroux (1816), p.316.
Melobesia pustulata ß *canellata* Kützing (1849), p.696. (L!).
Titanoderma verrucatum (Lamouroux) Y.Chamberlain (1986), p.201.

Key characters: Plants thin, lacking erect filaments, bisporangial conceptacles prominent, with distinct periphery often marked by a 'skirt' (Fig.41A). Overgrowing *Melobesia membranacea*.

Illustrations: Garbary (1978), figs 7, 8 (as *Dermatolithon*); Chamberlain (1978), figs 3, 7 (as *Dermatolithon litorale*); (1986), figs 4–10 (as *Titanoderma verrucatum*); (1991b), figs 205–207.

Thallus encrusting, adherent, to at least 3mm diameter, to 90μm thick, orbicular to confluent, confluences flat, thallus flat and sometimes imbricating, *margin* entire, bistratose, *surface* smooth, colour pinkish, greyish or violet when dry (no Methuen record for fresh plants); *bisporangial conceptacles* very prominent, hemispherical, distinct at periphery and often with a 'skirt' (Fig.41A), 250–400µm diameter, surface stepped (SEM) (Fig.41A), sometimes with ring of *trichocytes* at base.

Basal filaments unistratose, palisade, cells oblique and sinuate, 7–16µm long × 33–91µm diameter with contents concentrated at the top; *erect filaments* absent except in immediate vicinity of conceptacles; *epithallial cells* triangular; *trichocytes* occasional in basal cells.

Gametangial and tetrasporangial conceptacles not seen; *bisporangial conceptacle* chamber floors two to three cell layers below thallus surface, chambers hemispherical, 130–350µm diameter, 68–215µm high, with the roof 26–52µm thick, roof composed of three

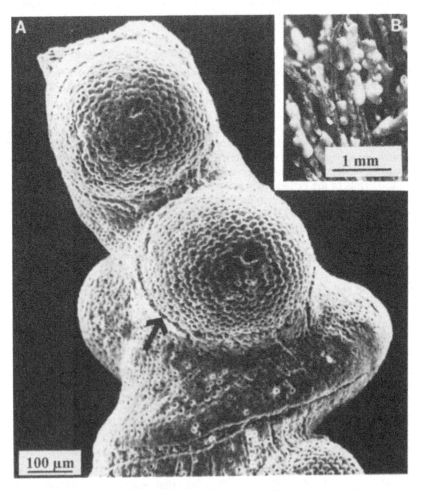

Fig. 41 *Titanoderma pustulatum* var. *canellatum*. (A) Bisporangial thallus growing on *Cladophora rupestris*; note skirt at conceptacle base (arrow) (YMC 78/142). (B) Thalli growing on *C. rupestris* (YMC 80/107).

cell layers, epithallial and inner cells small, middle cell taller and somewhat clavate, middle cell usually lengthening at pore and small papillae sometimes surrounding pore canal, columella slightly- to well-developed; *bisporangia* 60–75μm long × 28–36μm diameter.

Rare. Epiphytic, mainly on *Furcellaria* and turf algae such as *Cladophora* spp, *Jania rubens* and *Gelidium pusillum.*

Hampshire, Dorset, Co.Clare.
  Mediterranean.

Bisporangial conceptacles recorded throughout the year.

Since Lamouroux's (1816) original description of epiphytic *Melobesia verrucata* from the Mediterranean, the epithet has been misapplied to numerous crustose coralline taxa with

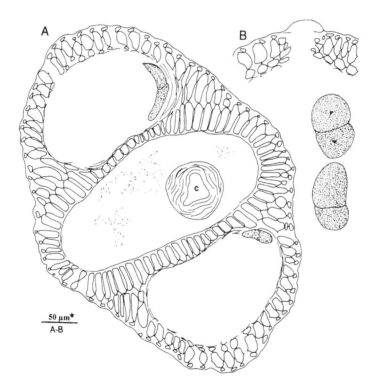

Fig. 42 Vertical sections of of *Titanoderma pustulatum* var. *canellatum*. (A) Two bisporangial conceptacles completely encircling a filament of *Cladophora rupestris* (c), two bisporangia have been extruded (YMC 80/107B). (B) Pore of bisporangial conceptacle (YMC 78/412).

small, thin thalli and minute but very prominent conceptacles, including species of *Fosliella* and *Pneophyllum*, as well as *Titanoderma* (Chamberlain, 1986).

Mediterranean plants of var. *canellatum* have been misidentified as *Lithophyllum* (*Dermatolithon*) *litorale* by Suneson (1943), which is now regarded as convarietal with *T. pustulatum* var. *confine* q.v. Nevertheless, the predominantly grey-violet colour of var. *canellatum* plants suggests a close relationship with var. *confine* and the distinction between var. *canellatum*, which lacks erect filaments, and minute plants of var. *confine* (q.v.), with some development of erect filaments, is somewhat arbitrary. Limited observations suggest that var. *canellatum* can also be distinguished by having a regularly stepped roof surface (SEM) as compared with an irregularly ridged one in minute plants of var. *confine* (Chamberlain, 1991b, figs 185, 205). Plants referable to var. *canellatum* are abundant as epiphytes in the Mediterranean, often occurring as minute thalli each bearing a conceptacle.

As in var. *confine* and var. *macrocarpum*, columella cells may become balloon-like and swollen cells sometimes develop from the underside of the conceptacle roof.

Titanoderma pustulatum var. **confine** (P.& H.Crouan) Y.Chamberlain (1991b), p. 50.
Figs 43–46

Lectotype: PC! (see Chamberlain, 1991b). Ex Herb. Crouan [presumably from Finistère].

Melobesia confinis P.& H.Crouan (1867), p.150.
Dermatolithon hapalidioides f. *confine* (P.& H.Crouan) Foslie (1900b), p.12, (as *confinis*).
Lithophyllum litorale Suneson (1943), p.39. (C!).
Dermatolithon litorale (Suneson) Hamel & Lemoine (1953), p.66.
Tenarea confinis (P.& H.Crouan) W.& P.Adey (1973), p.393.
Dermatolithon confine (P.& H.Crouan) W.& P.Adey in Parke & Dixon (1976) p.534 (comb.invalid, ICBN, Art. 33.2), (as *confinis*).
Titanoderma confine (P.& H.Crouan) J.Price *et al.* (1986), p.86, (as *confinis*).

Key characters: Plants bluish, thallus flat or imbricating, conceptacles rounded-conical with roof to 4 cells thick. Overgrowing *Melobesia membranacea* and *Pneophyllum* spp; about equal in competition with *Titanoderma corallinae*.

Illustrations: Rosenvinge (1917), figs 184, 185 (as *Lithophyllum macrocarpum* f. *intermedia);* Lemoine (1913a), p.137, fig.3 (as *Lithophyllum hapalidioides* f. *confinis);* Printz (1929), pl.LXXll, fig.7 (as *L.hapalidioides* f. *confinis);* Chamberlain (1978), figs 1, 2, 4, 5, 6, 8 (as *Dermatolithon litorale);* (1991b), figs 160–204.

Thallus encrusting, adherent to loosely attached, orbicular to confluent, to at least 30mm diameter, to 620 (900)μm thick, often extensive, adjacent plants overgrowing one another, individual thalli flat to extensively imbricating; *margin* entire, bistratose; *surface* chalky, smooth or patchy due to areas of regeneration, colour mauvish or bluish (Methuen greyish-violet, dull violet, dark violet), often becoming bleached; *bisporangial conceptacles* widely spaced to densely crowded, rounded-conical, mainly with a distinct periphery, often with a rim at base, 400–500μm diameter, surface usually ridged (SEM).

Basal filaments unistratose with cells palisade, oblique, usually sinuate, often very high, 11–22μm long × 25–179μm diameter; *erect filaments* absent or to 14 cells long with cells 14–93μm long × 9–24μm diameter; *epithallial cells* triangular, rectangular or domed with cell contents usually sparse and starch grains uncommon, new basal filaments often

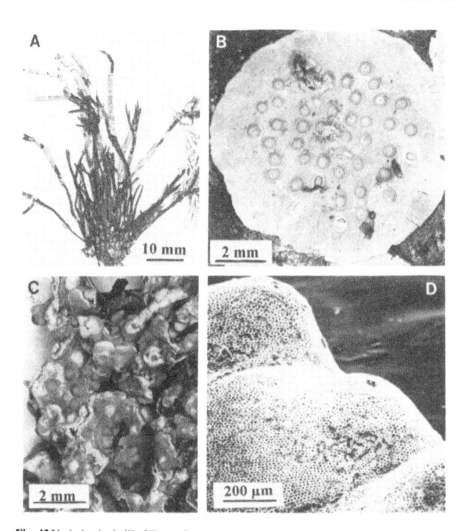

Fig. 43 Variation in thalli of *Titanoderma pustulatum* var. *confine*. (A) On *Furcellaria* (YMC 80/105). (B) Thin, flat thallus on *Palmaria* (YMC 86/153A). (C) On the base of *Corallina officinalis* (YMC 84/399). (D) Bisporangial conceptacles (YMC 79/208).

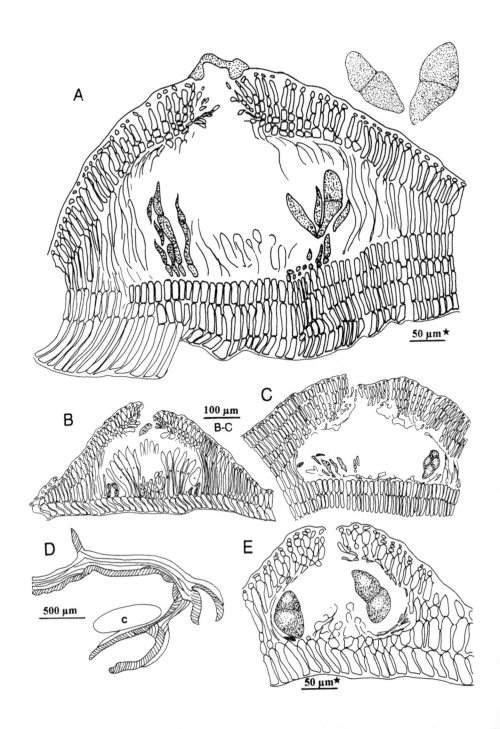

regenerating from erect filament cells to form thallus lobes; *trichocytes* occurring occasionally in basal or erect filaments.

Gametangial and tetrasporangial plants not seen; *bisporangial conceptacle* chamber floors two to three cell layers below thallus surface, chambers hemispherical to elliptical, 234–416μm diameter, 91–234μm high, roof 29–62μm thick with roof thickening to 3–5 cell layers near pore with cells irregularly sized, columella slightly- to well-developed; *bisporangia* 62–104μm long × 26–52μm diameter (occasional stunted tetrasporangia seen), to *c*.60 per conceptacle; old conceptacles either becoming infilled or buried in thallus. [Minute plants described below.]

Common. Epiphytic, mainly on *Corallina, Furcellaria, Gelidium pusillum* and *Laurencia pinnatifida*, or on rocks, stones or shells, mainly littoral but reported to 45m depth (Adey & Adey, 1973, fig.58, as *Tenarea*).

Cornwall to Hampshire, Yorkshire to Northumberland; Pembroke, Anglesey, Isle of Man, Argyll, Sutherland, Shetland Isles; Co.Clare, Co.Galway, Co.Down. [Minute plants seen in Hampshire and Dorset].

British Isles to Canary Isles, W. Baltic, W. Mediterranean; India, USA (California).

Bisporangia present throughout the year in some localities or present July-September elsewhere, prolific spore settlement seen in July.

The Crouan brothers (1867) described *Melobesia confinis* as being, among other things, pale violet with cells from four to six times as long as wide (in VS) and these characters have been consistently noted in descriptions of plants from the British Isles (Lemoine, 1913a, as *Lithophyllum hapalidioides* f. *confinis*; Adey & Adey, 1973, as *Tenarea confinis*; Chamberlain, 1978, as *Dermatolithon litorale* pro parte; 1991b). Suneson (1943, p.37, as *Lithophyllum litorale*) likewise mentioned a greyish violet colour in Swedish plants and observed that they could also be whitish, presumably due to bleaching.

The formation of overgrowing (imbricating) lobes by thallus regeneration was mentioned particularly by Foslie (1900b, as *D.hapalidioides* f. *confinis*), Lemoine (1913a), Rosenvinge (1917, as *L.macrocarpum* f. *intermedia*), Hamel & Lemoine (1953, as *D.hapalidioides* f. *confinis*) and Chamberlain (1991b). The description of tetra/ bisporangial conceptacles as being conical with rounded or somewhat flattened tops has also remained consistent. This steady perception of the characteristics of the taxon contrasts with the instability of its rank, reflected in the large number of synonyms; Chamberlain (1991b) resolved the difficulty by grouping all members of this complex in a single variable species with a number of varietal focal points.

Columella cells are sometimes balloon-like and similar swollen cells can develop from the under surface of the roof, but this feature does not seem to have taxonomic significance.

Fig. 44 Vertical sections of *Titanoderma pustulatum* var. *confine*. (A) Bisporangial conceptacle and extruded bisporangia (YMC 84/147). (B) Bisporangial conceptacle with swollen paraphyses (YMC 87/62). (C) Old bisporangial conceptacle becoming immersed in the thallus (YMC 84/416). (D) Diagrammatic representation of a continually regenerating thallus growing among *Corallina* (c) plants; diagonal stripes represent palisade basal filaments (YMC 84/200). (E) Bisporangial conceptacle of minute growth form (YMC 79/208).

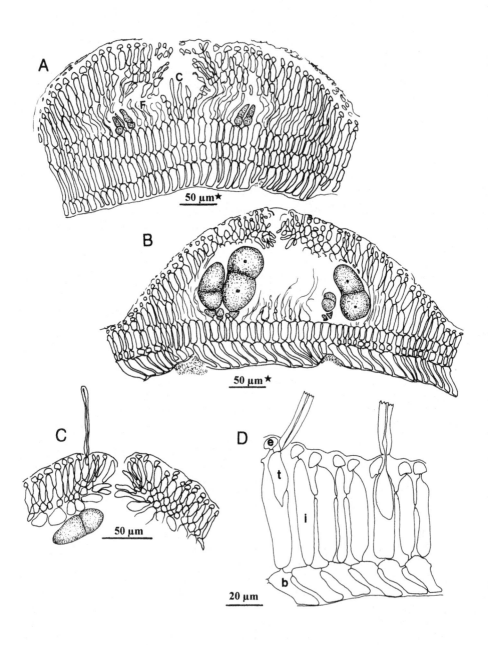

In Hampshire and Dorset *minute epiphytic plants* (Figs 44E; 46) occur which differ perceptibly in size from the main populations of var. *confine* among which they grow. Their thalli do not exceed five cells thick and bear hemispherical bisporangial conceptacles measuring 169–250μm internal diameter, 91–130μm high with roof 26–78μm thick (Chamberlain, 1991b).

Sequential field observations in Hampshire, Dorset and the Isle of Man (Chamberlain, 1991b) showed that the abundance of var. *confine* varies spasmodically both during the year and from year to year, suggesting that populations are maintained at a low level in associations with other species in algal turfs, from which they are able to colonize other substrata very rapidly under favourable conditions. In Hampshire (Chamberlain, 1991b), var. *confine* grows abundantly on *Furcellaria* together with *T. corallinae* which may be overgrown by var. *confine* or vice versa.

SEM characteristics are discussed by Chamberlain (1991b), the conceptacle roof being mainly ridged to somewhat stepped.

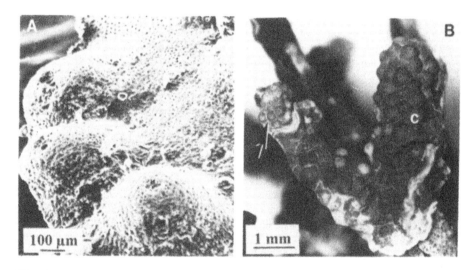

Fig. 46 Minute growth form of *Titanoderma pustulatum* var. *confine*. (A) Bisporangial conceptacles (YMC 79/208). (B) Thallus of typical var. *confine* (c) adjacent to a minute thallus (arrow) showing marked difference in size (YMC 80/104).

Fig. 45 Vertical sections of *Titanoderma pustulatum* var. *confine*. (A) Immature conceptacle showing bisporangial initials (shaded), columella (c) and interspersed filaments (f) (YMC 84/28B). (B) Mature bisporangial conceptacle (YMC 84/28B). (C) Bisporangial conceptacle roof bearing a trichocyte (YMC 84/28B). (D) Thallus cells bearing trichocytes (t), note basal filament cell (b), subepithallial initial (i) and epithallial cell (e) (YMC 87/62).

Titanoderma pustulatum var. *macrocarpum* (Rosanoff) Y.Chamberlain (1991b, p.33).
Figs 47–49

Lectotype: CHE! (see Chamberlain, 1986; 1991b). *Le Jolis*, Alg. Mar. Cherbourg no.276, France (Rochers des Flamands, Cherbourg), 19.iii.1863, on *Phyllophora crispa*.

Melobesia macrocarpa Rosanoff (1866), pp.74–75.
Melobesia simulans P.& H.Crouan (1867), p.149. (CO!).
Melobesia hapalidioides P.& H.Crouan (1867), p.150. (CO!).
Lithothamnion adplicitum Foslie (1897), p.17. (BM!).
Dermatolithon pustulatum f. *macrocarpum* (Rosanoff) Foslie (1898b), p.11, (as *macrocarpa*).
Dermatolithon hapalidioides (P.& H.Crouan) Foslie (1898b), p.11.
Lithophyllum hapalidioides (P.& H.Crouan) Heydrich (1901), p.537.
Tenarea hapalidioides (P.& H.Crouan) W.& P.Adey (1973), p.394.
Titanoderma hapalidioides (P.& H.Crouan) J.Price *et al.* (1986), p.86.
Titanoderma macrocarpum (Rosanoff) Chamberlain (1986), p.203.

Key characters: Plants reddish-orange, thallus flat or imbricating, bisporangial conceptacles conical, roof to 7 cells thick, basal cells (VS) often unusually short and vertical. Overgrowing *Melobesia membranacea*.

Illustrations: Rosanoff (1866), pl.IV, figs 4–6, 8, 11–20 (as *Melobesia macrocarpa*); Suneson (1943), figs 22, 23, pl.VIII fig.37 (as *Lithophyllum pustulatum*); Chamberlain (1986), figs 12–21 (as *Titanoderma macrocarpum*); (1991b), figs 82-119.

Thallus encrusting, adherent, to at least 50mm diameter, to 700µm thick, orbicular to confluent, flat to extensively imbricating; *margin* entire, bistratose; *surface* of thallus smooth, colour rose-, to salmon-, to red-, to orange-pink (Methuen greyish red *inter alia*), texture somewhat glossy; *bisporangial conceptacles* widely spaced to densely crowded, convex to hemispherical, varying from lacking a distinct periphery to having a well-defined rim at the base, 400–900µm diameter; conceptacle surface with horseshoe-like or stepped cells (SEM).
   Basal filaments unistratose with cells mainly palisade and oblique, but often with some areas of short, vertically orientated cells, 9–36µm long × 5–139µm diameter; *erect filaments* from absent to 16 cells long, cells 5–80µm long × 9–41µm diameter, often with particularly pronounced secondary pit connections; single *epithallial cells* flattened; cell contents usually sparse, but lower thallus cells frequently full of large (to 6µm diameter) starch grains; new basal filaments often regenerating from erect filament cells; *trichocytes* not seen.
   Gametangial and tetrasporangial conceptacles unknown; *bisporangial conceptacle* chambers low elliptical to hemispherical, 190–540µm diameter, 65–210µm high, with the roof 52–105µm thick, thickening to 3–7 cells thick at the pore canal that is sometimes lined with papillae; columella slightly to well-developed, *bisporangia* 50–117µm long × 21–65µm diameter, to *c.*80 per conceptacle; old conceptacles becoming buried in the thallus. [Imbricating plants described below].

Rare. Epilithic, epizoic, or rarely epiphytic particularly on *Phyllophora* and old maerl. Sublittoral to 33m (Adey & Adey, 1973, as *Tenarea hapalidioides*, fig.59), rarely in lower littoral pools.

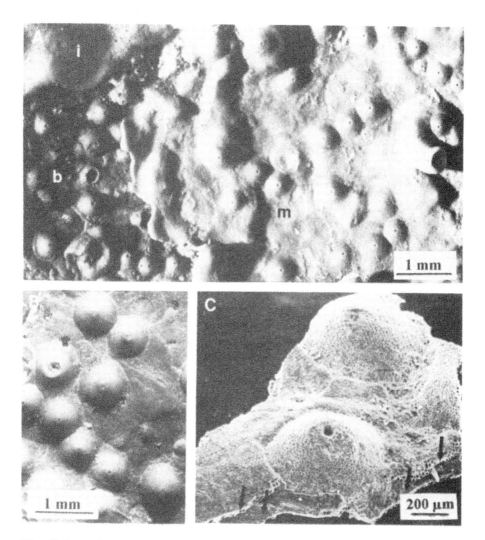

Fig. 47 *Titanoderma pustulatum* var. *macrocarpum*. (A) Epilithic thallus of bisporangial var. *macrocarpum* (m) overgrowing imbricating growth form (b). *Lithophyllum incrustans* (i) overgrows both thalli (YMC 86/135). (B) Type (bisporangial) of *Melobesia hapalidiodes* (CO, *P. & H. Crouan*, Brest). (C) Bisporangial conceptacles of imbricating growth form; arrows indicate thallus layers (YMC 86/135).

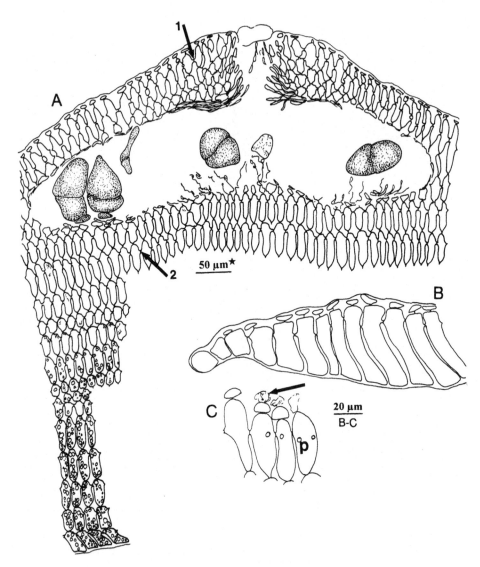

Fig. 48 Vertical sections of type of *Melobesia hapalidioides* now subsumed in *Titanoderma pustulatum* var. *macrocarpum*. (A) Bisporangial conceptacle and thallus cells; note tube-like secondary pit connections (1,2). (B) Bistratose thallus margin. (C) Epithallial shedding (arrow), secondary pit connections are seen in face view (p).

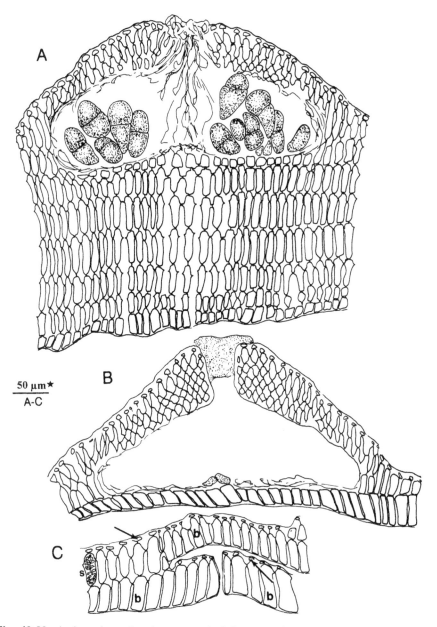

Fig. 49 Vertical sections showing anatomical features of *Titanoderma pustulatum* var. *macrocarpum*. (A) Bisporangial conceptacle (YMC 84/229). (B) Bisporangial conceptacle of imbricating growth form (YMC 83/162). (C) Imbricating thallus layers showing regeneration to produce thalli composed of basal (b) and epithallial (arrows) cells; all cells are full of starch grains (s) (YMC 78/199).

Northumberland, Sussex-Cornwall, Pembroke, Isle of Man; Channel Isles; Co. Clare, Co. Mayo.

British Isles to France, W. Baltic, E. Mediterranean. Reports for Norway (Rueness, 1977) and Italy (Bressan, 1974) not confirmed.

Bisporangia recorded throughout the year.

The reddish or orange plants of var. *macrocarpum* were noted by the Crouan brothers (1867, as *Melobesia hapalidioides* and *M. simulans*), Hamel & Lemoine (1953, as *Dermatolithon hapalidioides*) and Adey & Adey (1973, as *Tenarea hapalidioides*) and it has always been regarded as a mainly sublittoral, epilithic or epizoic taxon. The taxonomic significance of differences in roof structure was noted by Chamberlain (1986, 1991b) who classified epiphytic plants such as *Melobesia macrocarpa* Rosanoff (1866), and *Lithophyllum pustulatum* (as described by Suneson, 1943) in the same infraspecific taxon as epilithic plants generally attributed to var. *hapalidioides* because they all have a thickened bisporangial conceptacle roof with up to 7 cell layers.

As in var. *confine*, conceptacles filled with balloon-like cells occur sporadically and Rosanoff's drawing (1866, pl.IV, figs 11–15) show this feature in the type material, although it is no longer evident in herbarium specimens (CHE!).

A tendency for thallus regeneration to form imbricating lobes is common in epilithic plants of var. *macrocarpum*; some plants collected in Hampshire and Dorset lack erect filaments and are composed entirely of layers of bistratose, imbricating thalli (Figs 47C; 49B,C). Such a growth pattern recalls that seen in *Hydrolithon boreale, q.v.*, with all layers belonging to a single plant. Conceptacles in such imbricating plants are very prominently domed and the plants are salmon pink. None of the synonyms refers to imbricating plants as such, but a good example is to be seen on the shell that also bears the type specimen of *Lithothamnion adplicitum* (BM!).

Further evidence that var. *macrocarpum* forms thin lobes comes from the following observation. In November 1990 epiphytic thalli were found growing on *Cladophora* sp., at the air-water interface, in a laboratory tank at Hayling Island Marine Laboratory, Hampshire. These occurred as bistratose thallus lobes measuring about 1mm across and orientated at right angles to the host, together with thicker thalli bearing typical var. *macrocarpum* bisporangial conceptacles and forming a sleeve round the host. Trichocytes occurred in thallus and conceptacle roof cells. The source of these thalli was probably an epilithic thallus of this species that had remained in the laboratory tank for some years.

Sublittoral epilithic plants are fairly common in the British Isles and var. *macrocarpum* also appears to be a characteristic epiphyte on old maerl (C.Maggs, pers. comm.). Epiphytic plants are otherwise rare in the British Isles and France although the epiphytic type material, collected at Cherbourg in 1863, was prolific (Rosanoff, 1866). Suneson's (1943) epiphytic Swedish plants were growing on *Phyllophora* spp., various other algae and on a tunicate. The species may persist as long-lived, sublittoral, epilithic plants from which epiphytic plants can quickly be recruited when conditions are favourable.

MASTOPHOROIDEAE Setchell

by Yvonne M.Chamberlain

MASTOPHOROIDEAE Setchell (1943), p.134 (as Mastophoreae).

Type genus: *Mastophora* Decaisne (1842), p.365.

Thallus nongeniculate; some cells of contiguous vegetative filaments normally joined by cell fusions, secondary pit connections absent or rare; tetra/bisporangia lacking apical plugs and borne within uniporate conceptacles.

A group of genera, including *Mastophora*, was set aside as the tribe Mastophoreae by Setchell (1943) who defined them as: "differing from all other Corallinaceae by their lack of, or, at least, by their very feeble differentiation of tissues". It appears to have been Suneson (1945) who pinpointed the criteria (a combination of uniporate gametangial conceptacles and fusions between cells of contiguous vegetative filaments) which distinguish a group of genera now regarded as a subfamily.

Generic delimitation within the subfamily has undergone many changes in recent years. Characters used diagnostically have included whether the thallus was thin or thick, whether it was monomerous or dimerous, whether or not a coaxial medulla was present, and whether trichocytes occurred individually or in discrete groups (e.g. Dawson, 1960; Adey, 1970a; Johansen, 1981; Adey *et al.*, 1982; Chamberlain, 1983; Mendoza & Cabioch, 1986; Cabioch, 1988; Penrose & Woelkerling, 1988; Woelkerling, 1988). Additionally Chihara (1972, 1973a,b,c, 1974a,b), Notoya (1976a) and Chamberlain (1983; 1984) suggested that spore germination pattern could be a diagnostic character in the Corallinaceae. This character has maintained its significance with respect to distinguishing between *Pneophyllum* and the *Fosliella*-state of *Hydrolithon* and in the species taxonomy of the latter.

Penrose (1991, 1992a,b), Penrose & Woelkerling (1991, 1992) and Penrose & Chamberlain (1993) studied type specimens together with recently collected southern Australian material and showed that vegetative characters rarely form a sound basis for generic delimitation. These authors proposed generic definitions that depend mainly on diagnostic differences in habit and reproductive structures such as male conceptacle structure and pore structure in tetra/bisporangial conceptacles. A key to eight recognized genera, *Hydrolithon, Lesueuria, Lithoporella, Mastophora, Metamastophora, Neogoniolithon, Pneophyllum* and *Spongites* is provided in Penrose & Chamberlain (1993).

The subfamily is poorly represented in the northern Atlantic. Elsewhere, massive rock-encrusting species of the genera *Hydrolithon, Neogoniolithon* and *Spongites* form a conspicuous element of the nongeniculate coralline flora and foliose or taeniform plants are a characteristic element of Australian representatives. *Fosliella* and *Pneophyllum* (see Chamberlain, 1983) have been recorded for the British Isles but *Fosliella* has recently been subsumed in *Hydrolithon* (Penrose & Chamberlain, 1993).

Fosliella was a new name proposed by Howe (1920) for thin species with uniporate tetra/bisporangial conceptacles previously included in *Melobesia* (subgen. *Eumelobesia* Foslie) Foslie (1898b). A new name had to be chosen because Lamouroux (1812) had previously assigned the multiporate species *Corallina membranacea* Esper (1806) to this genus. Chamberlain (1983) revised the concept of *Fosliella* to include species with a 4-celled centre to the germination disc and trichocytes terminating basal filaments (cf. *Pneophyllum*). The complexities of historical concepts are summarised in the above publications and in Mason (1953) and Woelkerling (1988). Penrose & Chamberlain (1993) found that the *Hydrolithon* pore character described above occurred also in *Fosliella* and concluded that this genus must be subsumed in *Hydrolithon*. However, the thin dimerous thalli, with 4-celled germination discs, attributed to *Fosliella*, are generally rather distinct from thicker dimerous or monomerous plants traditionally classified as *Hydrolithon sensu stricto* (see Woelkerling, 1988). Penrose & Chamberlain (1993) proposed the retention of the concept of this form by referring to it as the *Fosliella*-state of *Hydrolithon* sensu lato.

Harlin *et al.* (1985) confirmed the presence of 4-celled germination discs in species attributed to *Fosliella* and further suggested that the pattern of surrounding cells could be used as a diagnostic species character. Examination of material from many parts of the world by Penrose & Chamberlain (1993) has supported this concept. These authors typified *Fosliella* with *Melobesia farinosa* Lamouroux (1816) which had a germination disc of four central cells flanked by twelve surrounding cells (Fig.3A). It was suggested (Cabioch, 1972, p.250) that taxa like *Fosliella*, with thin, fast-growing thalli that become reproductive at an early stage, are neotenic. This seems probable and the similarly 4-celled germination disc reported by Chihara (1974b, fig.5 M-R) in *Porolithon onkodes* (Heydrich) Foslie lends support to this suggested relationship. At present no further data are available to confirm this theory and it will probably be necessary to undertake culture studies to test this hypothesis.

HYDROLITHON Foslie

HYDROLITHON Foslie (1909), p.55, sensu lato, emend. Penrose & Woelkerling (1992), p.87.

Type species: *Hydrolithon reinboldii* (Weber van Bosse & Foslie in Foslie) Foslie (1909), p.55, see Penrose & Woelkerling (1992), p.83.

Heteroderma Foslie (1909), p.56 pro parte.
Porolithon Foslie (1909), p.57.
Fosliella Howe (1920), p.587.

Plants nongeniculate and attached or unattached; varying from flat, warty, lumpy, fruticose attached thalli to unattached forms; all conceptacle types uniporate.

Thallus occasionally unconsolidated, usually pseudoparenchymatous. Structure sometimes dorsiventral, dimerous and composed of a single layer of non-palisade basal filaments, with or without erect filaments that may become very long, and non-flared epithallial cells; sometimes monomerous and composed of medullary filaments running more or less parallel to the substratum giving rise to cortical filaments running more or less vertically; sometimes with monomerous protuberances with radial construction. Contiguous cells of adjacent filaments often joined by cell fusions, secondary pit connections unknown. Trichocytes common, occurring either terminally (Fig.50C) in basal filaments, or dorsally on basal cells individually or in discrete groups (fields) (Figs 52A; 53E;F), or terminally on erect or cortical filaments; also present on conceptacles (Fig.54B). Spore germination disc of the *Fosliella*-state of *Hydrolithon* (Figs 3A-C) with four-celled central element.

Gametangial plants monoecious or dioecious; simple spermatangial systems borne on conceptacle floor only; carposporangial conceptacles with central fusion cell bearing gonimoblast filaments peripherally; tetrasporangial conceptacles each arising from discoid, subepithallial primordium, conceptacle roof formed from filaments interspersed among the sporangia, pore canal surrounded by enlarged, vertically orientated cells; zonate tetrasporangia borne peripherally or across conceptacle floor; bisporangial conceptacles as tetrasporangial ones but bearing two uninucleate bispores.

Spore germination discs of the *Dumontia*-type (Chemin, 1937), spore usually dividing to give four central cells (Figs 3A-C) each giving rise to a single surrounding cell that then cuts off further surrounding cells, the number depending on the species. In *Hydrolithon boreale* the four surrounding cells produce no further surrounding ones but each central cell usually cuts off a small cell apically (Fig.3B) that divides no further.

Chamberlain (1983) recorded one species of *Fosliella* from the British Isles as *F. farinosa* (Lamouroux) Howe on the basis of the presence of a 4-celled germination disc and trichocytes terminating basal filaments. She further distinguished the unconsolidated f. *callithamnioides*. The taxonomic significance of the disposition of cells surrounding the germination disc, elucidated by Bressan *et al.* (1977) and Harlin *et al.* (1985), has since become evident. Penrose & Chamberlain (1993) re-examined the type material (CN!) of *Hydrolithon farinosum* and could not find the pattern previously attributed to *Fosliella farinosa* (Chamberlain, 1983, fig.7B) but two patterns were seen, one with eight surrounding cells as in *F. cruciata* Bressan in Bressan *et al.* (1977, fig.1; see Fig.3C) and a second one with twelve surrounding cells (Fig.3A). As the latter was more abundant and the former was already well established for *F. cruciata*, *Hydrolithon farinosum* was lectotypified with the second pattern. On the diagnostic basis of this pattern, a few plants seen by Chamberlain (1983, figs 20A,E,22B,23) in the British Isles pertain to *H. farinosum*. Some *H. cruciatum* (*q.v.*) plants have also subsequently been identified. Most British Isles plants have the pattern seen in Fig.3B (see Chamberlain, 1983, fig.7B) which is associated with the predominant *Hydrolithon* s.l. species that was cultured by Chamberlain (1977b) and from which culture the germination disc was illustrated. This species, which characteristically develops imbricating thallus layers, agrees well with the type (TRH!) of *Melobesia farinosa* f. *borealis* Foslie (1905a, p.96) and this entity is now renamed *Hydrolithon boreale* (Foslie) Y.Chamberlain. Finally *Pneophyllum sargassi* is now transferred to *Hydrolithon*. Although neither germination discs nor trichocytes have been seen, the presence of a *Hydrolithon* type of pore merits its reclassification and it is suggested that this is a thickened form of *H. farinosum* although culture studies are needed to test this theory. Penrose (1992a) showed that in Australian plants of *H. cymodoceae* (Foslie) Penrose, thin *Fosliella*-state thalli occurred on ephemeral seagrass leaves and thicker thalli developed on the longer-lived stems and it is possible that the *H. farinosum/H. sargassi* relationship is similar. This is reinforced by Masaki & Tokida's (1963, p.5, as *Melobesia*) observation that the type gathering of *Melobesia sargassi* (TRH) contained mainly *H. farinosum*. A further species, *H. samoënse* (Foslie) D.Keats & Y.Chamberlain, which does not pertain to the *Fosliella*-state, is newly recorded for the British Isles.

It should be noted that although germination disc pattern appears to provide a satisfactory diagnostic character for species delimitation in *Fosliella*-state species of *Hydrolithon*, considerable variation is seen (YMC obs.) in plants derived from these sporelings and it would be very difficult to delimit species on the basis of vegetative and reproductive anatomical features alone.

KEY TO SPECIES

1 Thallus multistratose, erect filaments present, germination disc unknown 2
   Thallus bistratose, erect filaments absent, germination disc with 4-celled centre 3
2 Thallus dimerous, epiphytic .... *H. sargassi*
   Thallus dimerous or monomerous, epilithic .... *H. samoënse*
3 Thallus forming imbricating layers, germination disc centre surrounded by 4 cells (Fig.3B) .... *H. boreale*
   Thallus not forming imbricating layers, germination disc centre surrounded by 8 or 12 cells 4
4 Germination disc centre surrounded by 8 cells (Fig.3C) .... *H. cruciatum*
   Germination disc centre surrounded by 12 cells (Fig.3A) .... *H. farinosum*

Hydrolithon boreale (Foslie) Y.Chamberlain comb.nov.
   Figs 3B; 50; 51; 66

Lectotype: TRH! (now selected) *M.H.Foslie*, Roundstone, Ireland, 18.iv. 1899, sample a) on *Gigartina*

Melobesia farinosa f. *borealis* Foslie (1905a), p.96.
Fosliella farinosa sensu Chamberlain (1983), p.343 pro parte.
Fosliella farinosa f. *callithamnioides* sensu Chamberlain (1983), p.351 (non *Hapalidium callithamnioides* P.& H.Crouan, 1859; nec *Melobesia callithamnioides* (P.& H.Crouan) Solms-Laubach (1881).

Key characters: Germination disc pattern with four surrounding cells (Fig.3B), thallus imbricating, bisporangia predominant. Overgrowing *Melobesia membranacea, Pneophyllum limitatum* and other *Pneophyllum* spp, overgrown by *Titanoderma* spp.

Illustrations: Chamberlain (1977b), figs 1–4,8–20; (1983), figs 21,22A,C,E,F,23D,F,G; (1984), figs 15,16,18; Garbary (1978), figs 15–18 (all as *F. farinosa*).

Thallus encrusting, adherent, to at least 5mm diameter, to 20μm thick (individual lobes), flat or sprouting imbricating layers, partly unconsolidated, or orbicular, or lobed, becoming confluent, confluences flat; *margin* entire, thin, usually with, sometimes without orbital rings;

surface smooth, cell surface *Pneophyllum*-type (SEM), colour mauve-pink (Methuen greyish ruby), texture grainy; *conceptacles* self-coloured with thallus, domed, pore rather small, not sunken, *carposporangial conceptacles* to 170μm diameter, *tetrasporangial* to 230μm, *bisporangial* to 250μm with a star-like ring of cells surrounding pore (SEM) (Garbary, 1978, fig.17, as *Fosliella farinosa*).

Thallus entirely bistratose, *basal filaments* single-layered, cells in surface view 14–29μm long × 5–17μm diameter, in VS to 20μm high, often regenerating to form thallus lobes; *epithallial cells* very conspicuous, rounded, often with shiny central spot; *trichocytes* single, terminating basal filaments, becoming immersed in thallus, 30–43μm long × 12–20μm diameter.

Gametangial plants monoecious or spermatangial only; *spermatangial conceptacles* (when monoecious) adjacent to carpogonial/carposporangial ones, chambers low-domed, 33–91μm diameter, 26–44μm high, spout pronounced; *carpogonial conceptacle* chambers domed; *carposporangial conceptacle* chambers domed, 62–110μm diameter, 52–83μm high, with the roof 10–16μm thick and composed of cells somewhat higher than wide with conspicuous epithallial cells, trichocytes not seen on roof surface, fusion cell thick and narrow, *gonimoblast filaments* borne from periphery, to four cells long including carposporangium; *tetrasporangial conceptacle* chambers domed, 65–94μm diameter, 52–78μm high, roof 10–20μm thick, roof structure as in carposporangial conceptacles with c.6 enlarged, vertically orientated cells surrounding the pore; *tetrasporangia* 36–65μm long × 23–39μm diameter, deeply invaginated when immature, c.8 per conceptacle; *bisporangial conceptacle* chambers domed, 166–208μm diameter, 45–125μm high, roof 18–39μm thick, roof mainly 3 cells thick, innermost cell squarish, middle cell tall, epithallial cell domed, pore structure as in tetrasporangial conceptacles; *bisporangia* 54–78μm long × 26–42μm diameter, deeply invaginated when immature, c.20 per conceptacle.

Spore *germination disc* (Fig.3B) with four surrounding cells and usually four small apical cells.

Common. Epiphytic on a wide range of algae, particularly *Laminaria* spp., *Fucus serratus*, *Cystoseira nodicaulis* (especially basal tophules), *Palmaria* and *Chondrus*, rarely on *Zostera*, quite commonly on stones, rocks and shells, mainly littoral but also sublittoral to c.8m depth.

South and south-west England, Jersey, Wales, Ireland, western Scotland, not seen in eastern Scotland or England.

British Isles to Canary Isles, Italy; Arabian Gulf, southern and western Australia, probably more or less cosmopolitan (germination disc pattern seen from many parts of the world).

Spermatangia throughout the year, carposporangia and tetrasporangia mainly in summer, bisporangia throughout the year but predominant in winter; extensive thalli sometimes lack conceptacles entirely.

Hydrolithon boreale initially forms thin thalli, but as plants age layer upon layer of imbricating thallus lobes, up to 7 layers deep (Figs 50A; 51E), sprout successively from the surface. Sometimes thalli that are partly unconsolidated occur: this growth form was named f. *callithamnioides* in Chamberlain (1983), but Solms's (1881) drawings of the type of this taxon shows it to bear triangular propagules which probably pertain to a different taxon. It now seems unnecessary formally to name sporadically occurring, unconsolidated growth forms as taxonomic entities.

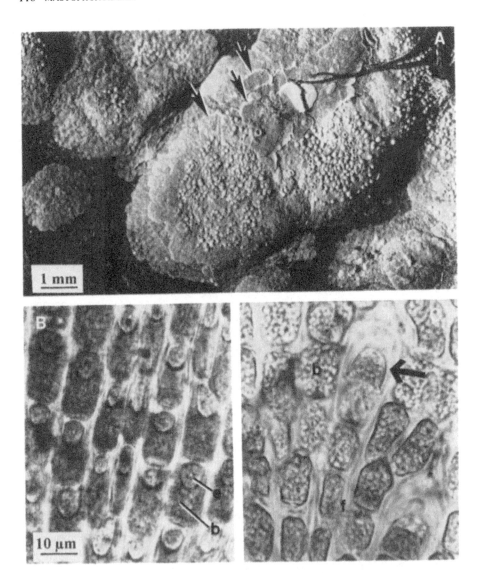

Fig. 50 *Hydrolithon boreale*. (A) Group of bisporangial thalli on *Palmaria*; note successively imbricating thallus lobes (arrows) (YMC 76/273). (B & C) Surface views of bistratose thalli; (B) is calcified and shows epithallial (e) and basal filament (b) cells; (C) is decalcified and focussed down to show basal filament cells (b), cell fusions (f) and a terminal trichocyte (arrow) (YMC 78/229).

Chamberlain (1977b, 1983, as *F. farinosa*) found the species throughout the year, often forming a bloom in autumn (September to November), usually together with *Pneophyllum limitatum*. In winter bisporangial plants predominate but spermatangial plants and monoecious plants with empty carposporangial conceptacles also occur.

Germination discs are rarely seen in field-collected material but bispores grown in laboratory culture (Chamberlain, 1977b, p.355) produced the characteristic disc

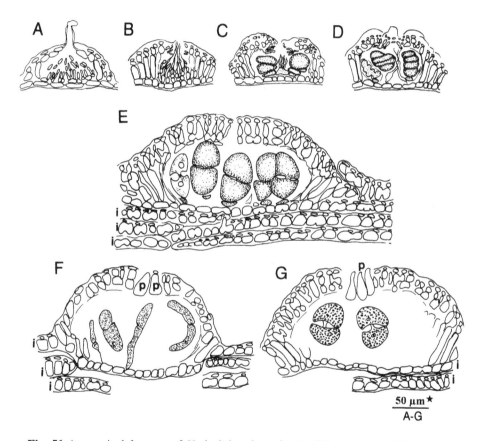

Fig. 51 Anatomical features of *Hydrolithon boreale*. (A) VS spermatangial conceptacle (YMC 76/281). (B) VS carpogonial conceptacle (YMC 78/17M). (C) VS carposporangial conceptacle (YMC 77/218). (D) VS tetrasporangial conceptacle (YMC 78/275). (E) VS mature bisporangial conceptacle (YMC 76/37). (F) VS bisporangial conceptacle from lectotype (TRH) showing pore cells (p) (Foslie slide 2). (G) VS young bisporangial conceptacle showing pore cells (p) (YMC 76/37). i = imbricating thallus lobes (E-G).

(Chamberlain, 1983, fig.7B) and further bisporangial conceptacles, thus demonstrating a self-perpetuating, bisporangial life history.

Balakrishnan (1947), Masaki & Tokida (1960b) and Ganesan (1971) made detailed studies of plants identified as *F. farinosa*. Unfortunately, no germination discs were illustrated. However, the imbricating habit mentioned by Ganesan (1971), Masaki & Tokida (1960b) and Dawson (1960) suggests that their concept conformed to the present one of *H. boreale*.

Hydrolithon cruciatum (Bressan) Y.Chamberlain comb. nov.
Figs 3C; 52; 53

Holotype: TSB (not seen). *G.Bressan*, Trieste (A52).

Fosliella cruciata Bressan in Bressan, Miniati-Radin & Smundin (1977), p.43.

Key characters: Germination disc pattern with eight surrounding cells (Fig.3C), trichocyte fields present, thallus cells relatively broad. Overgrowing *Pneophyllum fragile*.

Illustrations: Bressan *et al.* (1977), figs 1–7; Jones & Woelkerling (1983), figs 1–2; (1984), figs 1,2,4,20.

Thallus encrusting, adherent, to at least 2mm diameter, to 22µm thick, lobed, becoming confluent, confluences flat; *margin* entire, thin, with or without orbital rings; *surface* smooth, cell surface *Pneophyllum*-type (SEM), colour pink (Methuen colour not recorded), texture somewhat glossy; *trichocyte* fields more or less circular, 33–80µm diameter; *conceptacles* self-coloured with thallus, high-domed, pores small, not sunken, *carposporangial conceptacles* to 170µm diameter, *tetrasporangial* ones to 250µm.

Thallus entirely bistratose, *basal filaments* single-layered, cells in surface view 9–20µm long × 4–20µm diameter, in VS to 22µm diameter (i.e. high); *epithallial cells* conspicuous, rounded; single terminal *trichocytes* at thallus margin, 13–33µm long × 13–26µm diameter, *trichocyte fields* comprising 7–12 trichocytes borne on upper surface of basal cells, each composed of a single cell subtending a hair.

Gametangial plants monoecious, *spermatangial conceptacles* immediately adjacent to carpogonial ones, chambers domed, 23–34µm diameter, 20–26µm high, spout very pronounced, up to 40µm long; *carpogonial conceptacles* slightly raised, chambers elliptical; *carposporangial conceptacle* chambers domed, 91–104µm diameter, 44–62µm high, with the roof 26–29µm thick, roof composed of triangular cells with conspicuous epithallial cells at apex, roof thickening to elongate cells at pore, no trichocytes seen on roof, fusion cell narrow and thick, raised on small cells above floor, *gonimoblast filaments* borne from the periphery, composed of 2–3 small cells and a larger cell subtending a relatively large carposporangium; *tetrasporangial conceptacle* chambers domed, 104–130µm diameter, 52–91µm high, with the roof 17–39µm thick, roof structure as in carposporangial

---

Fig. 52 *Hydrolithon cruciatum*. (A) Thalli showing a field of trichocytes (arrow), individual trichocytes (arrowheads), a spermatangial conceptacle (m), a carposporangial conceptacle (c) and a tetrasporangial conceptacle (t) (YMC 88/81). (B) Population of thalli on *Zostera* (YMC 76/186).

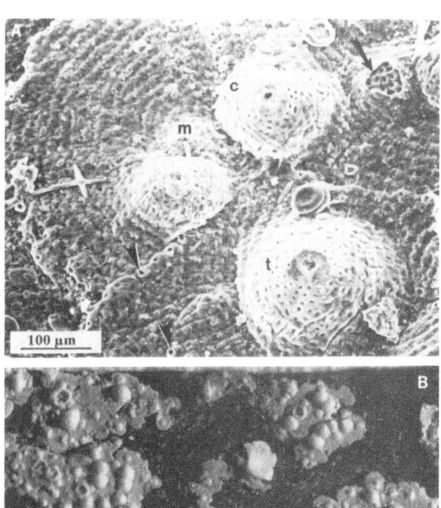

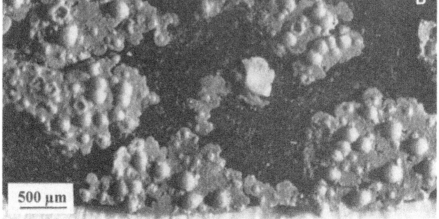

conceptacles; *tetrasporangia* relatively broad, 39–72μm long × 26–46μm diameter; *bisporangial* plants unknown.

Spore *germination discs* with eight surrounding cells (Fig.3C).

Rare. Epiphytic, mainly on *Zostera*, once on *Ulva*, sublittoral to 3m depth.

Dorset, Ayrshire, Ireland (*Bean*, BM!).

Confirmed from Sweden, Italy, Canary Isles, Arabian Gulf, southern and western Australia, probably widely distributed.

Spermatangia, carposporangia and tetrasporangia known to occur in June, August, September and November.

Three populations are known from the British Isles. Two herbarium specimens (BM!) have been seen: one from Sea Mill, Ayrshire, Scotland (*Holmes*, August 1890), the other from Ireland (*Bean*). Fresh material was collected from The Fleet, Dorset, where *H. cruciatum* is quite common on *Zostera hornemanniana* Tutin and reproductive plants have been collected from June-November (no data for December-May). At all times all reproductive types are present and single and field trichocytes occur. Spores from plants collected in September

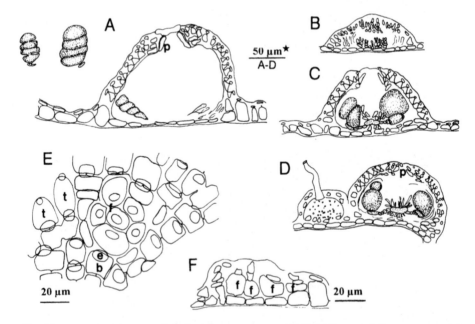

Fig. 53 Anatomical features of *Hydrolithon cruciatum* (YMC 88/81). (A) VS tetrasporangial conceptacle and released tetrasporangia; note pore cells (p). (B) VS carpogonial conceptacle. (C) VS carposporangial conceptacle. (D) VS spermatangial conceptacle (left) adjacent to carposporangial conceptacle; note pore cells (p). (E) Surface view of thallus showing basal cells (b), epithallial cells (e), a field of trichocytes (f) and individual trichocytes (t). (F) VS of a field of trichocytes (f).

settled and grew readily in laboratory culture (YMC unpublished), trichocytes were present in these cultures only until December. Jones & Woelkerling (1984) found that, in laboratory culture, field trichocytes occurred only at photon flux densities of 12–25µmol m⁻² s⁻¹ and that basal trichocytes were more abundant at these values than at 3–6.5µmol m⁻² s⁻¹.

Hydrolithon cruciatum has relatively broader thallus cells in surface view than other species in the British Isles. Bressan *et al.* (1977) noted that cell size varied throughout the year, being larger in summer (May-September) than in winter (October-April). They compared the plants with *Pneophyllum fragile* (as *F. lejolisii*) and *H. boreale* (as *F. farinosa*) finding at any one time the cells of *H. cruciatum* (as *F. cruciata*) to be longer and broader than those of other species. They suggested that it might be polyploid. Jones & Woelkerling (1983) found that the most favourable conditions for growth and reproduction lay within the range 15–22°C, 12–25µmol m⁻² s⁻¹ (i.e. summer conditions), and that sustained growth was better at 15°C than 22°C.

Bressan & Tomini (1980, 1981) cultured *H. cruciatum* (as *Fosliella*) at varying light and temperature levels and found responses similar to the above.

Hydrolithon farinosum (Lamouroux) D.Penrose & Y.Chamberlain (1993), p.295
Figs 3A; 54; 55

Lectotype: CN! (see Penrose & Chamberlain, 1993). Herb. Lamouroux, Mediterranean, on *Sargassum linifolium*.

Melobesia farinosa Lamouroux (1816), p.315.
Fosliella farinosa (Lamouroux) M.Howe (1920), p.587.

Key characters: Germination disc pattern with twelve surrounding cells (Fig.3A), single basal trichocytes conspicuous as dots in surface view, conceptacles with slightly sunken pores and ring of trichocytes at base. Overgrowing *Melobesia membranacea*.

Illustrations: Chamberlain (1983), figs 20A,E, 22B,D, 23C,E; Penrose & Chamberlain (1993), figs 2–29.

Thallus encrusting, adherent, to at least 2mm diameter, to 21µm thick, flat, more or less orbicular becoming confluent; *margin* entire, thin; *surface* smooth, cell surface *Pneophyllum*-type (SEM), colour pale pinkish, almost transparent (no Methuen colour recorded); texture floury, *trichocytes* often visible as dots throughout thallus; *conceptacles* pinkish, domed, with basal skirts, pores conspicuous and slightly sunken like ring doughnuts, often ringed by trichocytes, *carposporangial conceptacles* to 170µm, *tetrasporangial* to 240µm.

Thallus entirely bistratose, *basal filaments* single-layered, cells in surface view 13–24µm long × 5–16µm diameter, in VS to 21µm diameter (i.e. high); *epithallial cells* conspicuous, rounded; *trichocytes* single, terminal, becoming immersed in thallus, 26–33µm long × 13–18µm diameter.

Gametangial plants monoecious; *spermatangial conceptacles* adjacent to carpogonial ones, domed, chamber 36–53µm diameter, 26–38µm high, spout pronounced, *c.*40µm long; *carpogonial conceptacle* chambers domed; *carposporangial conceptacle* chambers domed, 85–104µm diameter, 36–52µm high, with the roof 20–23µm thick, roof composed of cells somewhat higher than wide with conspicuous epithallial cells, roof cells near pore

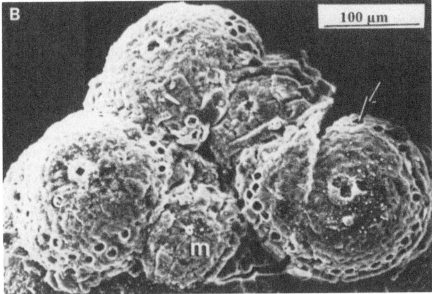

Fig. 54 *Hydrolithon farinosum.* (A) Thalli growing on *Cystoseira foeniculacea* (YMC 77/23, France: Brittany). (B) Gametangial thallus with spermatangial conceptacles (m) and carposporangial conceptacles (c); note trichocytes (arrow) at conceptacle base (YMC 77/115).

developing downward-pointing papillae, small pore collar present, trichocytes forming a ring round conceptacle base but not seen on roof surface, central fusion cell narrow and thick, attached to conceptacle floor by tall cells, *gonimoblast filaments* borne from side of fusion cell, composed of three small cells and a larger cell subtending a relatively large carposporangium; *tetrasporangial conceptacle* chambers domed, 65µm diameter, 34µm high, with the roof 14µm thick, pore surrounded by enlarged, vertically orientated cells; *tetrasporangia* relatively broad, 36–42µm long × 24–29µm diameter; *bisporangial* plants unknown.

Spore *germination disc* (Fig.3A) with four central cells being longer from apex to centre than wide, giving a diamond shape, twelve surrounding cells more or less-equally sized, four trichocytes nearly always present in positions shown (Fig.3A).

Epiphytic on branches of *Cystoseira foeniculacea* in a lower littoral pool.

Dorset.

British Isles to western France, Mediterranean, Canary Isles; Arabian Gulf, Red Sea, Kenya, South Africa, Indonesia, Florida (USA), southern and western Australia, probably widely distributed.

Spermatangia, carposporangia and tetrasporangia known to occur in July.

Only one small population has been seen in the British Isles but this species is an abundant epiphyte on various branching red and brown algae in littoral pools in northern France (YMC obs.). Now that the germination pattern has been characterised, it is possible that the species

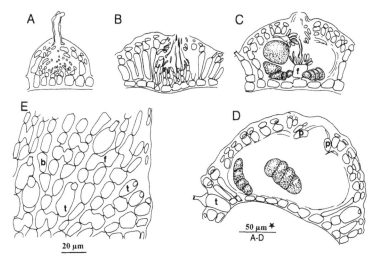

Fig. 55 Anatomical features of *Hydrolithon farinosum*. (A) Spermatangial conceptacle (YMC 77/115). (B) Carpogonial conceptacle (YMC 77/115). (C) Carposporangial conceptacle; note fusion cell (f) (YMC 77/115). (D) Tetrasporangial conceptacle; note pore cells (p) and trichocyte (t) (YMC 77/23, France: Brittany). (E) Surface view of thallus showing basal filament cells (b), cell fusions (f) and terminal trichocytes (t) (YMC 77/115).

will be identified from other southern parts of the British Isles. Rosanoff (1866, p.II, fig.3) illustrated a similar disc in material he identified as *Melobesia farinosa*. See *Hydrolithon sargassi* for possible relationships.

Hydrolithon samoënse (Foslie) D.Keats & Y.Chamberlain (1994), p.15.
Figs 56; 57

Lectotype: TRH! Satana, Savai'i, Samoa, *Reichinger*, July 1905.

Lithophyllum samoënse Foslie (1906), p.20.
Pseudolithophyllum samoënse (Foslie) Adey (1970a), p.13.
Spongites wildpretii Afonso-Carrillo, (1988), p.99.
Lithophyllum illitus Lemoine (1929b), p.54.
Neogoniolithon illitus (Lemoine) Afonso-Carrillo (1984), p.133.

Key characters: Thin, flat, dimerous/monomerous thalli, with low surface bumps, overgrown by most other corallines, roofs of all types of conceptacle more or less flush with thallus surface, tetrasporangial conceptacle roof composed of epithallial cell and a tall thin cell, sometimes with a small lower cell.

Illustrations: Printz (1929), pl.LIII, fig.19 (as *Lithophyllum*); Masaki (1968), pls XX, figs 3,4; XXIII, figs 1–4; LXI, figs 3–6; LXII, figs 1–5 (as *Lithophyllum*); Dawson (1960), pl.28, fig.2 (as *Lithophyllum*); Afonso-Carrillo (1988), figs 7–30 (as *Spongites wildpretii*).

Thallus encrusting, adherent, flat, to at least 5mm diameter, 156μm thick, orbicular becoming confluent, confluences flat; *margin* entire, adherent, bistratose, without orbital rings; *surface* with low bumps, colour pink (Methuen colour not recorded), texture matt; *conceptacles* varying from flat-roofed to slightly raised or slightly sunken.

Thallus initially dimerous, *basal filaments* single-layered, cells in surface view (Fig.57B) measuring 8–18μm long × 5–9μm diameter, in VS c.5–7μm square, cells of *erect filaments* squarish to shorter than wide, measuring 4–8μm long × 4–7μm diameter; *epithallial cells* in single layer, elliptical, measuring c.2.5μm long × 4–7μm diameter, older thalli with some monomerous development; *trichocytes* (Fig.57A) occasional, términating erect filaments, composed of basal cell and upper, hair-bearing cell together measuring c.12μm long × 8μm diameter.

Gametangial plants monoecious; *spermatangial conceptacle* chambers flask-shaped, measuring 65μm diameter, 33μm high, with the roof 16μm thick, spermatangial systems on floor only; *carposporangial conceptacle* chambers elliptical, measuring 78–98μm diameter, 39–46μm high, with the roof 13–18μm thick, roof composed of epithallial cell plus tall thin cell (Fig.57C), pore surrounded by few enlarged cells, fusion cell relatively wide and thin, senescent carpogonia borne on surface, *gonimoblast filaments* borne peripherally; *tetrasporangial conceptacle* chambers elliptical, measuring 69–96μm diameter, 34–49μm high, with roof 13–14μm thick, roof and pore structure as for carposporangial conceptacles; *tetrasporangia* about twice as long as wide, measuring 26–33μm long × 13–26μm diameter. *Bisporangial* plants not seen.

Old conceptacles shed, initially leaving craters, not becoming buried in the thallus. *Germination disc* unknown.

Known only as an epilith on stones in the lower littoral.

Norfolk (East Runton) and Co. Clare.

British Isles to Spain, Canary Isles; South and East Africa, India, Mexico, Samoa, Tahiti, Japan (probably widely distributed).

Syntypes of *Lithophyllum samoënse* from PC & TRH (see Woelkerling, 1993) were examined by Keats & Chamberlain (1994). Syntype characters agreed with present material and also with both the type of *Spongites wildpretii* Afonso-Carrillo (1988) and material from Spain and France (YMC obs.) confirmed as *S. wildpretii* by Dr Afonso-Carrillo. Descriptions of *Lithophyllum illitus* (Lemoine, 1929b; Afonso-Carrillo, 1984, as *Neogoniolithon*), and examination of slides made by Dr. Afonso-Carrillo from the type (C), led Keats & Chamberlain (1994) to consider that this species was also conspecific with *Hydrolithon samoënse*.

Collections (YMC) from Guéthary in France, near the border with northern Spain, showed that this species was very common together with *Lithophyllum orbiculatum*. Both have similar, flat thalli. The type of *Goniolithon subtenellum* Foslie (1898c) was collected at this locality and examination of the lectotype (TRH!) showed that only *Lithophyllum orbiculatum* was present (Chamberlain, 1993, p.112). However, in the protologue Foslie (1898c) noted that a mixture of species was present and records of *Goniolithon subtenellum* persistently refer to the presence of a large central pore surrounded by small pores (e.g. Hamel & Lemoine, 1953, p.98, as *Lithothamnion*). It is probable that such observations refer to the tetrasporangial pore of *Hydrolithon samoënse* which, in surface view, appears to be surrounded by pores that are, in fact, enlarged, *Hydrolithon*-type pore cells.

At present this species is known only from Norfolk and Co. Clare. Now that it is recognized it may prove to be more widely distibuted in the British Isles.

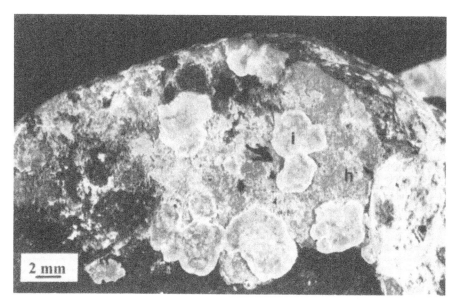

Fig. 56 *Hydrolithon samoënse* (h) overgrown by *Lithophyllum incrustans* (i) (YMC 93/41).

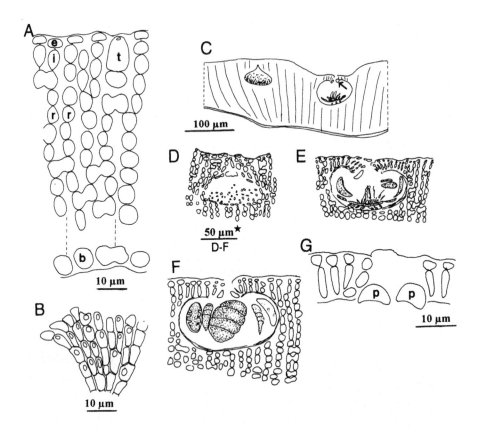

Fig. 57 Anatomical features of *Hydrolithon samoënse*. (A) VS dimerous thallus showing basal cells (b), cells of erect filaments (r), subepithallial initials (i), epithallial cells (e) and trichocyte (t) (YMC 93/41A). (B) Bistratose thallus margin in surface view (YMC 90/1). (C) Diagrammatic drawing of VS gametangial thallus with spermatangial conceptacle (left) and young carposporangial conceptacle (right), note pore cells (arrow) (YMC 93/41A). (D) VS spermatangial conceptacle (YMC 93/41A). (E) VS carposporangial conceptacle (YMC 93/41A). (F) VS tetrasporangial conceptacle and released tetrasporangia (YMC 90/1). (G) VS tetrasporangial conceptacle pore showing enlarged pore cells (p) (YMC 90/1).

Hydrolithon sargassi (Foslie) Y.Chamberlain comb. nov.
  Figs 58; 59

Holotype: TRH (not seen), (see Masaki & Tokida, 1963). Yendo, *Misaki,* Japan, on *Sargassum serratifolium,* April 1903.

Melobesia marginata f. *sargassi* Foslie (1904a), p.22.
Melobesia sargassi (Foslie) Foslie (1908c), p.6.
Heteroderma sargassi (Foslie) Foslie (1909), p.57.
Pneophyllum sargassi (Foslie) Chamberlain (1983), p.445.

Key characters: Erect filaments present, conceptacles rounded with sunken pore. Overgrowing *Melobesia membranacea*.

Illustrations: Masaki & Tokida (1963, as *Melobesia*), pl.IV, figs 5,6, pl.V, figs 4–9, pls IX, X; Chamberlain (1983, as *Pneophyllum*), figs 84–89.

Thallus encrusting, adherent, to at least 2mm diameter, to 200µm thick, flat, orbicular, becoming confluent; *margin* entire thin; *surface* smooth, cell surface *Pneophyllum/ Phymatolithon*-type (SEM), colour semi-transparent milky-grey (Methuen colour not recorded), texture grainy; *conceptacles* hemispherical with slightly sunken, dark centre and skirted base, *carposporangial conceptacles* to 170µm diameter, *tetrasporangial* to 210µm.
Basal filaments single-layered, cells at margin in surface view (often not visible) 10–14µm

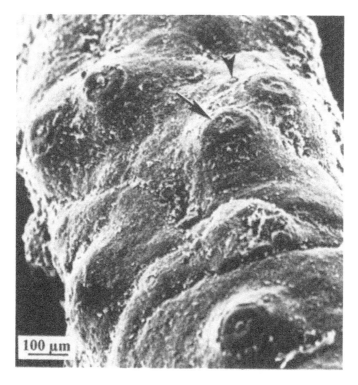

Fig. 58 *Hydrolithon sargassi* on *Halopithys* showing carposporangial (arrow) and spermatangial (arrowhead) conceptacles (YMC 78/253).

long × 3–8μm diameter, *erect filaments* to 12 cells long, cells squarish to elongate, 7–17μm long × 5–11μm diameter; *epithallial cells* flattened to domed, occasionally to three cells deep; *trichocytes* not seen.

Gametangial plants monoecious; *spermatangial conceptacles* immersed beside carpogonial ones, conceptacle chambers 28–59μm diameter, 19–44μm high, with conspicuous spout; *carpogonial conceptacles* immersed to somewhat raised, chambers flask-shaped, pore deeply invaginated and with a plug; *carposporangial conceptacle* chambers elliptical, 72–130μm diameter, 48–95μm high, with the roof 15–33μm thick, roof mainly composed of single layer of triangular cells topped by more or less domed epithallial cells, a number of small cells developing near pore, no trichocytes seen on roof, central fusion cell narrow and deep, borne above conceptacle floor on small cells which later disintegrate, *gonimoblast filaments* borne peripherally, composed of *c*.3 small cells and a larger cell

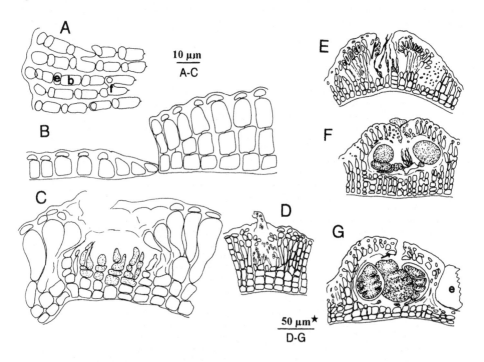

Fig. 59 Anatomical features of *Hydrolithon sargassi*. (A) Bistratose thallus in surface view showing basal cells (b), epithallial cells (e) and cell fusions (f) (YMC 77/215). (B) VS showing thallus with bistratose margin on left and thallus with thickened margin on right (YMC 77/215). (C) VS conceptacle with young carpogonial branches (shaded) (YMC 78/226). (D) VS spermatangial conceptacle (YMC 77/215). (E) VS gametangial thallus with carpogonial conceptacle and edge of spermatangial conceptacle (YMC 77/215). (F) VS carposporangial conceptacle (YMC 77/215). (G) VS tetrasporangial conceptacle; note pore cells (arrow); a leafy epiphyte (e) is attached to the conceptacle (YMC 77/215).

subtending a relatively very large carposporangium; *tetrasporangial conceptacle* chambers narrowly elliptical, 78–122µm diameter, 66–106µm high, with the roof 14–26µm thick, roof composed of triangular to squarish cells with somewhat flattened epithallial cells, pore surrounded by elongate vertically orientated cells, young conceptacle stages not deeply invaginated; *tetrasporangia* 39–65µm long × 26–50µm diameter, young ones deeply invaginated, *c.*8 per conceptacle; *bisporangial conceptacles* unknown.

Epiphytic, mainly on *Halopithys incurvus* (Hudson) Batters, once on *Palmaria.*

Hampshire, Dorset.
    British Isles to France (north & Atlantic), Italy; Japan.

Spermatangia, carposporangia and tetrasporangia recorded throughout the year, but plants apparently in best condition from September to December. At this time the thalli are thinner because they are growing faster areally, and the sporangia are clear and healthy-looking. In very cold periods and in summer the thalli are thicker, more heavily calcified, and spores are often degenerate or absent. Differences are also noted from year to year, *H. sargassi* was, for example, more abundant in 1978 than 1979.

As discussed under the genus description, it is possible that *H. sargassi* represents a thick growth form of *H. farinosum* which has been observed only once in the British Isles. The sunken conceptacle pore and similar conceptacle construction suggests such a relationship. Masaki & Tokida (1963, as *Melobesia*) observed considerable variation in the plants they described in Japan and noted that the type specimen (TRH) of *H. sargassi* comprises plants that they regarded as *Fosliella farinosa.* The true relationship between these various entities will probably be elucidated if data about germination discs or culture studies become available.

PNEOPHYLLUM Kützing

PNEOPHYLLUM Kützing (1843), p.385.

Type species: *P. fragile* Kützing (1843), p.385.

Hapalidium sensu P.& H.Crouan (1859), p.284, non Kützing.
Heteroderma Foslie (1909), p.56, pro parte.

Plants nongeniculate and attached, forming flat, crustose thalli, or, rarely, largely un-consolidated filamentous thalli; all conceptacle types uniporate. Thallus partly to entirely pseudoparenchymatous, or sometimes composed largely of unconsolidated filaments; structure dorsiventral, dimerous, composed of a single layer of basal filaments, with or without erect filaments up to *c.*20 cells long; epithallial cells occurring dorsally on basal cells or erect filaments. Cells of basal filaments shortish to tallish but not palisade, cells of erect filaments mainly squarish, contiguous cells sometimes joined by cell fusions, secondary pit connections unknown; trichocytes common, either intercalary or, very rarely, terminal in basal filaments, or apical on erect filaments; spore germination pattern comprising an 8-celled centre from which the thallus develops.
    Gametangial plants monoecious; spermatangial systems simple, borne on conceptacle

floor only; central fusion cell in carposporangial conceptacles usually quite wide and rather thin, bearing gonimoblast filaments from the periphery, from the upper surface at the periphery, or rarely from the entire surface; tetrasporangial conceptacles with pore canal surrounded by numerous papillae that, at least initially, are orientated more or less parallel to the conceptacle roof, roof developing from cells surrounding as well as cells interspersed among the sporangia, tetrasporangia borne peripherally, with central columella; bisporangial conceptacles as tetrasporangial ones but producing bisporangia with two uninucleate bispores.

The name *Pneophyllum* was largely forgotten after its description by Kützing (1843), the species now attributed to this genus being classified variously in *Melobesia*, *Fosliella* and *Heteroderma*; it was revived by Chamberlain (1983) for dimerous mastophoroid species with an 8-celled germination disc and terminal trichocytes in basal filaments. Accounts of the involved history of the taxon were given by Chamberlain (1983) and Woelkerling (1988).

Voigt (1981, as *Fosliella inexspectata* Voigt) found fossilised casts of plants, showing thallus and spore germination characteristics of *Pneophyllum*, from the Upper Cretaceous and thus about 75 million years old. The epiphytic thalli had been sandwiched between an algal host and a bryozoan and were preserved as casts on the underside of the bryozoan.

Chamberlain (1983) included 11 species in the British Isles flora. Of these, *P. sargassi* is now transferred to *Fosliella* and five species are reduced to synonymy, leaving five species. Discussion with Dr Woelkerling and Dr Penrose has indicated that, in southern Australia, there is a prolific flora of *Pneophyllum* spp. Many species show wide variation ranges and apparent links between 'species' with thalli lacking, and thalli bearing erect filaments. It is possible that when the Australian work is complete more British Isles species will have to be subsumed. For the time being, however, it seems advisable to maintain the seven species now recorded.

Species concepts (as in Chamberlain, 1983) are based mainly on the nature of the tetra/bisporangial conceptacle roof and pore structure. It has been found (YMC obs.) that, as with varieties of *Titanoderma pustulatum* (*q.v.*), populations observed over a number of years remain very stable in thallus structure but may differ from those at other localities. It is difficult to determine species limits in these circumstances. As has been noted, while spore germination pattern in the *Fosliella*-state of *Hydrolithon* appears to provide a reliable basis for differentiating species, size and structure of plants derived from similar discs may show considerable variation. In *Pneophyllum* no similarly diagnostic character has been perceived and at present it has not been found possible to determine species limits satisfactorily. Further experimental studies may clarify this situation.

Taxa now recognized, together with Chamberlain's (1983) synonyms are:

P. caulerpae (including *P. rosanoffii* and *P. zonale* sensu Chamberlain, 1983) with long, free, unicellular filaments surrounding the pore (Figs 60C,D);

P. confervicola, usually lacking specialized pore structures, with prominent conceptacles;

P. fragile (including *P. lejolisii* and *P. microsporum*) lacking specialized pore structures, with mainly flat conceptacles;

P. limitatum with a corona of fused filaments surrounding the pore (Fig.66C);

P. lobescens (including *P. plurivalidum*) with a multicellular, calcified dome surrounding the pore (Fig.68C);

P. myriocarpum (including *P. concollum*) with a raised hyaline collar round the pore (Fig.71C);

Pneophyllum sp. with pore surrounded by massive growth of very thin, upward-pointing filaments (Fig.73D); referred with some doubt to this genus as tetra/bisporangial development not seen.

KEY TO SPECIES

(Roofs and pore canals refer throughout to tetra/bisporangial conceptacle roofs and pore canals)

1 Roofs mainly flush with thallus surface or slightly raised 2
  Roofs prominent 3
2 Cells surrounding pore canal (Figs 65B-E) not specialised .... *P. fragile*
  Pore canal surrounded by filaments of small cells forming a calcified dome (Fig.68C)
  .... *P. lobescens*
3 Conceptacle (roof plus chamber) to 430µm high .... *P.*sp.
  Conceptacle (roof plus chamber) less than 200µm high.. 4
4 Pore canal surrounded by long filaments 5
  Pore canal simple or surrounded by a hyaline collar 6
5 Pore canal surrounded by corona of fused multicellular filaments (Fig.66C)
  .... *P. limitatum*
  Pore canal surrounded by free, unicellular filaments (Fig.60C,D) .... *P. caulerpae*
6 Pore canal surrounded by hyaline collar (Fig.71C)............................ *P. myriocarpum*
  Pore canal mainly simple (Figs 63D,E,G), occasionally with hyaline extrusions (Fig.63C)
  .... *P. confervicola*

Pneophyllum caulerpae (Foslie) P.Jones & Woelkerling (1984), p.184.
  Figs 60; 61

Holotype: TRH! Bay of Islands, New Zealand, *Setchell* no. 6080a, June 1905.

Melobesia caulerpae Foslie (1906), p.16.
Heteroderma caulerpae (Foslie) Adey (1970a), p.16.
Pneophyllum zonale (P.& H.Crouan) sensu Y.Chamberlain (1983), p.435.
Melobesia trichostoma Rosenvinge (1917), p.253. (C!).
Fosliella tenuis Adey & Adey (1973), p.398 - pro parte, type only (USNC!).
Pneophyllum rosanoffii Y.Chamberlain (1983), p.367. (BM!).
  [NOTE: Re-examination of the type (CO!) of *Hapalidium zonale* P.& H.Crouan (1859) has not confirmed the presence of diagnostic pore filaments. The status of this species is therefore uncertain and its basionym is removed from synonymy together with the following synonyms: *Melobesia zonalis* (P.& H.Crouan) Foslie (1898b) and *Fosliella zonalis* (P.& H.Crouan) Feldmann (1942).]

Key characters: Erect filaments present, conceptacles raised, free unicellular pore filaments present. Overgrowing *P. fragile;* overgrown by all large crustose corallines; more or less equal in competition with *P. myriocarpum*.

Illustrations: Chamberlain (1983), figs 11A-C,E,F,14,78–83 (as *P. zonale*); figs 12,33–37 (as *P. rosanoffii*); Jones & Woelkerling (1984), fig.18.

Thallus encrusting, adherent, flat, to at least 10mm diameter, to 130μm thick, orbicular to lobed, becoming confluent, confluences flat or sometimes overgrowing; *margin* entire, thin, without orbital rings; *surface* smooth, cell surface *Pneophyllum*-type (SEM), colour brownish to mauvish to pinkish (Methuen colour not recorded), texture matt; *conceptacles* raised with pore often looking shiny due to protruding filaments, *carposporangial* conceptacles to c.170μm diameter, *tetrasporangial* (epiphytic and epilithic) and *bisporangial* (epilithic) ones to c.300 μm.

Basal filaments single-layered, cells in surface view 5–17μm long × 4–14μm diameter, in VS 7–16μm diameter (i.e. high), *epithallial cells* elliptical; *erect filaments* absent in epiphytic plants, to 8 cells long in epilithic ones, cells 10–29μm long × 7–15μm diameter; *epithallial cells* terminating erect filaments, domed; *trichocytes* intercalary in basal filaments, larger than thallus cells, 10–30μm long × 8–14μm diameter.

Gametangial plants monoecious; *spermatangial conceptacles* adjacent to carpogonial ones, chambers flask-shaped, 25–50μm diameter, 19–46μm high, spout present; *carpogonial conceptacle* chambers more or less orbicular; *carposporangial conceptacle* chambers high elliptical, 84–156μm diameter, 45–117μm high, with the roof 13–52μm thick, roof 1 (epiphytic) to 3 (epilithic) cells thick plus epithallial cell, a profusion of downwardly orientated papillae surrounding pore canal base and long, free, terete or flattened filaments forming corona round pore at surface, these often giving criss-cross appearance in VS, fusion cell wide and quite thin, *gonimoblast filaments* borne peripherally, composed of up to 3 small cells and relatively large carposporangium; *tetrasporangial conceptacle* chambers elliptical, 104–240μm diameter, 70–164μm high, with the roof 13–34μm thick, roof structure as for carposporangial conceptacles, columella present; *tetrasporangia* borne peripherally, 41–78μm long × 26–46μm diameter, to c.20 per conceptacle (probably more in epilithic plants); *bisporangial conceptacle* chambers (epilithic plants only) low elliptical, 169–234μm diameter, 65–117μm high, roof 18–39μm thick, roof structure as for carposporangial conceptacles, columella present; *bisporangia* borne peripherally, 41–73μm long × 26–34μm diameter, to c.60 per conceptacle.

Quite common. Epiphytic, particularly on *Halopithys* together with *Hydrolithon sargassi*, on *Palmaria* with *Pneophyllum myriocarpum* and *Zostera* with *P. fragile*, epilithic on rock, stones, glass etc., also on shells, epiphytic plants mainly littoral, epilithic plants low littoral to 20m depth.

Hampshire, Dorset, Devon, Cornwall, Jersey, Anglesey, Co. Galway, Co. Clare.

Norway (Oslofjord) to Mediterranean, Madagascar, southern Australia. Probably widely distributed.

Epiphytic plants recorded most abundantly from September to December, all types of sporangium collected throughout the year; epilithic plants bearing mainly bisporangia and these collected throughout most of the year.

This species was previously recorded as *P. zonale* (Chamberlain, 1983). Subsequent examination of the lectotype (CO!) has failed to confirm the presence of pore filaments and it seems that *zonale* must be rejected as a specific epithet. Jones & Woelkerling (1984) discussed germination characteristics of *Pneophyllum caulerpae* and discussions with Deborah Penrose and Bill Woelkerling have confirmed that conceptacle and pore structure of *P. caulerpae* agree with that seen in *P. zonale* as described in Chamberlain (1983). There

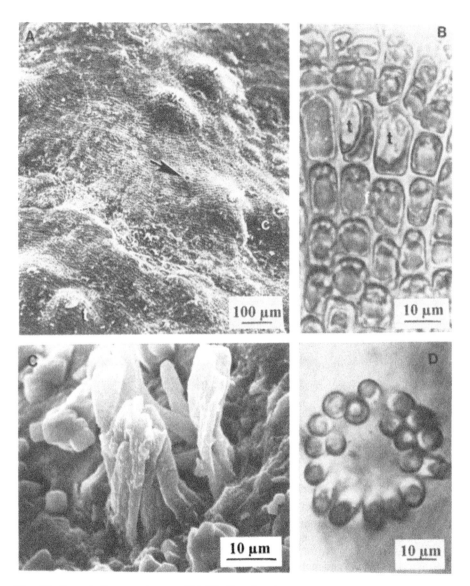

Fig. 60 *Pneophyllum caulerpae.* (A) Thallus showing spermatangial conceptacle pore (arrow), carposporangial conceptacle with pore filaments (c) and adjacent thallus with tetrasporangial conceptacle (t) (YMC 78/235). (B) Thallus in surface view with basal filaments, cell fusions (f) and intercalary trichocytes (t) (YMC 79/149). (C) Pore filaments (SEM) (YMC 78/235). (D) Pore filaments seen from above under light microscopy (YMC 79/149). Note that pore filaments are only seen in mature conceptacles.

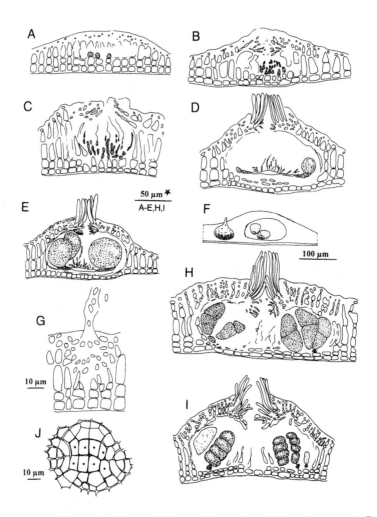

Fig. 61 Anatomical features of *Pneophyllum caulerpae* (vertical sections except J). A-E Stages in carposporangial conceptacle development: (A) Initiation (YMC 78/109). (B) Young carpogonia (shaded) (YMC 78/109). (C) Carpogonial branches (shaded) (YMC 78/109). (D) Young gonimoblast filaments (YMC 78/109). (E) Mature gonimoblast filaments (YMC 77/70). (F) Diagrammatic drawing of monoecious thallus with spermatangial (left) and carposporangial (right) conceptacles (YMC 77/70). (G) Spermatangial conceptacle (YMC 77/70). (H) Bisporangial conceptacle (YMC). (I) Tetrasporangial conceptacle (YMC 78/147). (J) Surface view of germination disc with eight central cells (dots) (YMC 81/48).

are differences between epiphytic and epilithic plants in the British Isles. Epiphytic ones lack erect filaments except in the vicinity of conceptacles, the roof of carposporangial and tetrasporangial conceptacles is mainly two cells thick, conceptacles tend to be smaller and bisporangial conceptacles have not been seen. Epilithic plants develop erect filaments, are mainly bisporangial, the roof of carposporangial, tetrasporangial and bisporangial conceptacles is up to five cells thick and equivalent conceptacles tend to be larger than in epiphytic plants. The differences can probably be attributed to the fact that epiphytic populations are generally short-lived due to the transient nature of the host, whereas epilithic plants can presumably persist for long periods.

Chamberlain, (1983, 1987) cultured spores released by a mixture of *Pneophyllum* thalli growing on a bottle collected from 8m depth in Devon. Bisporangial plants of *P. caulerpae* (as *zonale*) were obtained and bispores germinated to produce further bisporangial plants.

Spore germination characteristics were discussed in Chamberlain (1984, as *zonale*) and trichocyte periodicity in Chamberlain (1985b, as *zonale*) who showed that intercalary, basal filament trichocytes occur from about May-September and are absent from October-April.

Historical details regarding epilithic plants of *P. caulerpae* are given in Chamberlain (1983, p.438, as *zonale*). The epithet *zonale* has generally been applied in Europe to epilithic plants with thalli up to 7–8 cells thick and raised conceptacles to 300μm diameter (e.g. Feldmann, 1942; Hamel & Lemoine, 1953; Bressan, 1974). Unfortunately it is impossible to confirm records without examining specimens. Lemoine (1913a) recorded *Melobesia zonale* from Clew Bay, Ireland; this material (BM-K!, DBN!) is *Hydrolithon boreale*. *Fosliella tenuis* Adey & Adey (1973) appears, from the text description, to be based partly on *P. myriocarpum*, but the slide of the type specimen (USNC! 67–16,50–70E, Oslofjord) shows the characteristic *P. caulerpae* pore (Chamberlain, 1983, fig.82E, as *zonale*).

The diagnostic character of this species is now considered to be the presence of free filaments surrounding the pore. This was not appreciated previously (Chamberlain, 1983) when thick epilithic plants with this character were determined as *P. zonale*, and thin, epiphytic ones as *P. rosanoffii*. In the above description the dimensions of the two previously accepted taxa are aggregated and mention is made of characters that differ somewhat in epiphytic and epilithic plants.

Historical details regarding epiphytic plants of *P. caulerpae* were given in Chamberlain (1983, p.370, as *P. rosanoffii*). These epiphytic plants became confused with *P. fragile* (usually as *lejolisii*) which has flattened conceptacles and no pore filaments, although the first description (Le Jolis, 1863, as *Melobesia farinosa*) mentioned pores surrounded by hairs. Rosenvinge (1917, figs 158D, E, as *M. lejolisii*) illustrated pore filaments in some plants from Denmark.

Pneophyllum confervicola (Kützing) Y.Chamberlain (1983), p.385 (as *confervicolum*). Figs 62; 63

Holotype: L! 941.156.120 (see Chamberlain, 1983). Herb. Kützing, near Trieste, on *Conferva vasta*.

Phyllactidium confervicola Kützing (1843), p.295.
Hapalidium phyllactidium Kützing (1849), p.695. (L!).
Hapalidium callithamnioides P.& H.Crouan (1859), p.289. (CO!).
Melobesia minutula f. *typica* Foslie (1909), p.56. (TRH!).

Melobesia fosliei Rosenvinge (1917), p.249. (C!).
Heteroderma minutula (Foslie) Dawson (1956), p.47.
Fosliella minutula (Foslie) Ganesan (1963), p.38.
Melobesia minutula f. *lacunosa* Foslie (1905a), p.107 (nom. illeg. Art.63.1)
Pneophyllum confervicolum f. *minutulum* (Foslie) Chamberlain (1983), p.392.

Key characters: Erect filaments absent, tetra/bisporangial conceptacles hemispherical, rarely exceeding 100μm internal diameter, pore mainly simple, surface stepped (SEM). Overgrowing *Melobesia membranacea*; overgrown by *Pneophyllum myriocarpum*, *Hydrolithon boreale* and *Titanoderma* spp.

Illustrations: Rosenvinge (1917), figs 168–172 (as *Melobesia fosliei* and *M. minutula*); Suneson (1943a), figs 14–18, pl.V, figs 23,24 (as *M. minutula* f.*typica* & f.*lacunosa*); Chamberlain (1983), figs 43–48.

Thallus encrusting, adherent, flat, to at least 2mm diameter, to 10μm thick, partly unconsolidated, or orbicular, or lobed, becoming confluent, confluences flat; *margin* entire, thin, without orbital rings; *surface* smooth, cell surface *Pneophyllum*-type (SEM), colour from pale pink, almost transparent, to bright pink and solid-looking (Methuen colour not recorded), texture matt; *conceptacles* mainly hemispherical to somewhat raised, often bubble-like, from scattered to very crowded, surface usually stepped (SEM), *carposporangial and tetra/bisporangial* conceptacles to *c.*150μm diameter.

Thallus entirely bistratose, *basal filaments* single-layered, cells in surface view usually squarish, 5–14μm long × 4–16μm diameter, in VS to *c.*10μm diameter (i.e. high); *epithallial cells* in surface view wide and short, often almost lenticular; intercalary *trichocytes* common, similar in size to thallus cells.

Gametangial plants monoecious; *spermatangial conceptacles* adjacent to carpogonial ones, chambers flask-shaped, 30μm diameter, 23μm high (1 only measured), spout present; *carpogonial conceptacle* chambers domed; *carposporangial conceptacle* chambers domed to elliptical, 47–104μm diameter, 39–68μm high, with the roof 6–16μm thick, roof composed of one or two layers of irregularly arranged small cells, pore varying from unelaborated to somewhat fringed or beaked, central fusion cell quite narrow and thin, *gonimoblast filaments* borne peripherally, up to four cells long plus enlarged carposporangium; *tetrasporangial conceptacle* chambers domed to elliptical, 52–91μm diameter, 30–74μm high, with the roof 8–18μm thick, roof structure as for carposporangial conceptacles, small columella usually present; *tetrasporangia* borne peripherally, 26–46μm long × 15–28μm diameter, *c.*10 per conceptacle; *bisporangial conceptacle* chambers domed to elliptical, 68–107μm diameter, 52–65μm high, roof 9–10μm thick, roof structure as for carposporangial conceptacle, small columella usually present; *bisporangia* borne peripherally, 26–39μm long × 14–21μm diameter, 5–15 (fide Suneson, 1943a) per conceptacle.

Quite common. Epiphytic on a wide range of algal hosts, particularly *Cladophora* spp., *Palmaria* and *Phyllophora,* also on geniculate corallines, *Zostera,* hydroids and bryozoa; mainly littoral, also sublittoral to 7m depth.

Northwards to W.Inverness, Shetland Isles, eastwards to Norfolk, Hampshire, Channel Isles; Co. Antrim, Co. Down, Co. Wexford, Co. Cork, Co. Clare, Co. Galway.

Norway (Hordaland) to Mediterranean, Madeira; southern USSR, India, Pacific Mexico, central Pacific.

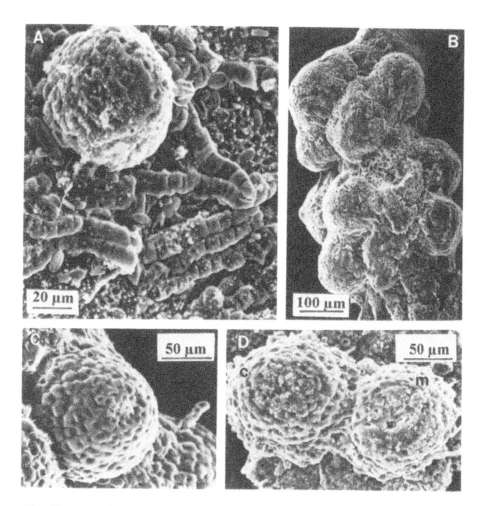

Fig. 62 *Pneophyllum confervicola.* (A) Unconsolidated thallus with conceptacle (YMC 76/73). (B) Thallus growing on *Cladophora* (YMC 78/142). (C) Spermatangial conceptacle (with spout) and carposporangial conceptacle (YMC 78/142). (D) Carposporangial (c) and spermatangial (m) conceptacles (YMC 78/134).

All reproductive types recorded throughout the year, but plants occurring most commonly from about May-November.

This species shows a wide range of structure as is indicated in the description and implied by the abbreviated list of synonyms provided. The type specimen (see Chamberlain, 1983, figs 43A, B) shows hemispherical, bubble-like conceptacles with a stepped surface (SEM) and characteristically short, wide thallus cells. This appearance is common in British Isles material and can, perhaps, be regarded as typical. However, some thalli have more elongate cells, and conceptacles vary from abruptly hemispherical to slightly raised. The pore may be entirely unelaborated or have a small spout or hyaline fringe (Fig.63C). At Kimmeridge, Dorset, populations of typical *P. confervicola* are common in summer and are accompanied by plants of *P. fragile* with entirely flat conceptacles. Suneson (1943a, as *Melobesia minutula*) also noted considerable variation in this species. In addition the thallus may be consolidated, or partially or entirely unconsolidated. The unconsolidated form was previously given infraspecific status, as f. *minutula* (see Chamberlain, 1983, p.392), but it is now considered that this is unnecessary. Suneson (1943a, as *Melobesia minutula* f. *lacunosa*) gave a detailed description of unconsolidated plants from Sweden.

Some of the plants now included in *P. confervicola* may belong to different species. They are so small that they are difficult to investigate satisfactorily, and in the British Isles are

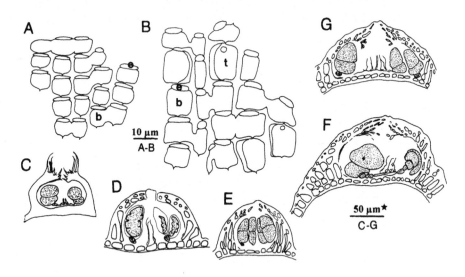

Fig. 63 Anatomical features of *Pneophyllum confervicola*. (A) Surface of thallus with smaller cells (YMC 76/137). (B) Surface of thallus with larger cells; note basal filament cells (b), short, wide epithallial cells (e) and intercalary trichocytes (t) (YMC 81/48). (C) Diagrammatic drawing of VS of carposporangial conceptacle with hyaline pore filaments (YMC 78/167). (D) Tetrasporangial conceptacle in VS (YMC 76/73). (E) Bisporangial conceptacle in VS (YMC 76/73). (F) Carposporangial conceptacle in VS (YMC 81/34). (G) Bisporangial conceptacle in VS (YMC 81/34).

often present in very small quantities mixed with other taxa. Two small species of *Pneophyllum* have recently been described from Pacific Russia as *P. elegans* and *P. japonicum* (Kloczcova & Demeshkina, 1987) on the basis of monospecific populations on *Zostera*. Hopefully, similar studies, combined with culture work, will eventually clarify the taxonomy of this difficult group.

Historical aspects of the classification of this species are discussed in Chamberlain (1983). Batters (1892, as *Melobesia callithamnioides*) recorded an unconsolidated form from the British Isles, and Rosenvinge confirmed the identification of typical plants (as *M. minutula* f. *typica*) in Knight & Parke (1931).

All plants seen have epithallial cells, but they are not always obvious. Harvey (1848) described a sterile plant as the new genus and species *Lithocystis allmanni*, based on the absence of epithallial cells. It had been found by Professor Allmann growing on *Lomentaria clavellosa* (Turner) Gaillon attached to an oyster he was eating for dinner. The name has haunted the synonymy of small coralline algae from the British Isles ever since, but it has proved quite impossible to find the specimen. It seems unlikely that we shall ever know the identity of this taxon, but *P. confervicola* is one possibility (see Chamberlain, 1983, p.301).

Garbary (1978, fig.20, as *Fosliella minutula*) first noted the stepped nature of the conceptacle roof; this structure occurs in many plants but others show different surface features (Chamberlain, 1983).

Pneophyllum fragile Kützing (1843), p.385.
Figs 64; 65

Holotype: L! no. 941.241.152. Herb. Kützing, Mediterranean, on *Sphaerococcus coronopifolius* Stackhouse (see Chamberlain, 1983, p.356, figs 26, 27).

Melobesia lejolisii Rosanoff (1866), p.62, pro parte.
Dermatolithon lejolisii (Rosanoff) Foslie (1898b), p.11.
Heteroderma lejolisii (Rosanoff) Foslie (1909), p.56.
Fosliella lejolisii (Rosanoff) M.Howe (1920), p.588.
Melobesia microspora Rosenvinge (1917), p.256.
Pneophyllum microsporum (Rosenvinge) Y.Chamberlain (1983), p.395.

Key characters: Conceptacles not, or slightly, raised, pore lacking elaboration. Overgrown by *Hydrolithon cruciatum*, *Pneophyllum caulerpae* and *P. myriocarpum*.

Illustrations: Rosanoff (1866), p.I, figs 4, 10, 12, pl.vii, figs 9–11 (as *Melobesia lejolisii*); Suneson (1937), figs 1–5 (as *M. lejolisii*); Masaki (1968), pls XII, XLIX, L (as *Fosliella lejolisii*); Chamberlain (1977a), figs 14–23 (as *F. lejolisii*); (1983), figs 28–32 (as *P. lejolisii*); Penrose & Woelkerling (1991), figs 1–27.

Thallus encrusting, adherent, flat, to at least 5mm diameter (40mm in epilithic plants), to 140µm thick, orbicular to lobed, becoming confluent, confluences flat; *margin* entire, thin to thick, without orbital rings; *surface* smooth, cell surface *Pneophyllum*-type, colour pale to deep pink to brownish (Methuen colour not recorded), texture matt or grainy; *conceptacles* from flattened and closely crowded to slightly raised and more widely spaced, periphery indistinct, rosette of cells surrounding pore (SEM).

Thallus varying from entirely bistratose except for tall cells in vicinity of conceptacles (e.g. on *Zostera*) to thickened; *basal filaments* single-layered, cells in surface view 3–12µm

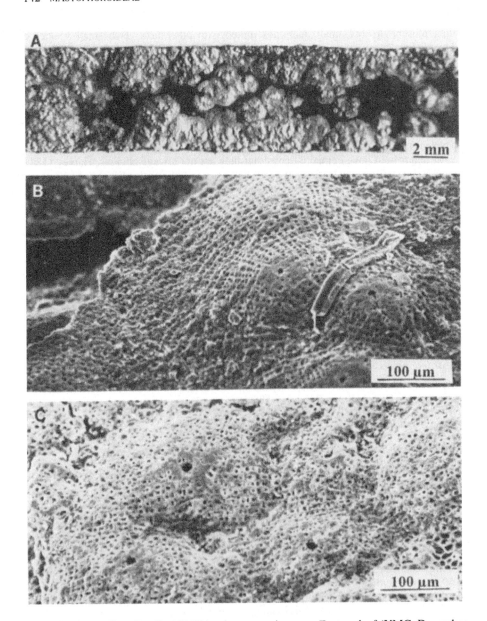

Fig. 64 *Pneophyllum fragile*. (A) Thin plants growing on a *Zostera* leaf (YMC, December 1975). (B) Conceptacles of thin plant growing on *Zostera* (YMC 78/300). (C) Thallus of thicker plants (previously known as *P. microsporum*) growing on *Chondrus* (YMC 79/8).

long × 6–12μm diameter, in VS 5–15μm long x 5–15μm diameter, *erect filaments* to 14 cells long, cells 5–21μm long x 5–14μm diameter; *epithallial cells* short and wide; intercalary *trichocytes* in basal filaments (Fig.65A) common, 11–16μm long × 8–14μm diameter, also seen terminally on erect filaments.

Gametangial plants monoecious or with carpogonial/carposporangial conceptacles only; *spermatangial conceptacles* immersed in thallus beside carpogonial ones, chambers low elliptical to flask-shaped, 13–49μm diameter, 12–34μm high, spout present; *carpogonial conceptacle* chambers low-elliptical; *carposporangial conceptacle* chambers low elliptical to elliptical, 45–104μm diameter, 22–78μm high, with the roof 6–31μm thick, roof to two cells thick with slight papillae near pore, central fusion cell narrow and moderately thick, *gonimoblast filaments* borne from periphery, to six cells long plus carposporangium; *tetrasporangial conceptacle* chambers low elliptical, 54–130μm diameter, 30–70μm high, with the roof 6–31μm thick, structure similar to carposporangial conceptacle, slight columella present; *tetrasporangia* borne peripherally, 22–70μm long × 18–49μm diameter, 4–12 per conceptacle; *bisporangial conceptacle* chambers low elliptical to elliptical, 63–130μm diameter, 33–65μm high, with the roof 7–20μm thick, structure as before; *bisporangia* 25–65μm long × 17–52μm diameter, 3–14 per conceptacle.

Common. Occurring in two forms: a thin form mainly epiphytic on *Zostera*, very occasionally on algae or epilithic, mainly littoral, also sublittoral to *c*.3m; and a thicker form (previously as *P. microsporum*) epiphytic on littoral plants of *Chondrus*, *Furcellaria* and other red and brown algae.

Throughout the British Isles in suitable habitats.

Norway (Nordland) to Canary Isles; Mediterranean; Baltic; Greenland, Canada (Newfoundland) to Caribbean; Black Sea; Pacific Mexico, Japan, Korea, Australia, New Zealand; probably more or less cosmopolitan.

All reproductive types recorded throughout the year. *Zostera* epiphytes seasonal from July to April, thalli persisting until some time between January and about April depending on the severity of the winter. Trichocytes abundant during warm, well-illuminated seasons, disappearing as winter arrives.

Plants of this species mainly occur in one of two forms: thin, seasonal plants (previously *P. lejolisii*, Chamberlain, 1983) are common on *Zostera* between about July and April, thicker plants (previously *P. microsporum*) grow on a range of red and brown littoral algae throughout the year and presumably represent a 'sink' from which more transient populations are recruited when *Zostera* leaves are at a suitable stage for spore settlement.

Chamberlain (1983) commented that *P. fragile* and *P. lejolisii* were probably conspecific but maintained them as separate species because the latter appeared to be restricted to seagrass hosts whereas the type specimen of *P. fragile* was growing on an alga. Since then Penrose & Woelkerling (1991) noted that thin and slightly thicker thalli occurred on the type specimen (L!) of *P. fragile* and have found a range of plants from ones lacking erect filaments to ones up to 700μm thick in southern Australia. On this basis they subsumed *P. lejolisii* in *P. fragile*. It is thus accepted that a wide range of thallus development can occur within this single species but very thick plants like those recorded in Australia (Penrose & Woelkerling 1991, figs 22,27) have not been seen in the British Isles.

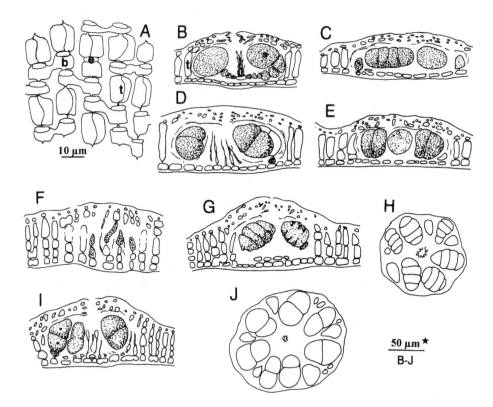

Fig. 65 Anatomical features of *Pneophyllum fragile*. A-E Thin plants growing on *Zostera*: (A) Surface view of thallus showing cells of basal filaments (b), epithallial cells (e) and intercalary trichocytes (t)(YMC 81/49). (B) VS carposporangial conceptacle; note tall cells at conceptacle periphery (t) (YMC 77/222). (C) VS tetrasporangial conceptacle with prostrate tetrasporangium (YMC 78/229). (D) VS bisporangial conceptacle with more or less vertical bisporangia (YMC 78/268). (E) VS bisporangial conceptacle with prostrate bisporangia (YMC 78/229). F-J Thicker plants of *Pneophyllum fragile*, with multicellular erect filaments, growing on *Halopithys* (YMC 78/307): (F) VS carpogonial conceptacle. (G) VS tetrasporangial conceptacle. (H) Surface view of tetrasporangial conceptacle. (I) VS bisporangial conceptacle. (J) Surface view of bisporangial conceptacle.

Many features of this species show considerable variability, for example in thin plants tetra/bisporangial plants may bear a mosaic of relatively small conceptacles or larger, widely spaced ones. Thallus cell size varies considerably throughout the growing season being relatively shorter and wider in summer, and longer and thinner in winter (Chamberlain, 1983).

A study was made of thin plants growing on *Zostera* in the littoral zone at Bembridge, Isle of Wight (YMC unpublished) where they are confined to host plants growing in pools. The species grows mainly on plants that are permanently submerged at the edges of clumps rather than on the surface of clumps where it would be more likely to suffer desiccation.

Pneophyllum limitatum (Foslie) Chamberlain (1983), p.376.
Figs 66; 67

Lectotype: TRH! (see Chamberlain, 1983, p.376). Herb. Foslie, Limfjorden, Lendrup Ron, Denmark, *L.K.Rosenvinge*, no.3807, 22 August 1892, on *Fucus vesiculosus*.

Melobesia lejolisii f. *limitata* Foslie (1905a), p.102.
Melobesia limitata (Foslie) Rosenvinge (1917), p.245.
Fosliella limitata (Foslie), Ganesan (1963), p.41.
Heteroderma limitata (Foslie) Adey (1970a), p.16.

Key characters: Erect filaments absent, conceptacles with multicellular pore filaments that are, in part at least, fused, often with further pore filament ring centrally. Overgrowing *Melobesia membranacea*; overgrown by *Hydrolithon boreale*.

Illustrations: Rosenvinge (1917), figs 163–167 (as *Melobesia*); Suneson (1937), figs 6–10, pl. IV, figs 19, 20 (as *Melobesia*); Chamberlain (1977a), figs 1–13 (as *Fosliella*); (1983), figs 39A,C,D, 40, 41, 42b-f.

Thallus encrusting, adherent, to at least 10mm diameter, to 20μm thick, orbicular becoming confluent, confluences flat, thallus flat; *margin* entire, thin, often with orbital rings; *surface* smooth, cell surface *Pneophyllum*-type, colour dull mauve-pink (Methuen colour not recorded), texture matt and surface often crazed; *conceptacles* conical with pore filaments appearing as pale central ring, *carposporangial conceptacles* to *c*.280μm diameter, *tetra/bisporangial* ones to *c*.300μm.

Thallus bistratose except in vicinity of conceptacles, *basal filaments* single-layered, cells in surface view 6–21μm long × 5–14μm diameter, in VS 15–20μm high; *epithallial cells* in surface view conspicuous, rectangular; intercalary *trichocytes* common, mainly larger than surrounding thallus cells, 14–26μm long × 10–17μm diameter.

Gametangial plants monoecious; *spermatangial conceptacles* immersed in thallus beside carpogonial/carposporangial ones, chambers low-elliptical to domed, 33–59μm diameter, 21–30μm high, spout present; *carpogonial conceptacles* somewhat raised, chambers low-elliptical; *carposporangial conceptacle* chambers elliptical, 117–169μm diameter, 52–109μm high, with the roof 8–26μm thick, roof composed of small, irregularly arranged cells to two cells thick, filaments in pore region united below, free above, forming a corona of fused multicellular filaments, lower filaments (Fig.67C) forming an inner ring of upwardly orientated filaments, central fusion cell wide and shallow, *gonimoblast filaments* borne peripherally or on upper surface at periphery, composed of up to 4 small cells and a relatively

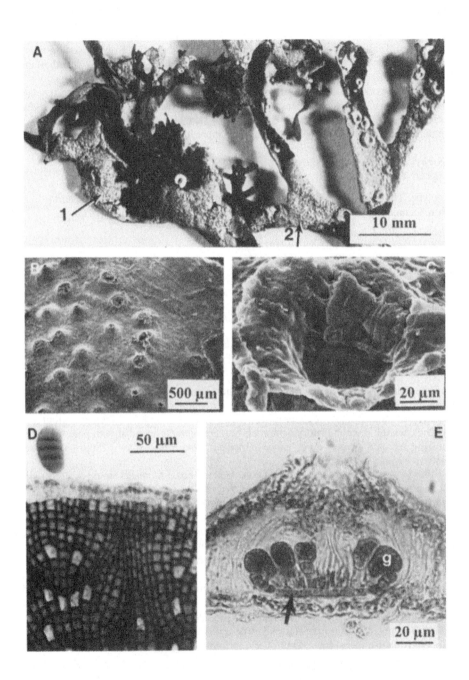

large carposporangium; *tetrasporangial conceptacles* conical, chambers elliptical, 151–229μm diameter, 68–117μm high, roof 10–26μm thick, roof and pore structure as for carposporangial conceptacles, small columella and scattered hyaline filaments present; *tetrasporangia* borne peripherally, 30–60μm long × 29–39μm diameter, *c*.30 per conceptacle; *bisporangial conceptacles* conical, chambers low-elliptical, 156–234μm diameter, 65–91μm high, with the roof 13–21μm thick, structure as for tetrasporangial conceptacles; *bisporangia* borne peripherally, 39–52μm long × 23–36μm diameter, *c*.30 per conceptacle.

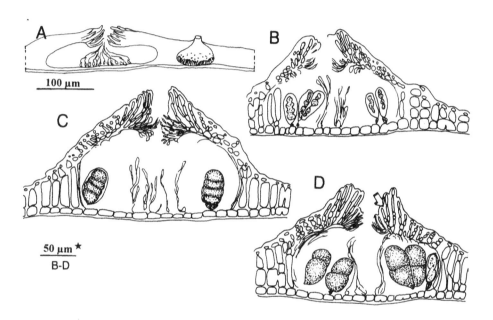

Fig. 67 Vertical sections of reproductive structures of *Pneophyllum limitatum* (YMC 77/217). (A) Diagrammatic drawing of gametangial thallus showing carpogonial conceptacle (left) and spermatangial conceptacle (right). (B) Young tetrasporangial conceptacle. (C) Mature tetrasporangial conceptacle. (D) Mature bisporangial conceptacle; note fused, segmented pore filaments (arrow).

Fig. 66 *Pneophyllum limitatum*. (A) *Chondrus crispus* frond covered with *Hydrolithon boreale* (1) and *P. limitatum* (2) thalli (fresh material). (B) Thallus with tetrasporangial conceptacles (YMC 79/290). (C) Corona of fused pore filaments on tetrasporangial conceptacle (YMC 79/290). (D) Thallus surface with intercalary trichocytes (pale cells); note extruded tetrasporangium. Photo 3.9.81. (E) VS of carposporangial conceptacle; note fusion cell (arrow) and gonimoblast filaments (g) (YMC 78/181E).

Quite common. Epiphytic, on wide range of littoral algae, particularly *Fucus serratus*, *Laminaria digitata*, *L. saccharina* and *Palmaria palmata.*; occasionally on *Zostera* and rarely on rock.

Throughout the British Isles except east coasts of Scotland and England.
Norway (Hordaland) to France (northern Brittany).

All reproductive types present throughout the year, bisporangial plants rare, plants and conceptacles most abundant and vigorous in autumn (September-November). At this time they usually occur together with *Hydrolithon boreale* on large algal hosts; as winter progresses *P. limitatum* gradually dies as *H. boreale* becomes more abundant. In some years these epiphytes are extremely abundant in autumn, occurring on large numbers of littoral algae; in other years they are sparse and difficult to find. Herbarium data (CHE! CO!) suggest that similar populations occur in northern France.

Thallus cells are larger in plants collected in autumn than at other times of year, trichocytes are also most abundant at this time and are absent between November and June. Chamberlain (1977a) noted that plants growing on *Zostera* bear more abundant trichocytes than those on algae.

It was noted (Suneson, 1937; Chamberlain, 1977a, 1983) that plants with flat-roofed bisporangial conceptacles sometimes occurred in *P. limitatum*. Suneson (1937) found such plants commonly among Swedish populations of *P. limitatum* and the same was observed in the British Isles (Chamberlain, 1977a, 1983) and France (Chamberlain, 1983). However, it was pointed out (Chamberlain, 1983) that these bisporangial plants were indistinguishable, except in occurrence, from those of *P. microsporum* (now subsumed in *P.fragile q.v.*). It is now concluded that they may pertain to this species. However, bisporangial plants of *P. fragile* normally occur in populations together with gametangial and tetrasporangial plants. It is possible that the occurrence solely of bisporangial *P. fragile* plants among *P. limitatum* populations indicates a relationship between the taxa, but no supporting data are at present available.

Pneophyllum lobescens Y.Chamberlain (1983), p.419.
Figs 68; 69

Holotype: BM! *Y.M.Chamberlain*, mouth of river Yealm, south Devon, 29 October 1977, YMC 77/228, 8m depth, type grown in laboratory culture from 77/228.

Pneophyllum plurivalidum Y.Chamberlain (1983), p.427.

Key characters: Erect filaments present, conceptacles immersed, with pale centre, calcified pore dome present (SEM). Overgrowing *P. myriocarpum;* overgrown by all large crustose corallines.

Illustrations: Chamberlain (1983), figs 8, 67–72 (as *P. lobescens*); figs 10B, 73–77 (as *P. plurivalidum*).

Thallus encrusting, adherent, flat to imbricating, to at least 60mm diameter, 100μm thick, initially lobed, becoming orbicular then confluent, with young thallus overgrowing old; *margin* entire, thin, without orbital rings; *surface* smooth, cell surface *Pneophyllum*-type, (SEM), colour brownish to purplish to bright rose-pink (Methuen greyish red), texture matt;

conceptacles flat to somewhat raised (especially when dry), with indefinite, pale centres, pore surrounded by a calcified dome (SEM), conceptacle periphery not visible externally.

Basal filaments single-layered, cells in surface view (at margin) squarish to elongate, 6–23μm long × 2–11μm diameter, *epithallial cells* rounded-elliptical to short and wide; *erect filaments* to 8 cells long, cells 7–23μm long × 7–16μm diameter; *epithallial cells* more or less domed; *trichocytes* in basal filaments intercalary, similar in size to basal cells, *trichocytes* in erect filaments terminal.

Gametangial plants monoecious; *spermatangial conceptacles* adjacent to or near carpogonial ones, chambers immersed, flask-shaped, 34–52μm diameter, 26–48μm high,

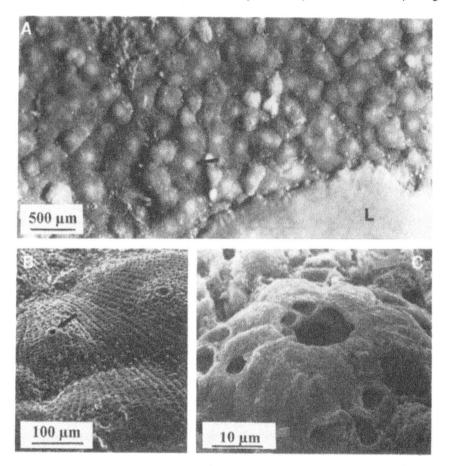

Fig. 68 *Pneophyllum lobescens*. (A) Thallus on rocks at Wembury, Devon, on 24.xi.1988, overgrown by *Lithophyllum crouanii* (L); note pale centred conceptacles (arrow) (YMC 88/118). (B) Bisporangial thallus grown in culture (from YMC 77/228). (C) Dome on bisporangial conceptacle in B.

spout present; *carpogonial conceptacle* chambers elliptical; *carposporangial conceptacle* chambers elliptical, 65–112μm diameter, 39–73μm high, with the roof 13–26μm thick, roof composed of large, irregularly arranged cells immediately above chamber, giving rise dorsally to upwardly orientated filaments up to 5 small cells long forming the pore dome, central fusion cell fairly wide and thick, *gonimoblast filaments* borne peripherally, composed of up to 6 small cells plus enlarged carposporangium; *tetrasporangial conceptacle* chambers elliptical, 104–138μm diameter, 65–91μm high, with the roof 15–26μm thick, roof structure as carposporangial conceptacles, columella present; *tetrasporangia* plump, borne peripherally, 43–55μm long × 24–42μm diameter, to 18 per conceptacle; *bisporangial conceptacle* chambers low elliptical, 82–138μm diameter,

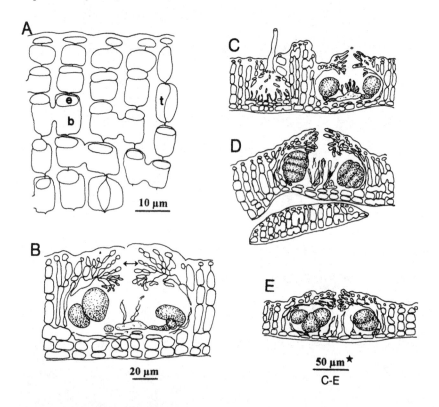

Fig. 69 Anatomical features of *Pneophyllum lobescens* (plants cultured from YMC 77/228, Yealm, Devon). (A) Bistratose thallus margin in surface view, showing basal cells (b), epithallial cells (e) and trichocytes (t). (B) VS carposporangial thallus to show detail of pore filaments (arrows) from which calcified domes (e.g. Fig. 68C) are derived. (C) VS gametangial thallus showing spermatangial conceptacle (left) and carposporangial conceptacle (right). (D) VS tetrasporangial thallus overgrowing another thallus lobe. (E) VS bisporangial thallus.

59–93µm high, with the roof 8–26µm thick, roof structure as carposporangial conceptacles, columella present; *bisporangia* borne peripherally, 36–51µm long × 19–35µm diameter, to 20 per conceptacle.

Quite common. Epilithic on rocks, stones, glass etc., also on shells; low littoral to 8m depth.

Hampshire, Dorset, Devon, Pembroke, Loch Sween, Shetland Isles.
British Isles to France (Normandy).

Carposporangia recorded for March, April, July, October and November, tetrasporangia for October, bisporangia for July and October.

Chamberlain (1983, 1987) cultured spores released by a mixture of *Pneophyllum* thalli found growing on a bottle at 8m depth in south Devon. Two rather different types of thallus with immersed conceptacles were observed and described (Chamberlain, 1983) as *P. lobescens* and *P. plurivalidum*. Gametangial, tetrasporangial and bisporangial plants were obtained in *P. lobescens* and gametangial and bisporangial ones in *P. plurivalidum*. The observed differences between the two species included squarish basal cells in surface view and rapid growth of bisporangial plants in *P. lobescens* as compared with very elongate basal cells and much slower-growing bisporangial plants in *P. plurivalidum*. The two taxa were very similar in having erect filaments, immersed conceptacles, and a characteristic dome round the pore. It is now considered that these last characters are diagnostic and the other features represent intraspecific variation. It is possible that *Fosliella valida* Adey & Adey (1973) is also conspecific: if sufficient type material of this species becomes available it will be examined to determine its qualitative characteristics.

The above culture study (Chamberlain, 1987) confirmed the presence of a *Polysiphonia*-type and a self-perpetuating (presumably diploid) bisporangial life history in this species.

Spore germination characteristics were discussed in Chamberlain (1984) and trichocyte periodicity was studied in Chamberlain (1985b), who showed that mature intercalary, basal trichocytes occur from about April-October, but overshadowing of thalli during this period can inhibit trichocyte production; immature trichocytes sometimes occur during the winter. Large, sac-like terminal cells resembling trichocytes were observed in October (see also *Pneophyllum myriocarpum*).

Two specimens (BM Box no. 505!) collected by Batters at Duke's Rock, Plymouth in 1897 and 1900, and identified by him as *Melobesia myriocarpa*, both pertain to *P. lobescens*.

Immersed conceptacles with pale, indefinite centres usually distinguish *P. lobescens* from other epilithic species of *Pneophyllum* such as *P. myriocarpum* and *P. caulerpae* that have raised to conical conceptacles, the three species often growing intermixed. Very often these species occur as scraps of thallus which makes identification difficult. Sometimes, however, excellent material becomes available: on 24 November 1988, low littoral rock surfaces at Wembury, south Devon, were found to be smothered with pale brownish pink (Methuen greyish red) plants of *P. lobescens*, easily identifiable by the abundant, flat-topped conceptacles with indefinite pale centres. Although not noticed on a previous visit to Wembury in December 1985, the plants had apparently been present for some time because crustose corallines such as *Lithophyllum crouanii* and *Phymatolithon laevigatum* had settled on their surface (Fig.68A) and grown to about 5mm and 12mm diameter respectively. A further bloom was reported by Jason Hall-Spencer at Loch Sween, Scotland.

Pneophyllum myriocarpum (P.& H.Crouan) Y.Chamberlain (1983), p.410.
Figs 70; 71

Lectotype: CO! (see Chamberlain, 1983, p.412, figs 61A,B). Herb. P.-L. & H.-M. Crouan, Rade de Brest, France, dredged, on porcelain, 12 December 1856.

Melobesia rosea sensu Rosanoff (1866), p.77, non *Hapalidium roseolum* Kützing.
Melobesia myriocarpa P.& H.Crouan (1867), p.150.
Melobesia zonalis f. *myriocarpa* (P.& H.Crouan) Foslie (1908b), p.4.
Fosliella zonalis f. *myriocarpa* (P.& H.Crouan) Bressan (1974), p.78.
Pneophyllum concollum Chamberlain (1983), p.402.

Key characters: Conceptacle pore surrounded by hyaline collar. Overgrowing *Melobesia membranacea*; overgrown by *Titanoderma* spp. and all large, crustose corallines; more or less equal in competition with *P. caulerpae*.

Illustrations: Chamberlain (1983), figs 55–60 (as *P. concollum*), figs 61–66 (as *P. myriocarpum*).

Thallus encrusting, adherent, flat, to at least 50mm diameter, 13μm thick (epilithic), or 75μm thick (epiphytic), orbicular to lobed, becoming confluent, confluences flat or overgrowing one another, thallus flat or with imbricating layers; *margin* entire, thin, without orbital rings; *surface* smooth, cell surface *Pneophyllum*-type, matt, colour mauve- to brown-pink (Methuen colour not recorded); *conceptacles* abruptly conical or domed (epilithic) to raised (epiphytic), *carposporangial and tetra/bisporangial conceptacles* to *c.*250μm diameter.

Basal filaments single-layered, cells in surface view 6–23μm long × 4–16 μm diameter, in VS to *c.*13μm in diameter (i.e. high); *epithallial cells* rectangular, *erect filaments* absent in epilithic plants, in epiphytic ones to 8 cells long, cells 4–29μm high, 6–15μm diameter, *epithallial cells* more or less domed; *trichocytes* intercalary in basal filaments of epilithic plants, of similar size to thallus cells.

Gametangial plants monoecious; *spermatangial conceptacles* immersed beside carpogonial/carposporangial ones, chambers more or less flask-shaped, 20–65μm diameter, 19–47μm high, spout present; *carpogonial conceptacles* slightly raised, chambers domed (epilithic) to elliptical (epiphytic); *carposporangial conceptacle* chambers domed (epilithic) to elliptical (epiphytic), 83–156μm diameter, 52–104μm high, with the roof (including collar) 13–31μm thick, roof 1–4 cells thick, cells small and irregularly arranged, filaments uniting into a hyaline collar surrounding the pore, sometimes extended into a spout, lower cells forming a downward orientated rim round bottom of pore, central fusion cell wide and very thin, *gonimoblast filaments* borne peripherally, composed of up to 6 gradually enlarging cells plus fairly large carposporangium; *tetrasporangial conceptacle* chambers domed (epilithic) to elliptical (epiphytic), 104–208 μm diameter, 39–143μm high, with the roof

Fig. 70 *Pneophyllum myriocarpum* (all except D as *P. myriocarpum* in Chamberlain, 1983). (A) Tetrasporangial thallus on *Rissoa* shell (YMC 79/147). (B) Young tetrasporangial plant (cultured from YMC 77/228). (C) Tetrasporangial conceptacles (cultured from YMC 77/228). (D) Tetrasporangial thallus (as *P. concollum* in Chamberlain, 1983) (YMC 76/134). (E) Thallus surface showing basal cells (b), epithallial cells (e) and intercalary trichocyte (t) (cultured from YMC 77/228). (F) Gametangial thallus with three conical carposporangial conceptacles and a spermatangial conceptacle (arrow) (cultured from YMC 77/228).

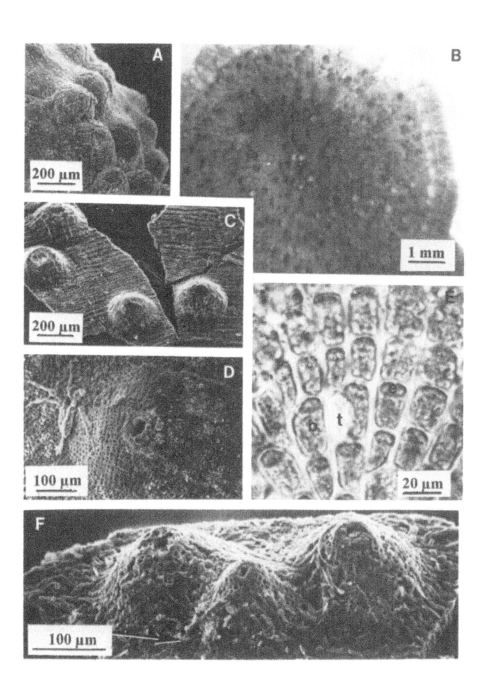

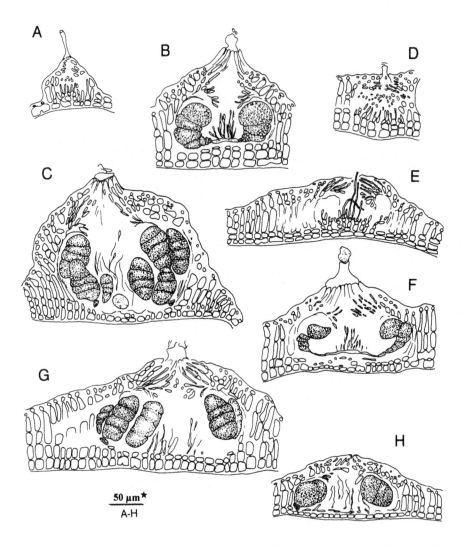

Fig. 71 Vertical sections of conceptacles of *Pneophyllum myriocarpum*. A-C recorded as *P. myriocarpum* in Chamberlain (1983). (A) Spermatangial conceptacle (cultured from YMC 77/228). (B) Carposporangial conceptacle on *Rissoa* shell (YMC 79/147). (C) Tetrasporangial conceptacle (YMC 77/228, original collection). D-H recorded as *P. concollum* in Chamberlain (1983). (D) Spermatangial conceptacle (76/100). (E) Carpogonial conceptacle (76/100). (F) Carposporangial conceptacle (YMC 77/149/6). (G) Tetrasporangial conceptacle (YMC 76/100). (H) Bisporangial conceptacle (from Shetland).

10–29µm thick, roof and pore structure as for carposporangial conceptacles, small columella present, balloon-like cells sometimes forming from roof and floor; *tetrasporangia* borne peripherally, 30–65µm long × 19–42µm diameter, *c*.8–20 per conceptacle; *bisporangial conceptacles* (1 only seen in epilithic, common in epiphytic plants) as tetrasporangial conceptacles in shape and structure, chambers 104–192µm diameter, 52–104µm high, roof 13–104µm thick; *bisporangia* 33–52 µm long × 15–42 µm diameter, 16–20 per conceptacle.

Common. Epilithic on rocks, stones, glass etc., also on shells, epiphytic plants on a wide range of algae, particularly *Furcellaria*, *Gracilaria*, *Chondrus* and *Palmaria*.

Kent, Hampshire, Dorset, Devon, Cornwall; Jersey; Co. Wexford, Co. Clare, Co. Galway. British Isles to France, Italy.

All reproductive types have been found in most months, sequential observations of plants growing, together with *P. caulerpae*, on *Palmaria* and other red algae at Beer, south Devon, showed that the plants occurred only from October-April (Chamberlain, 1983, p.408). At Kimmeridge, Dorset, *P. myriocarpum* grows on rissoid shells; studies by Wigham (1975) show that few individuals survive the winter so that *P. myriocarpum* must therefore grow and reproduce mainly in summer. It seems, therefore, that this species can grow and reproduce in all but the coldest seasons and that its actual occurrence depends on local conditions.

In laboratory culture (Chamberlain, 1983, 1987) gametangial and tetrasporangial thalli of epilithic plants were grown through several generations in normal, seasonal temperature and daylength conditions; plants were found to grow and reproduce throughout the year, but growth and maturation were quicker between April and October. A *Polysiphonia*-type life history occurs in *P. myriocarpum* (Chamberlain, 1987), plants becoming reproductively mature from 9–21 weeks after spore settlement. Spore segmentation characteristics are discussed in Chamberlain (1984). Intercalary trichocytes were shown to occur from May-October and large, sac-like, terminal structures resembling trichocytes, occurred in October (Chamberlain, 1985b).

P. myriocarpum was described by Chamberlain (1983) as a mainly epilithic species lacking erect filaments. Another, much thicker, mainly epiphytic species was newly described as *P. concollum* in the same work. It was pointed out (Chamberlain 1983, p.419) that although the two species appeared very distinct at their most characteristic, some qualitative features suggested that they might prove to be conspecific. It is now considered that this is the case. In the above description qualitative features that vary markedly in the two forms are mentioned separately, but dimensions are, for the most part, aggregated.

Further historical details were given in Chamberlain (1983, p.412).

Pneophyllum sp.
Figs 72; 73

Representative specimen: Fleet, Dorset, *W.F.Farnham*, 23.i.1989, on stones at 2m depth, YMC 89/6.

Key characters: Epilithic thalli with prominent, conical, often rostrate, bisporangial conceptacles measuring up to 430µm high.

Thallus encrusting, adherent, flat to imbricating, indefinite in diameter, up to at least 90µm thick; *margin* entire, bistratose, without orbital ridges; *surface* smooth, cell surface (SEM)

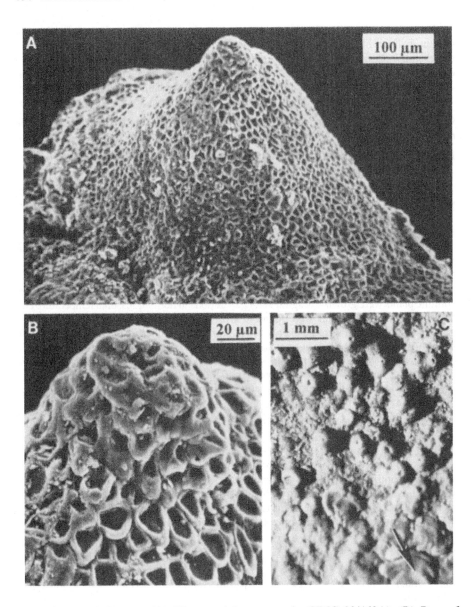

Fig. 72 *Pneophyllum* sp. (A) Bisporangial conceptacle (YMC 83/469A). (B) Pore of bisporangial conceptacle in A. (C) Bisporangial thallus partly overgrown by *Phymatolithon lenormandii* (arrow) (YMC 89/6).

Phymatolithon-type, colour mauvish to brownish pink (Methuen colour not recorded), texture matt; *conceptacles* very prominent, often leaning to one side, conical and often rostrate, surface (SEM) (Figs 72A,B) showing horseshoe-shaped calcification, conceptacle measuring up to 430μm diameter, up to 430μm high.

Basal filaments single-layered, cells elongate, in surface view 18–40μm long × 3–17μm diameter; *epithallial cells* in surface view rounded, *c*.10μm diameter; *erect filaments* to four cells long, cells 11–25μm long × 8–20μm diameter; *epithallial cells* flat to domed, 1–2 cells thick, imbricating thallus layers often developing; *trichocytes* not seen.

Gametangial and tetrasporangial plants not seen; *bisporangial conceptacle* chambers low elliptical, 247–312μm diameter, 78–130μm high, with the roof 88–117μm high, roof composed of vertically orientated filaments up to *c*.6 cells long, pore canal often extending into a rostrum, surrounded by long filaments of very thin cells that end in small papillae; small central columella, *bisporangia* borne across most of floor, 46–73μm long × 23–33μm diameter, *c*.70 per conceptacle.

Germination disc with four central cells.

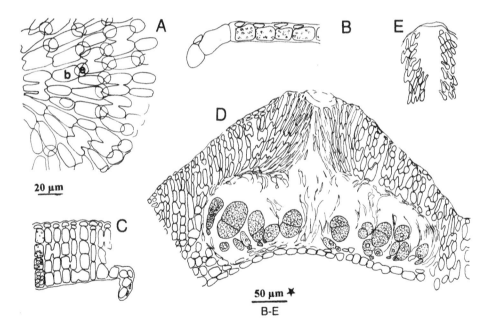

Fig. 73 Anatomical features of *Pneophyllum* sp. (A) Surface view of thallus margin showing basal cells (b) and epithallial cells (e) (YMC 89/6). (B) VS of bistratose thallus margin with unpigmented terminal initial and subtending cell (YMC 84/271). (C) VS of thickened dimerous thallus with bistratose regeneration on the right; note enlargement of starch grains down the thallus in left hand filament (YMC 84/271). (D) VS bisporangial conceptacle (YMC 89/6). (E) VS pore canal of bisporangial conceptacle lined with papillae (YMC 84/271).

Rare. Epilithic on rocks and stones in littoral channels or shallow sublittoral to *c*.3m depth. Usually in company with *Phymatolithon lenormandii* by which it is overgrown.

Dorset, Hampshire.
Unknown elsewhere.

Bisporangia recorded in January, June, November and December, probably fertile throughout the year.

This plant has cell fusions and lacks secondary pit connections, it bears bisporangia in uniporate conceptacles. These characters place it in the Mastophoroideae. It is crustose in habit and has a pore canal surrounded by numerous small papillae orientated more or less parallel to the conceptacle surface. This places it in one of the following genera: *Neogoniolithon*, *Spongites* or *Pneophyllum* (see key in Penrose & Chamberlain, 1993). In the absence of male conceptacles and the early stage of bisporangial conceptacle development, it is not possible to determine generic placement further.

In laboratory culture (YMC obs.) a germination disc with a four-celled centre was obtained. This resembles discs of the *Fosliella*-state of *Hydrolithon*, but pore structure shows that this species does not belong to *Hydrolithon*.

This species is fairly common in the Fleet, Dorset and in littoral channels at Bembridge, Isle of Wight. The prominent conceptacles resemble gametangial conceptacles of *Phymatolithon lenormandii* and bisporangial conceptacles of *Titanoderma pustulatum* agg. Subfamily characteristics can be used to eliminate misidentification.

No published description of a species attributable to the Mastophoroideae has been found to resemble this species.

MELOBESIOIDEAE Bizzozero

by Yvonne M. Chamberlain and Linda M. Irvine

MELOBESIOIDEAE Bizzozero (1885), p.109.

Type genus: *Melobesia* Lamouroux (1812), p.186.

Nulliporoideae Harvey (1849c), p.94, nom.illeg. (as Nulliporeae, ICBN, Art.19.1; see Woelkerling & Irvine, 1986a).
Lithothamnioideae Foslie (1908a), p.19.
Sporolithoideae Setchell (1943), p.134 (as Sporolitheae).

Thallus nongeniculate; some cells of contiguous vegetative filaments joined by cell fusions, secondary pit connections absent or rare; tetra/bisporangia with apical plugs and borne within multiporate conceptacles.

Most of the genera within this subfamily are distinguished by having either a discontinuous, irregularly-shaped fusion cell or several small fusion cells; sometimes a true fusion cell is apparently lacking while *Exilicrusta* has a continuous, central fusion cell. Different developmental types of spermatangial conceptacle occur, some of them involving an origin below the intercalary initials (see Johansen, 1981).

The type genus *Melobesia* and the genus *Exilicrusta* have a dimerous structure but in all other genera only monomerous thalli are known. Useful identifying characters include the size and shape of epithallial and outer cortical cells and the detailed structure of the tetra/bisporangial pore plate. These two features are illustrated on comparative plates (Figs 74; 75) rather than being discussed in detail in the synopses. Cabioch & Giraud (1978b) concluded that there was no fundamental difference between flared epithallial cells in *Lithothamnion* and domed ones in *Phymatolithon* as had been suggested by Adey (1964), but that the different shapes represented different stages in initiation, development, maturation and shedding. However, we find such a predominance of flared cells in *Lithothamnion* and domed ones in *Phymatolithon* that we regard the character as reliably diagnostic.

Of the nine genera recognized by Woelkerling (1988) *Lithothamnion*, *Melobesia*, *Mesophyllum* and *Phymatolithon* are known to occur in the British Isles. The genus *Leptophytum*, considered doubtful by Woelkerling, is also recorded and has recently been shown to be well-founded (Chamberlain, 1990). A further genus, *Exilicrusta* Y.Chamberlain (1992) has recently been described. Some species of *Leptophytum* and *Mesophyllum* have specialized cells lining the pores in the tetrasporangial roof (Chamberlain, 1990; Woelkerling & Irvine, 1986b) but the taxonomic significance of this feature is probably at species rather than genus level.

Lebednik (1978) studied spermatangial systems in melobesioid genera and concluded that

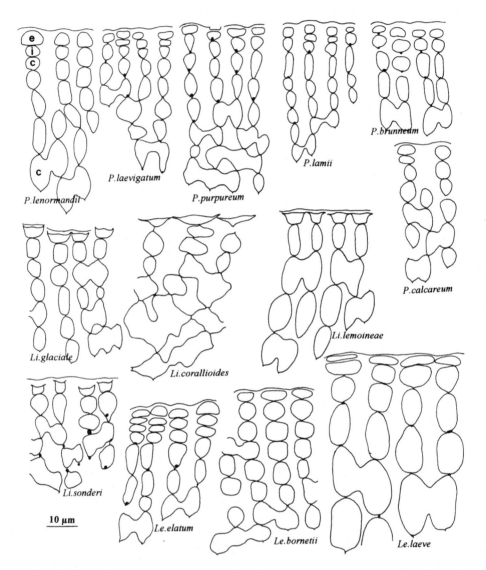

Fig. 74 Vertical sections to show comparative structure of upper thallus cells in the melobesioid genera *Leptophytum* (*Le.*), *Lithothamnion* (*Li.*) and *Phymatolithon* (*P.*). e = epithallial cell, i = subepithallial initial, c = cortical cells (one labelled exemplar is shown)

Fig. 75 Vertical sections to show comparative structure of tetra/ bisporangial conceptacle pore plates in the melobesioid genera *Exilicrusta* (*Ex.*), *Leptophytum* (*Le.*), *Lithothamnion* (*Li.*), *Melobesia* (*Mel.*), *Mesophyllum* (*Mes.*) and *Phymatolithon* (*P.*).

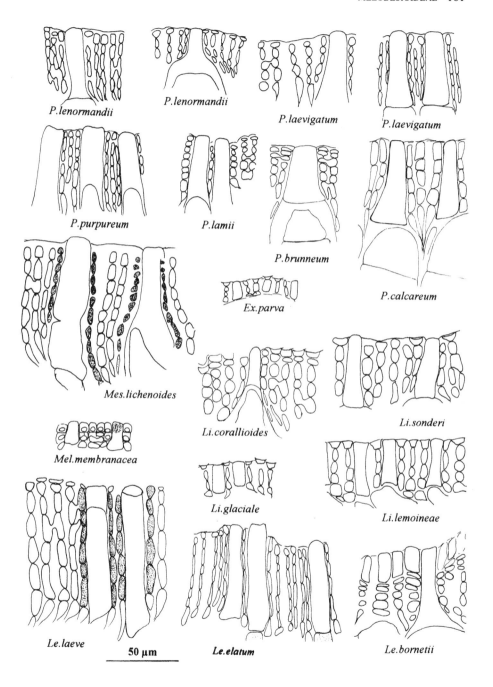

P.lenormandii

P.lenormandii

P.laevigatum

P.laevigatum

P.purpureum

P.lamii

P.brunneum

P.calcareum

Ex.parva

Mes.lichenoides

Li.corallioides

Li.sonderi

Mel.membranacea

Li.glaciale

Li.lemoineae

Le.laeve

50 μm

Le.elatum

Le.bornetii

these could yield diagnostic characters. We concur with this and find: 1) simple spermatangial systems all round conceptacle surfaces in *Melobesia* and *Mesophyllum*; 2) simple systems on the floor only in *Exilicrusta*; 3) dendroid systems on the floor and simpler systems on walls and roof in *Lithothamnion* and *Phymatolithon*; and 4) simple systems all round conceptacle surfaces except for a tuft of more or less dendroid ones in the centre of the floor in *Leptophytum*.

Records of *Clathromorphum* for the British Isles (Johnson & Hensman, 1899; Lemoine, 1913a; Cotton, 1912) have not been confirmed. Lemoine (1913a, as *Lithothamnion*) recorded *C. compactum* (Kjellman) Foslie from Clew Bay, Ireland. Examination of the specimen (BM!) on which this record is based shows it to be *Phymatolithon lamii*.

EXILICRUSTA Y. Chamberlain

EXILICRUSTA Y.Chamberlain (1992), p.185.

Type species: *Exilicrusta parva* Y.Chamberlain (1992), p.186.

Plants nongeniculate, attached, encrusting, flat, thin; gametangial conceptacles uniporate, tetrasporangial conceptacles multiporate.

Thallus pseudoparenchymatous, dimerous, composed of a single layer of basal filaments and erect filaments, subepithallial cells variable in length, epithallial cells with a flared outer wall; cell fusions common, secondary pit connections unknown.

Gametangial plants monoecious, with male and female gametes in different conceptacles on same thallus; spermatangial systems simple, borne only on the floor of the male conceptacle; carpogonial conceptacles with carpogonia and hypogynous cells borne singly on supporting cells; carposporangial conceptacles with central fusion cell bearing gonimoblast filaments peripherally; tetrasporangial conceptacles multiporate, not shedding cortical discs; bisporangial conceptacles not known. Trichocytes unknown. Spore germination characteristics unknown.

This genus was described by Chamberlain (1992) on the basis of plants collected in Hampshire, Dorset and Pembroke. It differs from all other British Isles melobesioid genera in having simple spermatangial systems confined to the conceptacle floor and a central discoid fusion cell with gonimoblast filaments borne peripherally.

Exilicrusta parva Y.Chamberlain (1992), p.186.
Figs 75–77

Holotype: BM! Princess's Shoal, Isle of Wight, Hampshire, *S.West & N.Thomas*, 15 May 1984, on a stone at 11m depth, YMC 84/212.

Key characters: Small thin plants with tiny multiporate conceptacles and flared epithallial cells. Usually in company with *Lithothamnion sonderi* and *Phymatolithon lamii*, overgrown by both of these species.

Illustrations: Chamberlain (1992), figs 1–12.

Fig. 76 *Exilicrusta parva* type specimen (BM) (YMC 84/212). (A) Holotype specimen (arrow). (B) Surface of type specimen. (C) Surface to show uniporate (single arrow) and multiporate (ring of arrows) conceptacle roofs on the same thallus.

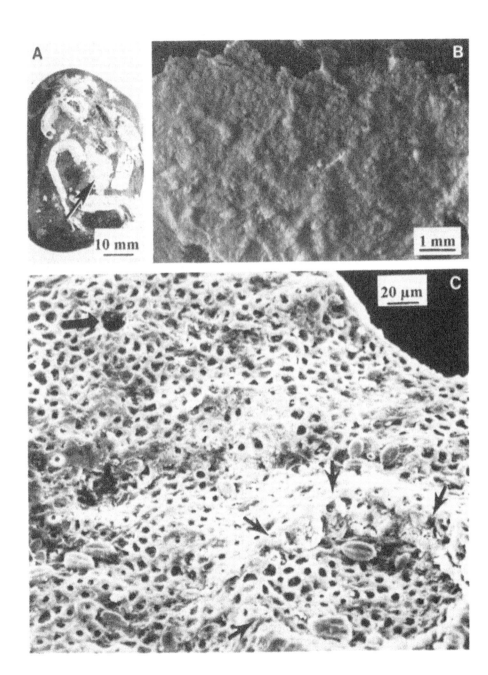

Thallus encrusting, strongly adherent, flat, to at least 30mm diameter, 312µm thick, becoming confluent; *margin* entire, thin, without orbital ridges; *surface* undulating and with microscopic bumps, cell surface (SEM) *Phymatolithon*-type, colour pale mauve to pink (Methuen purplish white), texture slightly glossy; *conceptacles* difficult to see, roofs flush with thallus surface; *gametangial conceptacles* with a thin white rim, white centre round the pore and pinkish area between, c.180µm diameter; *tetrasporangial conceptacles* with very slightly raised rim, pore plate with up to 20 pores, up to 180µm diameter.

Basal filaments unistratose, cells in surface view at bistratose margin 10–18µm long × 7–12µm diameter, in VS 5–6µm diameter (i.e. high); *erect filament* cells elongate above to relatively short and wide below, 3–11µm long × 3–9µm diameter, cell fusions common; *epithallial cells* wide and short with flared outer wall, in single layer.

Gametangial plants monoecious; *spermatangial conceptacles* adjacent to carpogonial ones, chambers flask shaped, 62–78µm diameter, 26–39µm high, simple spermatangial systems borne only on floor; *carpogonial conceptacles* with carpogonial branches

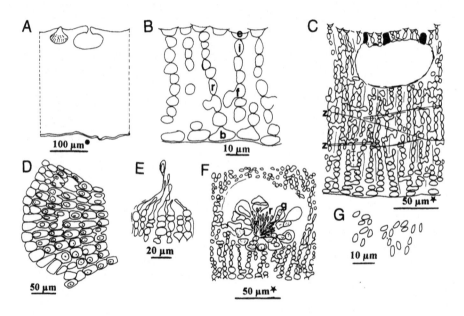

Fig. 77 Anatomical features of *Exilicrusta parva* type specimen (YMC 84/212). (A) Diagrammatic drawing of VS of gametangial thallus with spermatangial conceptacle (left) and empty presumed carposporangial conceptacle (right). (B) VS thallus showing basal filament (b), erect filaments (r), subepithallial initials (i) and flared epithallial cells (e); cell fusions (f) are present. (C) VS thallus with multiporate conceptacle; note zoning (z) of thallus. (D) Surface view of bistratose thallus margin showing basal cells subtending epithallial cells. (E) VS of carpogonial branches. (F) Oblique VS of carposporangial conceptacle showing carpogonial remnants (r) on surface of fusion cell and gonimoblast filaments (g) developing from periphery. (G) Released spermatangia.

comprising a supporting cell, hypogynous cell and single carpogonium extending into a trichogyne; *carposporangial conceptacle* chambers elliptical with central fusion cell and *gonimoblast filaments* borne peripherally, chambers 83–104μm diameter, 46–80μm high, with the roof 5–20μm thick; *tetrasporangial conceptacle* chambers elliptical, 85–91μm diameter, 52–59μm high, with the pore plate 12–14μm thick, pore plate composed of 1–2 cells plus epithallial cell; *bisporangial conceptacles* not seen.
   Old conceptacles of all types apparently becoming buried in the thallus.

Rare. Epilithic on stones in the sublittoral to at least 11m depth.

Hampshire, Dorset, Pembroke.
   Unknown elsewhere.

Gametangia recorded in May, tetrasporangia in February and May, no other collections available.

This small, obscure species is known only from a few specimens; it is difficult to find and may prove to be more widespread. Conceptacles are difficult to see but the type specimen (Fig.76C) appears to have uni- and multiporate conceptacles on the same thallus.

LEPTOPHYTUM Adey

LEPTOPHYTUM Adey (1966b), p.323.

Type species: *L. laeve* (Foslie in Rosenvinge) Adey (1966b), p.324.

Lithothamnion subgenus *Leptophytum* (Adey) Cabioch (1972) p.232.

Plants nongeniculate, attached, crustose, thin; gametangial conceptacles uniporate, tetra/ bisporangial conceptacles multiporate.
   Thallus pseudoparenchymatous, dorsiventral and monomerous; medulla composed of varying numbers of filaments subparallel to substratum with cells usually elongate; cortex composed of distal portions of medullary filaments curving towards thallus surface with cells increasing in length downwards from usually very short subepithallial initials; epithallial cells domed to flattened, not flared; fusions between cells of contiguous filaments common, secondary pit connections unknown; cell surface mainly *Leptophytum*-type, partly *Phymatolithon*-type (SEM); trichocytes unknown; spore germination characteristics unknown.
   Gametangial plants dioecious; spermatangial systems dendroid in centre of conceptacle floor, simpler peripherally and on walls and roof; carposporangial conceptacles with a central, branched fusion cell bearing scattered gonimoblast filaments peripherally; tetrasporangial conceptacles multiporate, primordium developing one or two cells below thallus surface, conceptacles never shedding a thick cortical disc, sometimes a thin veil; tetrasporangia zonate, borne across conceptacle floor, each terminating in an apical plug in conceptacle roof, pore containing plug lined by cells distinct from other roof cells at least when seen from roof surface (SEM); bisporangial conceptacles as tetrasporangial ones but producing bisporangia with two uninucleate bispores.

For an account of the genus see Chamberlain (1990). After erecting the genus (Adey, 1966b), Adey (1970) transferred eight further species of *Lithothamnion* to *Leptophytum,* six from the southern hemisphere, all having flat, thin crustose thalli; some were transferred on the basis of having raised conceptacles without their placement being confirmed by anatomical study.

Lebednik (1977b) suggested that the type and distribution of spermatangial systems was a diagnostic generic character in the Melobesioideae and that *Leptophytum* had mainly simple systems on all conceptacle surfaces with the exception of a tuft of somewhat dendroid ones in the centre of the conceptacle floor. Woelkerling (1988) included *Leptophytum* as a Holocene genus requiring further evaluation, pointing out that the generitype specimen had not been found and that many of the generic characters cited by Adey, particularly the raised tetra/bisporangial conceptacles, had proved notoriously unreliable in other melobesioid genera. There remain problems with respect to the distiction between *Phymatolithon* (*q.v.*) and *Leptophytum,* particularly as regards southern hemisphere populations. As they are distinct in the British Isles, however, the genera are maintained for the present. See further discussion under *Leptophytum laeve* and the generic description of *Phymatolithon.*

Chamberlain (1990) described *L. elatum* from the British Isles and, like all other known species, it has a rather flat thin thallus. She concluded that the characters first proposed by Adey (1966b) were sufficient grounds to consider this genus distinct from other melobesioid genera and also observed that, under SEM, thallus cells had a smooth, calcified top with a small central hole, all surrounded by a thin ridged wall (Fig.8D). This type of surface was not seen in *Phymatolithon* or *Lithothamnion* but occurs in *Clathromorphum* (Garbary & Scagel, 1979). This genus, however, differs from *Leptophytum* in spermatangial characteristics.

KEY TO SPECIES

1 Tetra/bisporangial conceptacles with markedly raised rim .... *L. bornetii*
   Tetra/bisporangial conceptacles with low rim 2
2 Tetra/bisporangial conceptacles mainly more than 600µm external diameter, lumen of roof cells 6–12µm diameter .... *L. laeve*
   Tetra/bisporangial conceptacles mainly less than 500µm diameter, lumen of roof cells 0.7–2µm diameter .... *L. elatum*

Leptophytum bornetii (Foslie) Adey (1966b), p.324.
Figs 74; 75; 78; 79

Holotype: TRH!, (see Adey & Lebednik, 1967; Chamberlain, 1990). *Ed.Bornet,* murs d'entrée du Port Militaire, Cherbourg, 30 Novembre 1853. Isotype: PC!

Lithothamnion bornetii Foslie (1898c) p.9.

Key characters: Conceptacle rim highly raised, pore plate cells relatively very short and wide. Overgrowing characteristics unknown.

Illustrations: Chamberlain (1990), figs 1,6–11,17–24,27,31,37.

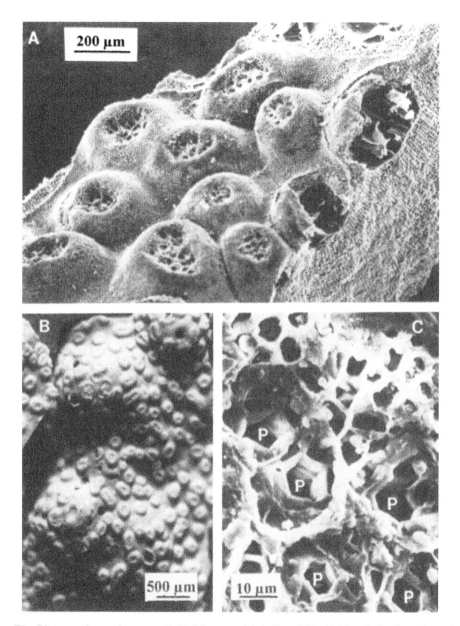

Fig. 78 *Leptophytum bornetii*. (A,B) Bisporangial thallus (BEL 4055). (C) Surface view of bisporangial conceptacle pore plate showing recessed pores (P) (YMC 83/413).

Thallus encrusting, adherent, diameter unknown, to lmm thick, irregularly shaped, flat to lumpy, without branches; *margin* entire, thin, without conspicuous orbital ridges; *surface* smooth, cell surface *Leptophytum*-type (SEM); colour reddish with bisporangial conceptacles whitish or reddish when alive, thallus and conceptacles dull pink when dried (Methuen colour not recorded), texture slightly glossy; *bisporangial conceptacles* crowded, mainly orbicular, 290–400µm external diameter, to 600µm if fused, lacking cortical discs or veils, raised above thallus surface with very prominent rims and recessed pore plates with up to *c*.50 recessed pores, pores surrounded by *c*.7 cells (SEM); old conceptacles shedding to leave fairly deep, cup-like depressions that are finally infilled.

Medulla thin, comprising less than 25% thallus thickness, filaments subparallel to substratum, cells elongate, 10–30µm long × 8–14µm diameter; *cortical filaments* in fairly distinct vertical files with cells heavily calcified, cells near thallus surface flattened, becoming

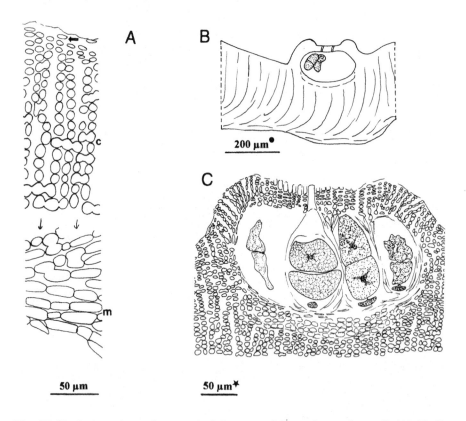

Fig. 79 Vertical sections of anatomical features of *Leptophytum bornetii*. (A) Thallus showing medulla (m), cortex (c) and short uppermost thallus cells (arrow) (BEL 4055). (B) Diagrammatic drawing to show position of bisporangial conceptacle (BEL 4055). (C) Bisporangial conceptacle (YMC 83/413).

rounded further down, (2)5–16µm long × (5)8–14µm diameter, cell fusions frequent from second to third cell down; *epithallial cells* thin-walled, domed, often absent. *Trichocytes* unknown. *Spore germination* characteristics unknown.

Gametangial and tetrasporangial plants unknown; *bisporangial conceptacle* chambers elliptical, 200–300µm diameter, 120–150µm high, pore plate 20–50µm thick, comprising *c.*5-celled filaments of very short cells, 2–5µm long × 5–8µm wide, topped by a tall (up to 16µm) thin-walled cell that is shed as conceptacle matures, rim composed of filaments of long, thin cells, pore plugs may or may not stain strongly in aniline blue, cells of filaments adjacent to pore plugs not staining differentially; *bisporangia* 80–100µm long × 50–80µm diameter.

Rare. Epilithic in lower littoral pools and sublittoral to 1m.

Confirmed from south Devon, Pembroke, Ayrshire (Loch Sween), Co.Down, possibly more widespread.

France (Cherbourg).

Gametangial and tetrasporangial plants unknown, bisporangia recorded for October and December.

The type specimen (TRH!), of which the isotype (PC!) is clearly part, is very small, but the distinctive external features (crowded, raised, high-rimmed 'tetrasporangial' conceptacles) are obvious although not well described by Foslie (1898c). Lemoine (in Hamel & Lemoine, 1953) recorded the species (as *Lithothamnium*) from a number of localities in both Atlantic and Mediterranean France, but the supporting specimens (PC!) are not correctly identified, the fertile ones, at least, being *Phymatolithon purpureum*. Specimens (BM!, PC!) from Glamorgan, coll. McLean, which Lemoine also identified as *L. bornetii*, are also *P. purpureum*. The identity of one of the specimens with raised conceptacles that Adey & Adey (1973, p.347) suggested might be this species (USNC!, Wembury, south Devon) has been confirmed. A full description of *L. bornetii* is given in Chamberlain (1990).

Leptophytum elatum Y.Chamberlain (1990), p.189.
Figs 8D; 74; 75; 80; 81

Holotype: BM! West Angle, Pembroke, *Y.M.Chamberlain*, 18 Feb 1984, on rock in lower littoral, YMC 84/38.

Key characters: Conceptacle roof raised, lacking raised rim, pore plate cells very narrow. Overgrowing characteristics unknown.

Illustrations: Chamberlain (1990), figs 4,25,28,33,34,36,38,40–44.

Thallus encrusting, rather loosely adherent, to at least 5cm diameter, to 900µm thick, flat or somewhat undulating, without branches or protuberances, irregularly orbicular, no confluences seen; *margin* entire, thin, with some orbital ridges; *surface* smooth, cell surface mainly *Leptophytum*-type, partly *Phymatolithon*-type (SEM), colour reddish (Methuen violet brown to reddish grey) with whitish conceptacles, texture slightly glossy; *spermatangial conceptacles* (observed once on otherwise tetrasporangial plant) flat-topped, to 430µm diameter; *tetrasporangial conceptacles* crowded, mainly orbicular in surface view, 280–570µm diameter, lacking cortical disc or veil, raised above thallus surface, rim apparent but only slightly raised, pore plate with up to *c*.100 pores, pores surrounded by 7–8 more or less hexagonal cells (SEM), pore plates of adjacent conceptacles fusing occasionally; old conceptacles shedding to leave shallow depressions that are finally infilled.

Medulla thick, composing up to 50% thallus thickness, filaments subparallel to substratum, with elongate cells 8–40μm long × 3–11μm diameter; *cortical filaments* in vertical files with cells well calcified, somewhat flattened near surface, becoming rounded further down, 1–12μm long × 4–12μm diameter, cell fusions common from about fifth cell down; *epithallial cells* domed, frequently absent.

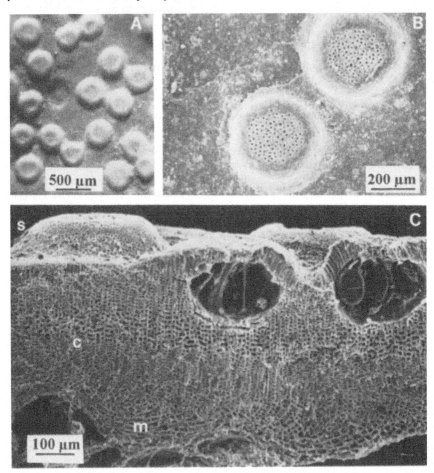

Fig. 80 *Leptophytum elatum*. (A) Bisporangial conceptacles in surface view (YMC 84/38-type). (B) Bisporangial conceptacles (YMC 84/38-type). (C) Vertical fracture of thallus showing medulla (m), cortex (c), bisporangial conceptacles in vertical fracture and side view (s) (YMC 84/62).

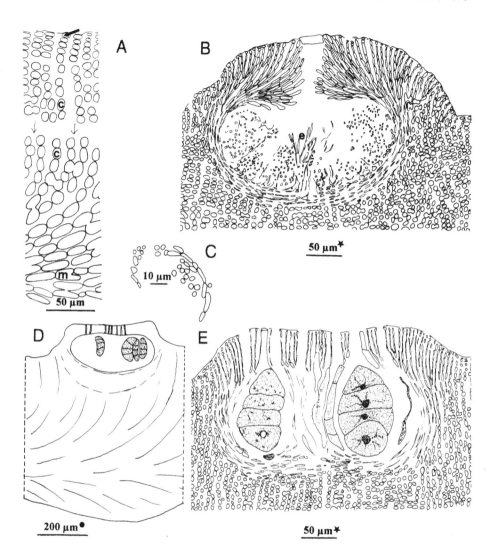

Fig. 81 Vertical sections of anatomical features of *Leptophytum elatum*. (A) Thallus showing medulla (m), cortex (c) and short uppermost thallus cells (arrow) (YMC 84/38-type). (B) VS spermatangial conceptacle; note more elaborate central spermatangial systems (e) (YMC 89/1). (C) Simpler spermatangial systems from B. (D) Diagrammatic drawing to show position of tetrasporangial conceptacle (YMC 84/38-type). (E) Tetrasporangial conceptacle (YMC 84/62).

Carpogonial/carposporangial plants unknown; *spermatangial conceptacle* chambers elliptical, mainly immersed in thallus, 234–286μm diameter, 130–156μm high, with the roof raised above the thallus surface, 73–78μm thick, roof composed of dense mass of upwardly orientated, branched filaments, floor composed of thick layer of flattened cells, spermatangial systems dense and more or less dendroid in centre of conceptacle floor but sparse and slightly branched peripherally and on walls and roof; *tetrasporangial conceptacle* chambers elliptical, 285–365μm diameter, 100–170μm high, with the roof (pore plate) 35–65μm thick, pore plate composed of 6-celled filaments of long thin cells, 4–13μm long × 0.7–2μm diameter with rounded epithallial cell, rim composed of similar cells, pore plugs fairly strongly staining with aniline blue, cells of roof filaments adjacent to pore plugs not staining differentially; *tetrasporangia* relatively long and thin, 70–105μm long × 30–55μm diameter.

Rare. On rock or shell debris in low littoral pools or shallow sublittoral.

Dorset, Pembroke.
Unknown elsewhere.

Spermatangia recorded in January, tetrasporangia in January, February and December.

This species can be difficult to distinguish from forms of *Phymatolithon lenormandii*, *q.v.*, with raised multiporate conceptacles. The most helpful identifying characters are the *Leptophytum*-type thallus surface, the long, thin pore plate cells (VS), and the lack of a veil over young conceptacles.

Leptophytum laeve (Foslie in Rosenvinge) Adey (1966b), p.324.
Figs 8E; 74; 75; 82; 83

Holotype: Missing (see Foslie, 1895, p.174). Recorded from Iceland (Eyrarbakki, on intertidal rocks and pebbles).

Lithophyllum laeve Strömfelt (1886), p.21, nom.illeg. (ICBN, Art. 64.1), non Kützing (1847), p.33. (= *Lithophyllum* sp., L!).
?Lithothamnion tenue Rosenvinge (1893), p.778. (type not seen).
Lithothamnion stroemfeltii Foslie (1895), p.173, nom. superfl.
Lithothamnion laeve Foslie in Rosenvinge (1898), p.14.

Key characters: Bisporangial conceptacles large (to 1mm), pore cells differentially staining. Overgrowing *Lithothamnion sonderi*; overgrown by *L. glaciale*.

Illustrations: Strömfelt (1886), pl.1, figs 11,12; Rosenvinge (1917), figs 129–132; Adey (1966b), figs 21,22,35–37,39,40,53,60–90; Rueness (1977), pl.VI, fig.3; Woelkerling (1988), fig.35; Chamberlain (1990), figs 5,26,29,35,39,45–47.

Thallus encrusting, flat, strongly adherent, to at least 5cm diameter, to 220(650)μm thick, irregularly orbicular to confluent, confluences flat; *margin* entire, thickish, with low, widely-spaced orbital ridges that spread across much of the thallus; *surface* of young plants smooth, old plants with low mounds; cell surface mainly *Leptophytum*-type, partly *Phymatolithon*-type (SEM), young plants deep pink (drying to Methuen greyish ruby/red/rose), with similarly coloured conceptacles that become white when old, older plants becoming less intensely pink, texture blotchy, matt; *tetra/bisporangial conceptacles* 500–1000μm diameter, scattered across thallus, shedding very thin cortical veil, raised above thallus

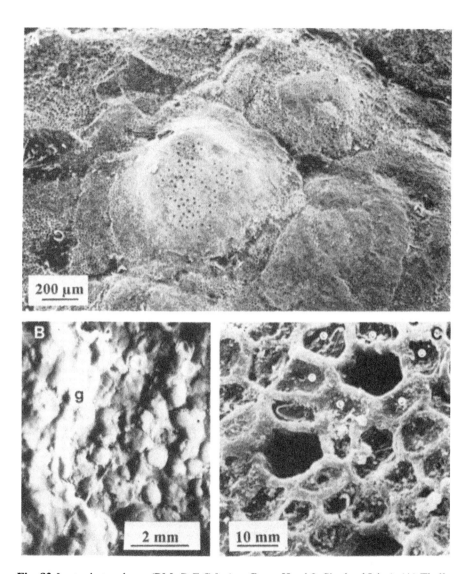

Fig. 82 *Leptophytum laeve* (BM. *D.E.G.Irvine*, Green Head 8, Shetland Isles). (A) Thallus with bisporangial conceptacles. (B) Thallus (right) being overgrown by *Lithothamnion glaciale* (g). (C) Surface view of pore plate showing bisporangial pore surrounded by 6 pore cells (white Os).

surface, flat topped without a distinct rim, usually with sides sloping from wider base to narrower pore plate, pore plate with up to 150 pores, each surrounded by up to 7 cells (SEM), rarely fusing with adjacent conceptacles; old conceptacles shed to leave shallow depressions that are finally infilled by regeneration.

Medulla 2–4 filaments thick, not exceeding 33% of the thallus thickness, filaments subparallel to substratum, cells elongate, 8–36μm long × 5–l0(12)μm diameter, fusions common; *cortical filaments* in distinct but (when decalcified) loosely arranged vertical files, cells well calcified, somewhat flattened near thallus surface then rounded to elongate, 3–15μm long × 5–14μm diameter, cell fusions frequent from about fourth cell down; *epithallial cells* flattened, often absent.

Gametangial plants not seen; only one *tetrasporangial conceptacle* seen; *bisporangial conceptacle* chambers elliptical, 312–520(680)μm diameter, 198–250(370)μm high with roof 62–78(90)μm thick, conceptacle rim and pore plate composed of c.7-celled filaments

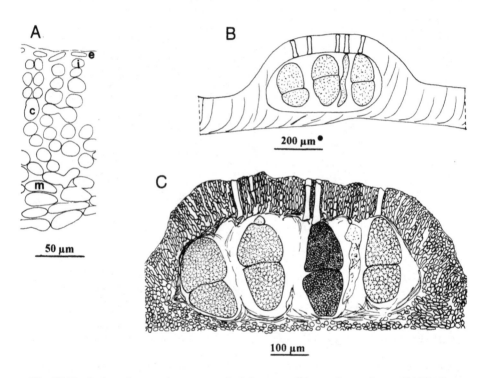

Fig. 83 Vertical sections to show anatomical features of *Leptophytum laeve*. (A) Thallus to show medulla (m), cortex (c), subepithallial initials (i) and flattened epithallial cells (e) (BM. *D.E.G.Irvine*, Green Head 8, Shetland Isles). (B) Diagrammatic drawing to show position of bisporangial conceptacle in the thallus (BM. *D.E.G.Irvine*, 192, Dales Voe, Shetland Isles). (C) Bisporangial conceptacle from same thallus as B (note that these conceptacles are too large to draw at the same scale as * conceptacles; cf. *Mesophyllum lichenoides* Fig. 99).

of squarish to elongate cells, 5–18μm long × 6–12μm diameter, often with cell fusions, pore plugs often staining strongly in aniline blue, surrounded by filaments of differentially staining cells that are noticeably thinner than adjacent cells, 3–4μm diameter; *bisporangia* 140–208μm long × 52–104μm diameter.

Epilithic and on shells (e.g. *Modiolus*), especially common on small stones, apparently intolerant of water movement and favouring bay and fjord situations in deep water, recorded only from 14–46m. Adey (1966c) found greatest relative abundance at 60–70m depth at 1–2°C in the NW Atlantic. Hooper (pers. comm.) reports that it grows to over 100m depth in Labrador. Adey *et al.* (1976) recorded this species as abundant from mid to deep water near the lower end of the photic zone in N. Japan.

Northwestern British Isles (including Rockall) to Bute.

Arctic Norway to British Isles; Iceland, Faroes, Denmark (Kattegat), arctic eastern Canada to USA (Connecticut), Greenland; Pacific Mexico, Japan, Antarctica.

Tetrasporangia and bisporangia recorded in July.

Rosenvinge (1917) found only bisporangial plants in Denmark. Tetrasporangia as well as bisporangia were recorded in the Gulf of Maine by Adey (1968, 1971) and in Iceland and Norway by Adey & Adey (1973). Gametangial conceptacles are not recorded for British Isles, but details of their structure are given for Gulf of Maine material in Adey (1966b) and Icelandic and Norwegian material in Adey & Adey (1973, pp.378 and 382); the conceptacles of both spermatangial and carpogonial/carposporangial plants are markedly raised-conical.

Adey (1966b,c) found that 76% of plants collected in the north-western Atlantic were fertile. Tetra/bisporangia, spermatangia and carpogonia/carposporangia occurred in about equal numbers whilst tetrasporangia were about twice as common as bisporangia and sometimes occurred in the same conceptacle. A cyclic pattern of development has not been demonstrated; Hooper (pers. comm.) reported spores only in winter in eastern Canada but they occur in July in western Scotland. Adey (1971) found the species grew well up to 12°C at low light intensities; he postulated that a requisite minimum winter temperature below 5–6°C would explain the recorded geographical distribution.

Adey (1966b) established *Leptophytum* to accommodate *Lithothamnion laeve* (Strömfelt) Foslie and *L. foecundum* Kjellman. He chose *L. laeve* as type species without examining the type specimen which cannot be found (Foslie, 1895; Woelkerling, 1988, p.218; Chamberlain, 1990). For the present *L. laeve* is interpreted according to Adey's (1966b) concepts. It is hoped that either the type will be found and evaluated, or a neotype comprising plants with all reproductive conceptacle types will be established. Such material is not available in the British Isles.

Adey & Sperapani (1971) described an adelphoparasite, *Kvaleya epilaeve,* growing on *L. laeve* from Labrador to Maine and in Norway; it has not been observed in the British Isles.

In the British Isles, Batters (1896) first recorded *L. laeve* from Bognor and Plymouth as *Lithothamnion stroemfeltii* Foslie but the specimens on which this record is based are *Titanoderma pustulatum* var. *macrocarpum* (*q.v.*) (BM Box 573!). Material distributed by Holmes as *Algae Britannicae Rariores Exsiccatae* no.295 from Bute (Cumbrae) as *Lithothamnion sonderi* is partly *L. laeve* (BM!). Records from Ireland (Johnson & Hensman, 1899; Batters, 1902; Adams, 1908; Lemoine, 1913a) are unconfirmed.

Because it is strictly sublittoral and has very large (to 1mm) multiporate conceptacles, this species is unlikely to be confused with any others. The name *Lithothamnion stroemfeltii* was used by Foslie (1895) as a substitute name for *Lithothamnion tenue* Rosenvinge (1893) which he mistakenly considered a homonym of *Lithophyllum tenue* Kjellman: *L. stroemfeltii* is thus a superfluous name.

LITHOTHAMNION Heydrich

LITHOTHAMNION Heydrich (1897b), p.412, nom.cons. (see Woelkerling, 1985b).

Type species: *L. muelleri* Lenormand ex Rosanoff (1866), p.101, pl.VI, figs 8–11.

Plants nongeniculate and attached or unattached; varying from flat, crustose thalli to bushy plants composed entirely of branches; gametangial conceptacles uniporate, tetra/bisporangial conceptacles multiporate.

Thallus pseudoparenchymatous; structure dorsiventral and monomerous in crustose parts, more or less radial and monomerous in protuberances and branches. Medulla of crustose thallus comprising varying numbers of filaments subparallel to substratum with cells of filaments usually elongate; cortex composed of distal portions of medullary filaments curving towards thallus surface, subepithallial initials elongate, cells scarcely elongating further below initials; epithallial cells with outer walls predominantly flared; fusions between cells of contiguous filaments common, secondary pit connections unknown; trichocytes rare; spore germination characteristics poorly known.

Gametangial plants normally dioecious; spermatangial systems dendroid on concpetacle floor, simple to arcuate on walls and roof; carposporangial conceptacles lacking fusion cells and gonimoblast filaments borne individually; tetrasporangial conceptacles multiporate with primordium developing in subepithallial initials and conceptacles not shedding a cortical disc or veil; tetrasporangia zonate, borne across conceptacle floor, each terminating in an apical plug in conceptacle roof, plug not surrounded by differentiated cells; bisporangial conceptacles as tetrasporangial ones, producing bisporangia with two uninucleate bispores.

For an account of the genus see Woelkerling (1988, p.169). Five species are recorded for the British Isles of which *L. sonderi* is reported by Adey & Adey (1973) to be the commonest shallow sublittoral crustose coralline species round our coasts. *L. corallioides, L. glaciale* and, to a small extent, *L. lemoineae* occur both attached and unattached, contributing significantly to the maerl. *L. apiculatum* Foslie, *L. hauckii* Rothpletz and *L. tophiforme* Unger have been reported for W. Ireland (Johnson & Hensman, 1899; Lemoine, 1913a; Adams, 1908) but material has not been available for confirmation. Lemoine (1913a) noted that *L. hauckii* is a Mediterranean species and the reports are probably misidentifications of *Phymatolithon purpureum, q.v. L. foecundum* was reported for Antrim (Adams, 1908) and is also probably a misidentification.

EXCLUDED SPECIES

L. norvegicum (Areschoug) Kjellman (1883), based on *L. calcareum* var. *norvegicum* Areschoug (1875) from Norway (Haugesund), has been regarded as a synonym of

L. glaciale, or *L. corallioides*, or *Phymatolithon calcareum*, *q.v*. From the original description, it is clearly based on maerl material and its correct assignation could be resolved by an examination of the type by someone familiar with the anatomy of maerl forms of species occurring in Norway. *L. norvegicum* has been reported for W. Ireland and southwest Scotland by Cotton (1912) and Lemoine (1913a) but the material in BM supporting these records is inadequate for confirmation.

Lithothamnion fruticulosum (Kützing) Foslie was first reported for the British Isles by Johnson & Hensman (1899) from Co.Galway (Roundstone). Material from Cornwall (Falmouth, 1906) was distributed by Holmes as *Algae Britannicae Rariores Exsiccatae* nos 292, 262. Examination of the specimens in BM upon which the reports for the British Isles are partly based has shown the Roundstone material to be *Lithophyllum fasciculatum*, and the Falmouth material is probably *Phymatolithon calcareum*, *q.v.*. Woelkerling (1985b) selected *Lithothamnion fruticulosum* as the type of the genus *Spongites* (Mastophoroideae) which is distinguished by having cell fusions between contiguous filaments, uniporate tetra/bisporangial conceptacles and a monomerous structure. No specimens pertaining to *Spongites fruticulosus* have been found in the British Isles.

KEY TO SPECIES

1   Plants almost entirely comprising unattached, branched thalli..................................... 2
    Plants epilithic, crustose 3
2   Branches fragile, younger thallus surface with low mounds, surface cells (SEM) as in
    Fig.8A, cortical cells as in Fig.74.. *L. corallioides*
    Branches hard and tough, younger thallus surface smooth or somewhat flaky, not
    consistently with low mounds, surface cells (SEM) as in Fig. 8B, cortical cells as in
    Fig.74 .... *L. glaciale*
3   Plants exclusively sublittoral, reddish in colour with conical white gametangial
    conceptacles .... *L. sonderi*
    Plants littoral to sublittoral, conceptacles not white 4
4   Thallus deep pink, speckled, matt, medulla to 7 filaments thick.................. *L. glaciale*
    Thallus brown-pink, uniform, somewhat glossy, medulla to 17 filaments thick ....
    .... *L. lemoineae*

Lithothamnion corallioides (P.& H.Crouan) P.& H.Crouan (1867), p.151, pl.20, figs 8–10.
Figs 8A; 74; 75; 84; 85; Table 4

Neotype: BM! Isoneotypes: CO!, PC, Roscoff (see Cabioch, 1966b). *Algues marines du Finistère* no.242, *P.-L. & H.-M. Crouan*, Brest, France.

Corallium pumilum album ... Ellis (1755), pl.76, pl.27 C pro parte, non *Millepora calcarea* Pallas (1766), p.265 (see Woelkerling & Irvine, 1986a).
Spongites corallioides P.& H.Crouan (1852), no.242. (CO!).
Lithothamnion solutum Foslie (1908a), p.214. (type not seen).
Lithophyllum solutum (Foslie) Lemoine (1915), p.13.

Key characters: Plants brown-pink, composed almost exclusively of infertile, unattached, slender, branched thalli, branches to 2mm diameter, thallus surface at branch ends beset with low mounds, cortical cells extensively fused. Epilithic crustose state overgrowing *Phymatolithon lenormandii*.

Illustrations: Cabioch (1966c, 1970); Adey & McKibbin (1970), figs 4,5,7; Bosence (1976), pl.53, text fig.10; Cabioch & Giraud (1978b), figs 6,10,11, pp.25–29,33.

Occurring either as attached crustose plant (rare) or unattached branching systems (common).

Crustose thallus closely adherent, to at least 6cm diameter, to 200µm thick, flat at periphery, simple or branched protuberances with knob-like apices soon developing, older branches thus formed to 5mm long × 2mm diameter individually (sometimes fasciated), confluent with confluences slightly overlapping or slightly crested; *margin* indefinite, thin on rock, or somewhat thickened and with orbital ridges when overgrowing other corallines; *surface* irregularly beset with numerous low mounds, sites of damage, white patches and lumps and pits where old conceptacles have broken out; cell surface (SEM) as in Fig.8A, colour brown-pink (Methuen greyish red) becoming more purplish (Methuen greyish red to violet brown) when dry, texture slightly glossy; *tetrasporangial conceptacles* densely crowded, mainly on protuberances, sometimes on basal crust, to *c*.350µm diameter (570µm if fused), somewhat domed with whitish pore plate and reddish periphery, pore plate with 15–25 pores, each surrounded by rosette of five cells (SEM); old conceptacles breaking out to leave high-rimmed crater before being infilled.

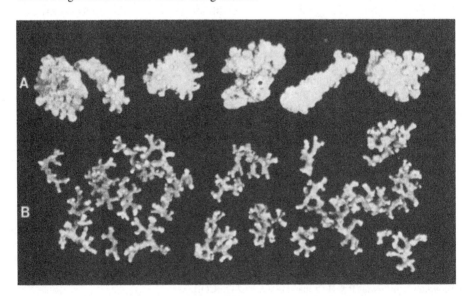

Fig. 84 *Lithothamnion corallioides*. (A) Crustose plants growing on shell fragments (*J.Cabioch*, Brittany, YMC 86/86) ×1. (B) Unattached, branched plants (*W.F.Farnham*, Falmouth, YMC 88/53) ×1.

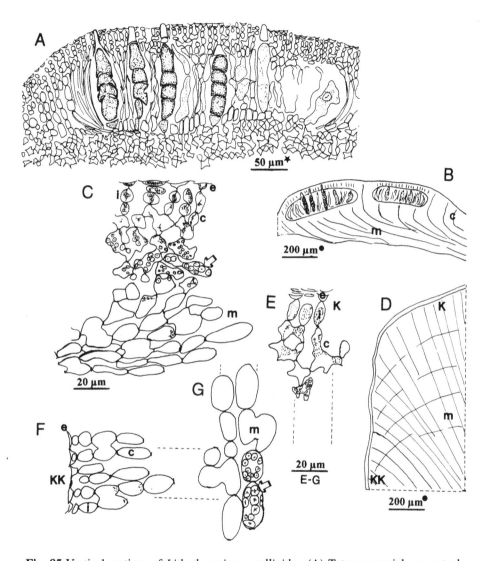

Fig. 85 Vertical sections of *Lithothamnion corallioides*. (A) Tetrasporangial conceptacle (YMC 88/45). (B) Diagrammatic drawing of crustose thallus showing position of tetrasporangial conceptacles (YMC 88/45). (C) Detail of crustose thallus (YMC 88/45). (D) Diagram of LS through half a branch (YMC 86/102). (E) Detail of cells at K on Fig. 85D. (F) Detail of cells at KK on Fig. 85D. (G) Detail of cells at m on Fig. 85D. c=cortex, m=medulla, i=subepithallial initials, e=flared epithallial cells, arrow=starch grains.

Unattached plants comprising single branches or subglobular to flattened branching systems to about 5cm diameter, sparsely to densely branched, branches sometimes confluent, generally rather slender, usually terete with knob-like apices, young branches *c.*1mm diameter with older parts to 2mm, thallus surface, at least in young parts, covered with low, cushiony mounds representing locally organized areas of growth, branch apices somewhat clubbed, often comprising a dead core covered with regenerating thallus, colour range as for crustose plant, surface somewhat glossy; *conceptacles* not seen but reported in French plants by Cabioch (1966c) and Spanish plants by Adey & McKibbin (1970).

Crustose thallus monomerous; *medulla* to 5–6 cells thick, to *c.*50µm thick (or more in crevices) with filaments loosely parallel to substratum, medullary cells 10–30µm long × 5–12µm diameter, usually full of floridean starch grains; *cortical cells* rounded to elongate and densely staining near thallus surface, cells becoming enlarged, fusiform and extensively fused further down, 5–35µm long × 5–10µm diameter, primary pit connections occupying entire width of end cell walls and fused groups of cells often star-like; *epithallial cells* mainly flattened and somewhat flared, often absent or encysted to two or three cells deep.

Unattached plants monomerous, in LS comprising a central core of filaments (*medulla*) composed of cells resembling the medullary cells of crustose plants, cells elongate, 6–24µm long, 5-10µm diameter, with fusions quite common and contiguous cells sometimes laterally aligned, cells either empty or packed with starch grains; *cortical filaments* radiating to periphery, outer cells rounded to elongate, inner cells becoming fusiform, extensively fused and star-like, these cells and epithallial cells similar to equivalent cells in crustose plant; entire branch appearing zonate with alternating arcs of longer empty cells and shorter, lightly staining cells, each pair possibly representing a year's growth.

Gametangial and bisporangial plants unknown; *tetrasporangial conceptacle* chambers only known on crustose plants, low-elliptical, 234–380µm diameter × 83–117µm high with roof 29–39µm thick; *tetrasporangia* long and thin, 78–86µm long × 21–26µm diameter.

Quite common. Unattached and crustose plants occur in sublittoral maerl beds (usually together with *Phymatolithon calcareum, q.v.*) in fairly sheltered areas; the beds in some areas (e.g. Falmouth) comprise deep, extensive deposits of dead and sub-fossil plants which develop a ridge and furrow system with relatively few living plants occurring at the surface, mainly in the furrows; in other areas (e.g. W. Ireland) flat maerl beds develop on the sea floor in various communities (Bosence, 1980; Blunden *et al.*, 1981). Crustose plants grow on pebbles and, more rarely, shells. An extensive epiflora (see Cabioch, 1969b; Blunden *et al.*, 1981), and epifauna (see Bosence, 1979) develops on maerl, predominantly in summer, and *Titanoderma pustulatum* var. *macrocarpum* (*q.v.*) is a characteristic coralline epiphyte occurring on old maerl thalli.

Confirmed from Dorset, Cornwall, west coast of Ireland. Records from Scotland (e.g. Adey & Adey, 1973; Maggs, 1986) not confirmed; plants examined have proved to be *Lithothamnion glaciale*.

Norway (Hordaland) to Canary Isles, Mediterranean (mainly recorded as *L. solutum*) . There is some doubt about the northern limit of this species: Rueness (1977) recorded it for Norway but Adey (1971) said it is both present (p.46) and absent (p.51) there. Reports for Cape Verde and Mauritania not confirmed.

Gametangial and bisporangial conceptacles unknown, the single epilithic, fertile plant from the British Isles had tetrasporangial conceptacles in July. Cabioch (1969b, 1970) reported that tetrasporangial conceptacles occurred mainly in winter in Brittany but Adey & McKibbin (1970) recorded a fertile plant in August in Spain.

Cabioch (1969b) noted that the proportions of *L.corallioides* and *P.calcareum* were continually changing in Breton maerl beds, possibly accounted for by phasic reproduction with peaks of fertility occurring about every six years. Cabioch (1969b, p.157; 1970) considered that recruitment to maerl populations is predominantly from branches shed from crustose plants and that vegetative propagation from unattached plants is rare. This may be the case in populations from Brittany, Madeira and Tenerife where crustose plants are frequent but crustose plants are so rare in the British Isles that unattached plants must presumably be almost entirely vegetatively propagated.

Field growth studies by Adey & McKibbin (1970) in Spain showed a maximum growth rate of 3.5µm/day in summer falling to nil in winter; this correlated with laboratory experiments giving 3.5µm/day at 10°C, 14h light: 10h dark daylength and 640 lux. They showed that maximum growth occurred at 10–12°C but that growth ceased if illumination fell below 170 lux. Adey (in Adey & Vassar, 1975, p.62) calculated an average marginal growth rate of 0.2mm per month in tanks in Norway.

Two fundamentally different growth forms occur in this species, the crustose, attached form and the secondarily derived, unattached, branched form. As only two crustose plants have been found in the British Isles, in Dorset and Devon, Cabioch's data (1966c), and our observations on crustose plants from Brittany, N.W.France, leg. J.Cabioch, are included in the above description. The branched form varies in appearance from simple, individual branches to lightly or densely branched, spheroidal, ellipsoidal or discoid masses. The Crouan brothers (1867) described *L.corallioides* as a distinct species, but Foslie (1905c) relegated it to a form of *L.* (now *Phymatolithon*) *calcareum* since when plants from northern France and the British Isles have usually been confused with this species (e.g. Newton, 1931). Mediterranean plants, however, became known as *L. solutum* Foslie (see Huvé, 1956) or *Lithophyllum solutum* (Foslie) Lemoine (1915). Cabioch (1966c) analysed populations of *L. corallioides* in Brittany, segregating it from *L. calcareum* and subsuming *L. solutum*. She considered that there were two varieties: var. *corallioides* having plants 1–5cm diameter, with branches exceeding 1mm diameter and clearly zoned in LS, and var. *minima* having plants under 1–2cm, with branches under 1mm diameter and a relatively homogeneous thallus structure. Within var. *corallioides* she included formae *corallioides*, *subsimplex* Batters (now reidentified as *Phymatolithon calcareum*, q.v.), *subvalida* Foslie, *australis* Foslie, *flabelligera* Foslie and *globosa* Cabioch, pointing out that parallel forms occurred in *Phymatolithon calcareum;* Bressan (1974) applied Cabioch's scheme to Italian populations.

Heydrich (1911b) discussed plants identified as *Lithothamnium rubrum* Philippi (1837) from Roscoff and concluded that they were probably conspecific in part with *Spongites corallioides* P.& H. Crouan. He erected five formae: *L. rubrum* f. *primigenia,* f. *corall[i]oides* (P.& H.Crouan), f. *minuta* (Foslie), f. *gracilis* (Philippi), and f. *crassa* (Heydrich). His drawings (pl.II, figs 1a,2) of f. *primigenia* show fertile, crustose plants strongly resembling plants of *L. corallioides* collected recently from Roscoff. It should be noted that the type of *L. rubrum* is an *Amphiroa* (Woelkerling, 1983a, p.172). The whereabouts of Heydrich's Roscoff specimens is unknown and their identity cannot therefore be confirmed.

It is not within the scope of this work either to examine all the type specimens pertaining to these formae or to unravel all the complexities resulting from the earlier concepts of the conspecificity of *L. corallioides* and *P. calcareum*. A new approach to the problem of form range within these plants was established by Bosence (1976, 1980) when he studied maerl beds at Mannin Bay, Co.Galway, on the west coast of Ireland. He interpreted the habitat in terms of varying degrees of exposure to wave action, substratum factors (sandy, muddy, etc.) and light penetration; at the same time he recognized three basic unattached thallus forms: spheroidal, ellipsoidal and discoidal, each with varying degrees of branching; he then proposed a scheme relating the resulting thallus forms to the habitat factors (Bosence, 1976, pl.53 and text fig.10), pointing out, like Cabioch, the existence of parallel growth forms in *L. corallioides* and *P. calcareum*. The forms of growth thus established can be related to the previously described formae (e.g. Cabioch, 1966c, pl.1B,C) and Bosence's scheme seems to offer a practical and meaningful frame of reference within which to consider the wide variation range seen in unattached thalli of *L. corallioides* and other species with a similar life-style. It seems preferable to use Bosence's descriptive terms rather than to regard the forms as having taxonomic status.

There are a considerable number of features by which *L. corallioides* and *P. calcareum* may be distinguished; these are summarised in Table 4. Cabioch (1966c, 1972) distinguished the two species as follows: in *P. (as L.) calcareum* branches are formed from perithallial filaments (defined as having ovoid cells), whereas they are composed of secondary hypothallial filaments (defined as having rectangular cells) in *L. corallioides,* differences in cell size and shape being valuable in distinguishing the species. Features that may be used to distinguish *L. corallioides* from *L. glaciale,* which possibly replaces it as a maerl component in the north of the British Isles, are shown in Table 4.

The basionym *Spongites corallioides* P.& H. Crouan (1852) is validated by the quoted description of *Corallium pumilum album* Ellis (1755); a lectotype should ideally be selected from his material (see ICBN Art.432.2), unfortunately now lost (Woelkerling & Irvine, 1986a). Included in Ellis's protologue was a reference to maerl from Falmouth, known to be a mixture, and from which Woelkerling & Irvine (1986a) selected one of the constituent species to neotypify *Phymatolithon calcareum* (Pallas) Adey & McKibbin, *q.v.* Microscopic examination of *Algues marines du Finistère* no.242 (CO, BM!) has shown that the Crouans effectively selected a different species, which also occurs in the Falmouth maerl, as the basis of their *Spongites corallioides.* In the absence of Ellis's original material, the specimens of no.242 in *Algues marines du Finistère* in BM are proposed as the neotype of *Spongites corallioides* P.& H. Crouan (see also Cabioch 1966c). Arguably, both Ellis (1755) and Pallas (1766) based their species on the same Falmouth sample, but even so this would not invalidate the above argument (see ICBN Art.7B.5).

An early reference to maerl beds was made by Ray (1690) as follows: "Corallium album pumilum nostras. Small white coral. A Corallio albo Lob.Ger.Park. differt. It is found plentifully in the ouze dredged out of Falmouth Haven to manure their lands in Cornwal: it is not unlikely that it may also grow on the Rocks about St.Michael's Mount, though we observed it not there". Maerl is still harvested at Falmouth (also more extensively in Brittany and western Ireland) and is a popular fertiliser in these days of "organic gardening". Blunden *et al.* (1981) investigated Falmouth maerl and found that *L. corallioides* predominated down to 6m and *P. calcareum* from 6–10 m. Chemical analysis of this maerl showed that it contained 32.1% $CaCO_3$ and 3.1% $MgCO_3$ (dry weight). A small percentage (10%)of total $CaCO_3$ occurred as aragonite, the remainder as calcite.

Cabioch & Giraud (1978b) studied *L. corallioides* under TEM, finding that the epithallial region strongly resembled *L. sonderi* in usually being one cell thick, having cells with flared upper edges (forme de carène) and appearing to be encysted, whilst subjacent cells were strongly secretory. They found, however, that sometimes the epithallial cells more closely resembled those of *Phymatolithon calcareum* (as *Lithothamnion*) in being domed, although the latter, in contrast, sometimes showed a flared aspect.

Adey & McKibbin (1970) reported that staining bodies are absent in *L. corallioides*.

Lithothamnion glaciale Kjellman (1883), p.123 (reprint p.93), pls 2,3.

Figs Frontispiece; 8B; 74; 75; 86; 87; Table 4

Lectotype: UPS (not seen) (no.241, provisionally selected by Adey, 1970a). Spitzbergen.

Lithothamnion colliculosum Foslie (1891), p.43 (TRH!).
Lithothamnion flabellatum f. *granii* Foslie (1895), p.98.
?*Lithothamnion roseum* Batters (1893), p.20.
Lithothamnion battersii Foslie (1896), p.l.
Lithothamnion granii (Foslie) Foslie (1900c), p.ll.

Key characters: Plants deep pink, hard in texture, often with profuse, clavate protuberances or branches, bisporangial conceptacles more or less immersed, mainly at base of branches, pore plate chalky, *c*.20–40 pores. Overgrowing *Pneophyllum* spp., *Lithothamnion sonderi*, *Phymatolithon laevigatum*, *P. lenormandii*; usually overgrowing *P. purpureum*.

Illustrations: Printz (1929), pls XVIII,XXI,XXIII,XXIV; Suneson (1943), figs 6,7; Rueness (1977), p.VI, fig.4; Adey (1966b), figs 9–14,19,7–30,51 (as *Lithothamnion* "a"); Adey & Adey (1973), figs 7–29; Cardinal *et al.* (1979), pl.I, figs 4,6, pl.II, figs 1,3,5, pl.III, fig.2.

Thallus encrusting, closely adherent, to at least 20cm diameter, initially thin, later thick and extensive to 4mm thick, initially flat, but usually developing simple to compound, cylindrical, often prolific branches to 15mm long, 4mm diameter with apices mainly rounded to somewhat pointed, also occurring as unattached plants ranging from individual branches to subspherical, knobbly to branched masses, orbicular to confluent with confluences somewhat crested; *margin* entire, thin and indistinct, without orbital ridges; *surface* smooth or with scattered low mounds; cell surface *Phymatolithon*-type (SEM); colour reddish to deep pink with a tinge of violet (drying to Methuen greyish red), usually with many small, white speckles, texture very hard, matt; *spermatangial conceptacles* hemispherical, to *c*.500μm diameter, *carposporangial conceptacles* conical, to *c*.600μm diameter, *tetra/bisporangial conceptacles* round branch bases or on branches, slightly raised, to *c*.380μm diameter, with small rim that is not raised, pore plate appearing chalky, with 20–40 pores, adjacent pore plates often fusing.

Crustose structure monomerous; *medulla* to 17 cells thick, to 50μm thick with cells 7–26μm long × 3–8μm diameter; *cortical cells* rounded to elongate, 8–12μm long × 4–8μm diameter, with extensive fusions, initials noticeably elongate; *epithallial cells* flared and relatively wide and shallow, sometimes up to 3 cells deep; *trichocytes* (megacells) reported rarely (Walker, 1984).

TABLE 4. Features contrasting and comparing *Lithothamnion corallioides, L. glaciale* and *Phymatolithon calcareum.* (Detached branches of these species are common in maerl beds/ deposits)

CHARACTER	*L. corallioides*
Colour tendency (fresh)	Brownish pink
Thallus surface	Covered with low mounds Cells (SEM) as Fig. 8A
Thallus texture	Slightly glossy
Branch hardness	Brittle
Branch size	Mainly <1mm diameter
Crustose plants in VS	
Epithallial cells	Mainly flared
Cortical cells (l)×(d)	Fusiform (5–35) × (5–10)μm
Cortical fusions	Very extensive
Groups of fused cells	Star-like
Primary pits	Occupying entire end wall
Medullary filaments	5–6 layers
Medullary cells (l)×(d)	(10–30) × (5–12)μm
Branch anatomy in LS	
Medullary cell distribution	Tiered
Medullary cells (l)×(d)	Rectangular (20–35) × (8–12)μm
Medullary cells pits	Bilenticular primary pits
Cortical cells (l)×(d)	(5–35) × (5–10)μm
Tetra/bisporangial conceptacle	
Shape	Without rim, pore plate convex
Ext. diam.	To c.350μm
Pores (SEM)	Ringed by 5 cells (SEM)
Chamber (VS) d	234–380μm
h	83–117μm
Roof thickness	29–39μm
Tetra/bisporangia	
Shape	Long and thin
Length	78–86μm
Diameter	21–26μm
Old conceptacles	Not known

L. glaciale	*P. calcareum*
Reddish to deep pink with violet tinge	Mauvish brown
Mainly smooth, some scattered low mounds	Some lowish mounds, frequently flaky areas
Cells (SEM) as Fig. 8B	Cells (SEM) as Fig. 8C
Matt	Somewhat chalky
Hard	Quite hard
Variable	Mainly >1mm diameter
Mainly flared	Mainly domed
Elliptical (8–12) × (4–8)μm	Elliptical (5–10) × (3–5)μm
More localised	More localised
Bead-like	Bead-like
Occupying only centre of end wall	Occupying only centre of end wall
To 17 layers	5–6 layers
(7–26) × (3–8)μm	(10–15) × (4–5)μm
Tiered	Not tiered
Elliptical to rectangular (4–18) × (3–11)μm	Elliptical (5–18) × (3–10)μm
Not known	Not bilenticular primary pits
(8–12) × (4–8)μm	(8–18) × (7–10)μm
With narrow, non-raised rim, pore plate level	Either with rim and prominent, or without rim and immersed, pore plate level
To c.380μm	To c.450μm
Ringed by 6 cells (SEM)	Ringed by 6 cells (SEM)
150–360μm	230–350μm
110–180μm	117–130μm
5–40μm	30–40μm
Elliptical	Plump
65–96μm	90–125μm
23–47μm	49–73μm
Usually becoming buried	Becoming buried

Branches secondary, structure monomerous, comprising a central core of filaments (*medulla*) composed of cells varying in LS from elliptical to rectangular, 4–18µm long × 3–11µm diameter, often with fusions, cells empty or with starch grains, filaments radiating to the periphery where the cells resemble *cortical* and *epithallial* cells of crustose thallus; entire branch conspicuously tiered in successive arcs of laterally aligned cells.

Gametangial plants dioecious; *spermatangial conceptacle* chambers domed, 295–430µm diameter × 100–250µm high with roof 20–70µm thick, spermatangial systems dendroid on conceptacle floor, simple or arcuate on walls and roof; *carpogonial conceptacle* chambers flask-shaped, 295–325µm diameter × 160–225µm high with roof (neck) 135–200µm thick; *carposporangial conceptacle* chambers domed, 385–520µm diameter, 135–320µm high with roof 90–270µm thick, fusion cell fragmentary and *gonimoblast filaments* attached individually, mainly at periphery but also over the conceptacle floor (Adey & Adey, 1973, figs 22–25); *tetra/bisporangial conceptacle* chambers elliptical, 150–360µm diameter × 110–180µm high with roof 5–40µm thick; *bisporangia* 65–96µm long × 23–47µm diameter; old conceptacles mainly becoming buried in the thallus, sometimes being infilled.

Common. Epilithic on bedrock and pebbles and a common colonizer of stabilized detritus; lower littoral in shady pools and runnels, sublittoral to 34m, well represented on pebbles, etc. in sheltered habitats with some water movement; branches can detach and contribute to maerl deposits (see Table 4).

Throughout northern British Isles south to north Devon (Lundy) and Yorkshire; in Ireland south to Co. Galway and Co. Down.

Arctic Russia to British Isles, Iceland, Faroes, W.Baltic, records for France (Hamel & Lemoine, 1953) unconfirmed, Arctic Canada to USA (Mass.), Greenland; Japan, China. The

Fig. 86 *Lithothamnion glaciale*: unattached plants (*J.Hall-Spencer*, W.Scotland, YMC 93/42) ×1.

species is abundant in boreal sublittoral areas and Adey (1966a, as *Lithothamnium* "a") showed that it formed 38.4% of total coralline cover from low tide level down to 17m in the Gulf of Maine. Adey *et al.* (1976) recorded this species as one of the principal deep-water crustose corallines of northern Japan.

Bisporangial conceptacles appear to be present throughout the year, ripe ones were reported, for example, in June/July by Rosenvinge (1917) and Suneson (1943) and were collected at Berwick in January. Adey (1966b) found that more plants from the north-west Atlantic were mature in midwinter. Growth rates have not been assessed but R. Hooper (pers. comm.) found that this species can completely cover a bottle with a mosaic of thalli up to *c*.5cm diameter in about five years.

The appearance of this species shows considerable variation according to age and habitat from initially thin flat crustose thalli, to knobbly ones, to extravagantly branched attached thalli. Unattached plants vary from individual branches to hedgehog-like structures with

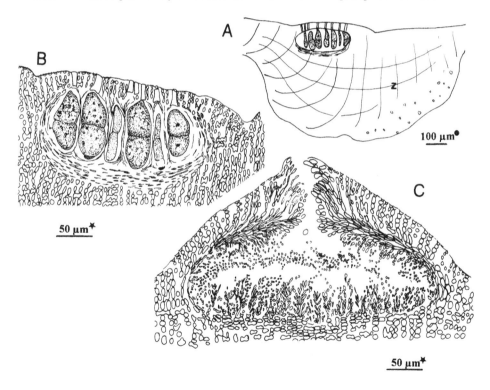

Fig. 87 Vertical sections to show anatomical features of *Lithothamnion glaciale*. (A) Diagrammatic drawing of bisporangial thallus; note conspicuous zoning (z) (YMC 83/12). (B) Bisporangial conceptacle (YMC 83/12). (C) Spermatangial conceptacle. Note: Procarpic and carposporangial features are illustrated by Adey & Adey (1973, figs 13–21).

branches in all planes and can also form sleeves round terete red algal hosts in some areas. As noted by Suneson (1943), branches are best developed on plants growing in narrow passages with streaming water. This variation has led to the proliferation of synonyms such as *L. battersii*, *L. colliculosum*, *L. granii* and *L. roseum*, plus various combinations of these at infraspecific levels (see Foslie, 1905a; Rosenvinge, 1917; Suneson, 1943). This species was distributed under the name *L. rosaceum* (instead of *L. roseum*) by Holmes in *Algae Britannicae Rariores Exsiccatae* no.142.

Adey (1970b) recorded growth rates up to 13μm/day at 13.5°C, 360 lux, 14h light:10h dark daylength which is rapid for coralline algae. He suggested that this might account for the species being the most abundant arctic North Atlantic crustose coralline in the photic zone.

Adey & Adey (1973) described procarp structure (figs 13–21) and reported (p.355) that staining bodies are absent and pit bodies very small.

Cardinal *et al.* (1979) found *L. glaciale* and *L. lemoineae* (*q.v.*) abundant in the littoral and sublittoral in Quebec and used discriminant analysis to establish significant epithallial and hypothallial differences between the two species.

Walker (1984) noted, for the first time, the occurrence of rudimentary trichocytes in cultured thalli.

The vivid, reddish pink, knobbly plants of littoral *L. glaciale* usually stand out very distinctly from all other crustose corallines except perhaps *L. lemoineae*.

It seems possible that *L. glaciale* replaces *L. corallioides* as a companion to *P. calcareum* in maerl beds in the northern British Isles and it has been found recently in the Scottish sea lochs Torridon and Sween (Sue Hiscock, Jason Hall-Spencer pers. comm.). Individual branches of *L. glaciale* are not easy to distinguish from *L. corallioides*. The harder thallus and the absence of numerous mounds on the surface of *L. glaciale* is a possible guide but the most reliable character is the difference in cortical cell structure (Fig.74; Table 4).

Lithothamnion lemoineae Adey (1970b), p.228.
   Figs 74; 75; 88; 89

Holotype: USNC (not seen). *Adey* 61–41 1–6B, 3–5m depth, Maine, 2 Nov. 1961.

Key characters: Plants brown-pink, with broad, somewhat tapering protuberances, bisporangial conceptacles more or less immersed, scattered across thallus surface, pore plate appearing silvery, *c.*20–40 pores. Overgrown by *Lithophyllum crouanii*.

Illustrations: Adey (1966b), figs 14–18,31–34,52 as *Lithothamnium* "b"; Cardinal *et al.* (1979), pl.I, figs 2,5,7, pl.II, figs 2,4,6, pl.III, figs 1,3.

Thallus encrusting, closely adherent, to at least 12cm diameter, initially thin, becoming thick, to *c.*2mm thick, usually with low protuberances developing into simple branches, widely or densely spaced, broad at base, to 2mm long with rounded to pointed apices, also occurring as unattached branching masses, orbicular, becoming irregular but confluences not seen; *margin* entire, thin, without orbital ridges; *surface* smooth or with low mounds and regenerating thallus covering old dead thallus, cell surface *Phymatolithon*-type (SEM); colour mainly uniform brown-pink (drying to Methuen dull red), texture slightly glossy; *bisporangial conceptacles* densely crowded across entire plant, slightly domed, to *c.*400μm diameter, rim surrounding pore plate distinct but not raised, pore plate silvery, with 20–40 pores, adjacent pore plates often fusing.

Crustose structure monomerous, *medulla* to 38 cells thick, to 220µm thick with cells 3–36µm long × 2–8µm diameter; *cortical cells* rounded to elongate, 5–16µm long × 4–8µm diameter with extensive cell fusions; *epithallial cells* flared, squarish to slightly wider than long, to three cells deep.

Gametangial and tetrasporangial plants unknown; *bisporangial conceptacle* chambers elliptical, 180–250µm diameter, 104–156µm high with roof 13–30µm thick; *bisporangia* 82–132µm long × 39–75µm diameter; most old conceptacles becoming buried in the thallus.

Rare. Epilithic on midlittoral rocks in shady places.

Northumberland (near Berwick-upon-Tweed), Orkney.

Eastern Canada (Labrador) to USA (Gulf of Maine) (South & Tittley, 1986). Adey *et al.* (1976) reported a similar species (as *Lithothamnion "pac-lem"*) in shallow waters to 15m in northern Japan.

Bisporangia recorded in January.

10 mm

Fig. 88. Bisporangial thallus of *Lithothamnion lemoineae* (YMC 83/3) (photo P.J.Mullock).

This species occurs in maerl beds in Orkney (Peter Sansom, obs.) where it forms branching masses, either surrounding small stones or without an inorganic centre. It forms a small proportion of massive *L. glaciale* maerl beds.

Adey (1966b) described this species (as *Lithothamnion* "b") from the Gulf of Maine and later (Adey, 1970b) published a formal description and useful keys based on various characteristics. Cardinal *et al.* (1979) discussed populations from Quebec and used discriminant analysis to distinguish *L. lemoineae* from *L. glaciale* (*q.v.*). Hooper & Whittick (1984) recorded the species from Labrador. *L. lemoineae* is a common species in the Gulf of Maine where Adey (1966b) showed that it formed 11.4% of total coralline cover from lower littoral to 17m sublittorally, occurring mainly on larger boulders. In this locality it varied greatly in degree of branching but tended to have a thicker crustose thallus and less dense branching than the more abundant *L. glaciale*. Adey & Adey (1973) did not find *L. lemoineae* in the British Isles and only one or two littoral plants have been found so far; however, it is probable that it will be found more commonly in both the littoral and sublittoral. Adey (1966b, p.331) reported pit bodies larger than those of *L. glaciale*.

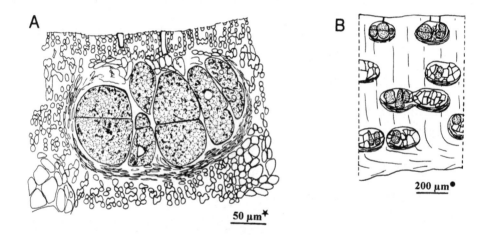

Fig. 89 Vertical sections to show anatomical features of *Lithothamnion lemoineae* (*R.Walker*). (A) Bisporangial conceptacle. (B) Bisporangial thallus with numerous buried conceptacles.

Lithothamnion sonderi Hauck (1883), p.273, pl.III. fig.5.
    Figs 74; 75; 90–92.

Holotype: L! Herb. F. Hauck, Helgoland.

Key characters: Plants reddish, thin and flat, tetrasporangial conceptacles raised, pore plate with up to *c*.80 pores. Overgrowing *Pneophyllum* spp., *Exilicrusta parva*, *Phymatolithon lenormandii* and *Corallina officinalis* basal crust; about equal in competition with *Phymatolithon lamii*, *P. laevigatum*; overgrown by *Lithophyllum nitorum*.

Illustrations: Hauck (1883), pl.III, fig.5; Heydrich (1900a), tab.II, figs 20–22; Printz (1929), pl.IV, figs 5–7; Suneson (1943), figs 4,5, pl.II, figs 5–7, pl.VIII, fig 35; Masaki (1968), pl.IX, fig.3, XI, XLVII, XLVIII; Kornmann & Sahling (1977), abb.122A-F; Chamberlain (1992), figs 13–41.

Thallus encrusting, closely adherent, to at least 5cm diameter, thin or to *c*.3mm thick, flat or with low protuberances or simple branches, orbicular becoming confluent, confluences forming small crests; *margin* entire, thin, wavy and indistinct, without orbital ridges; *surface* smooth or with slight mounds; cell surface *Phymatolithon*-type; colour deep rose red (Methuen greyish red) to slightly brownish, with white speckles (mainly in summer), drying to mauvy pink (Methuen greyish rose), quite glossy; *spermatangial conceptacles* subconical becoming apiculate, to 450μm diameter; *carposporangial conceptacles* subconical becoming apiculate, to 450μm diameter; *tetra/bisporangial conceptacles* raised with flat to slightly arched roof, to 520μm in diameter, pore plate with up to 80 pores (100 in type), without apparent rim at periphery, pores somewhat recessed and surrounded by rosette of 6–8 cells (SEM); all conceptacles shedding thin, white veil, initially same colour as thallus but soon becoming white; old conceptacle roofs being shed to leave persistent white-rimmed craters before being infilled.

Structure monomerous; *medulla* to *c*.6 cells thick, to 50μm thick with cells 10–21μm long × 3–8(9)μm diameter; *cortical cells* heavily calcified, mainly short to rounded, with extensive fusions from second cell down, 2–10μm long × (3)5–10μm diameter, conspicuous pit bodies usually present; *epithallial cells* flared, shallow and varying from wide to quite narrow.

Gametangial plants dioecious; *spermatangial conceptacle* chambers subconical, (160)276–312μm diameter, (70)104–109μm high with roof (45)88–130μm thick, spermatangial systems dendroid on conceptacle floor, simple to arcuate on walls and roof; *carpogonial conceptacles* raised, conical with chambers flask-shaped, 145–170μm diameter × 35–130μm high, and pore canal 70–115μm (Adey & Adey, 1973); *carposporangial conceptacle* chambers elliptical, *c*.395μm diameter, *c*.125μm high, with roof (pore canal) *c*.182μm thick, canal surrounded by papillae, no single fusion cell evident in VS, 2 celled *gonimoblast filaments* borne discontinuously among senescent carpogonia; *tetrasporangial conceptacle* chambers elliptical, 143–442μm diameter, 90–140μm high, roof (10)13–26μm thick; *tetrasporangia* relatively long and thin, 104–160μm long × 25–78μm diameter; *bisporangial conceptacles* (only one seen) similar to tetrasporangial ones.

Very common. Epilithic on bedrock and pebbles and on shells, worm tubes and dead maerl, not found unattached. This species is very characteristic of shallow, sublittoral habitats in sheltered places with streaming water where it forms thin, red-pink, irregularly shaped crustose thalli on stones among various animals; a good illustration of its habit is seen in Erwin & Picton (1987, p.113 top). According to Adey & Adey (1973, p.353) it is the primary deep water crustose coralline in the British Isles comprising, for example, up to 97% of crustose coralline cover in the 28–37m sublittoral zone at the Isle of Man, Belfast and Loch Broom. Clokie *et al.* (1979) reported *L. sonderi* at deep water photic limits down to 40m in western Scotland.

Throughout the British Isles.

Norway (Nordland) to N.Spain, W.Baltic, Mediterranean, Canary Isles; Japan.

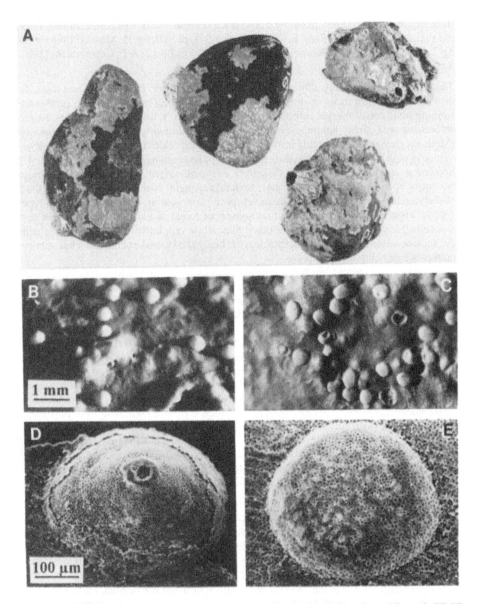

Fig. 90 *Lithothamnion sonderi*. (A) Thalli on cobbles collected in the sublittoral (YMC 89/32) ×1. (B) Thallus with carposporangial conceptacles (YMC 83/441). (C) Thallus with bisporangial conceptacles, scale as B (YMC 83/441). (D) Spermatangial conceptacle (YMC 83/441). (E) Bisporangial conceptacle, scale as D (YMC 83/441).

Conceptacles are found, apparently with fertile contents, throughout the year. Rosenvinge (1917), Suneson (1943) and Masaki (1968) recorded gametangia and tetrasporangia from May - October in Scandinavia and Japan respectively. Recent observations suggest that it is not fertile in Helgoland (type locality) in summer (Petra Leukart, pers. comm.). Suneson (1943) gave a full description of carpogonial conceptacles and the present account adds carposporangial conceptacle details.

The rather vivid, thin, glossy red thalli with white conceptacles or residual rims make mature plants quite easy to recognize. Suneson (1943) found that old plants had a distinctively rough, warty surface but young sterile plants were smooth with a rounded, zonate margin, making them virtually indistinguishable from forms of *L. glaciale* and *Phymatolithon lenormandii*. *L. glaciale* has no staining bodies and the medulla is better developed. Spermatangial plants cannot be distinguished from carpogonial/carposporangial plants without examining conceptacle contents since young conceptacles of both types are subconical with mature ones becoming apiculate and all are in the same size range. Adey & Adey (1973) reported plants with small branches.

This species is not found in Iceland, northern Norway or the western Atlantic. Adey (1971) suggested that this distribution is due to an inability to withstand winter temperatures below 6–7°C.

Cabioch & Giraud (1978b) made a study of *L. sonderi* under TEM noting, in particular,

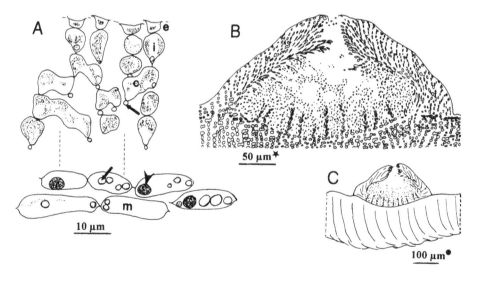

Fig. 91 Vertical sections showing anatomical features of *Lithothamnion sonderi*. (A) Thallus showing flared epithallial cells (e), subepithallial initials (i), cortical cells with wide fusions (c) and medullary cells (m); note oily globules at primary pit connections and in medulla (arrow) and densely-staining bodies (arrowhead) (YMC 84/37). (B) Spermatangial conceptacle (YMC 83/439). (C) Diagrammatic drawing to show position of spermatangial conceptacle in thallus (YMC 83/439).

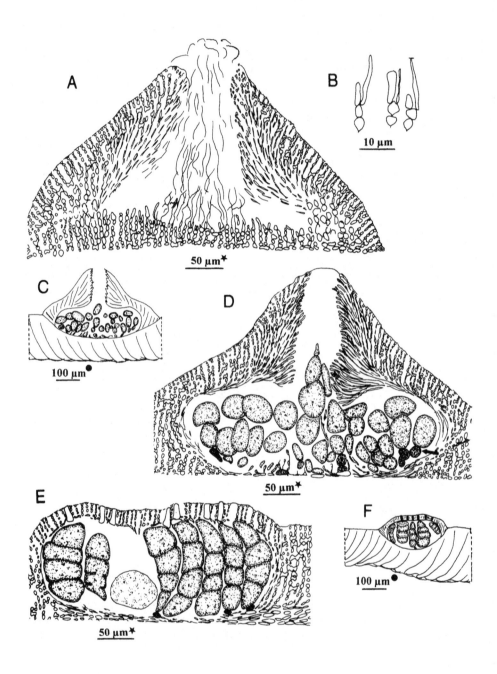

the thick cell walls, large fusions, and method of epithallial shedding. They found that large, siderophilic globules occurred commonly and floridean starch grains were abundant in the basal region of the thallus.

Adey & Adey (1973, p.355) reported abundant staining bodies.

MELOBESIA Lamouroux

MELOBESIA Lamouroux (1812), p.186.

Type species: *M. membranacea* (Esper) Lamouroux (1812), p.186 (see Chamberlain, 1985a).

Hapalidium Kützing (1843), p.385.
Epilithon Heydrich (1897b), p.408, nom. superfl.(ICBN, Art. 63.1).

Plants nongeniculate and attached; plane or undulate but not with protuberances or branches; calcification usually very light; gametangial conceptacles uniporate, tetra/bisporangial conceptacles multiporate.

Thallus pseudoparenchymatous, crustose; structure dorsiventral and dimerous, with a single layer of basal filaments; erect filaments up to about 8 cells long; epithallial cells non-flared; cells of basal filaments squarish to elongate; cells of erect filaments squarish to elongate, with fusions occurring between cells of contiguous filaments, secondary pit connections unknown; cell surface *Pneophyllum*-type (SEM); trichocytes unknown.

Gametangial plants monoecious or dioecious; spermatangial systems somewhat dendroid, borne over entire conceptacle chamber surface; in carposporangial conceptacles fusion cell apparently lacking but gonimoblast filaments developing only from peripheral supporting cells; tetrasporangial conceptacles multiporate with primordium developing dorsally on basal filaments, upper thallus layers shedding; tetrasporangia zonate, borne across conceptacle floor, each terminating in an apical plug in conceptacle roof, lacking specialised cells surrounding the plug; bisporangial conceptacles known only in *M. van-heurckii,* structure poorly known.

For an account of the genus, see Woelkerling (1988, p.186) and Wilks & Woelkerling (1991, p.508). Two species have been recorded from the British Isles. *Melobesia membranacea,* the type of the genus, is a distinctive, usually epiphytic species common throughout the British Isles. *M. van-heurckii* is only known to occur on hydroids; it has been recorded only from Jersey (the type locality) in the British Isles, but Cabioch (pers. comm.) reported it as common in northern Brittany and she has described its structure in some detail (Cabioch, 1972). Dixon & Irvine (in Parke & Dixon, 1976) suggested that it was merely a form of *M. membranacea,* but examination of the type material (AWH) has shown clear differences: it

---

Fig. 92 Vertical sections showing anatomical features of *Lithothamnion sonderi.* (A) Carpogonial conceptacle (*J.Hall-Spencer,* 1). (B) Carpogonial branches from conceptacle in A. (C) Diagrammatic drawing showing position of carposporangial conceptacle in thallus (YMC 83/439). (D) Carposporangial conceptacle; note gonimoblast filaments (arrow) (YMC 83/439). (E) Tetrasporangial conceptacle (YMC 83/439). (F) Diagrammatic drawing to show position of tetrasporangial conceptacle in thallus (YMC 83/439).

is uncalcified and has neither cell fusions nor secondary pit connexions. Its taxonomic position is thus uncertain, but we are provisionally retaining it in *Melobesia* pending further information.

KEY TO SPECIES

Plants calcified, cell fusions present, gametangia and tetrasporangia recorded, bisporangia unknown, occurring on a variety of hosts or on rocks . *M. membranacea*
Plants uncalcified, cell fusions absent, tetra- and bisporangia recorded, gametangia unknown, recorded only on a hydroid .... *M. van-heurckii*

Melobesia membranacea (Esper) Lamouroux (1812), p.186.
Figs 7B; 39A; 75; 93; 94

Neotype: CN! (see Chamberlain, 1985a). Herb. Lamouroux, France.

Corallina membranacea Esper (1806), tab. *Corallina* Tab.XII.
Melobesia corticiformis Kützing (1849), p.696. (L!).
Hapalidium coccineum P.& H.Crouan (1859), p.285. (CO!).
Hapalidium hildenbrandtioides P.& H.Crouan (1860), p.4. (CO!).
Epilithon membranaceum (Esper) Heydrich (1897b), p.408.
Lithothamnion membranaceum (Esper) Foslie (1904b), p.19.

Key characters: Thallus thin, remaining attached to substratum even when dry; conceptacles characteristically dark-centred. Overgrown by all other coralline algae.

Illustrations: Kylin (1928), figs 38–40; Hamel & Lemoine (1953), pl.XXII, figs 6, 7; Kornmann & Sahling (1977), abb.115D,E, abbs 116–117; Garbary (1978), figs 27,28; Stegenga & Mol (1983) pl.73, figs.6–10; Chamberlain (1984), figs 17,19, (1985), figs 1–4; Wilks & Woelkerling (1991), figs 1–22.

Thallus encrusting, closely adherent, lightly calcified, very thin, to at least 5cm diameter, to c.90µm thick, flat, orbicular, becoming confluent, confluences flat; *margin* entire, thin, without orbital ridges; *surface* smooth, red-violet, translucent, whitish when thickened, becoming wrinkled and not falling off substratum when dry; *spermatangial conceptacles* conical, to c.250µm diameter, *carposporangial conceptacles* conical, to c.280µm diameter, *tetrasporangial conceptacles* raised, to c.280µm diameter with dark-coloured, concave pore plate to c.150µm diameter.

Basal filaments unistratose with cells in surface view 5–16µm long × 2–6µm diameter, in VS 7–23µm diameter (i.e. high), *erect filaments* absent to 7 cells long with cells 2–13µm long × 3–10µm diameter, *epithallial cells* triangular or elliptical; *trichocytes* not known.

Gametangial plants monoecious or dioecious, conceptacles in monoecious plants either widely spaced, or spermatangial ones closely adjacent to carpogonial/carposporangial ones; *spermatangial conceptacle* chambers domed, 52–104µm diameter, 26–52µm high with roof 18–26µm thick; *carpogonial conceptacles* developing on basal filaments and shedding upper layers of thallus, domed; *carposporangial conceptacle* chambers domed, 109–143µm diameter, 65–83µm high with roof 52–59µm thick, *gonimoblast filaments* borne peripherally

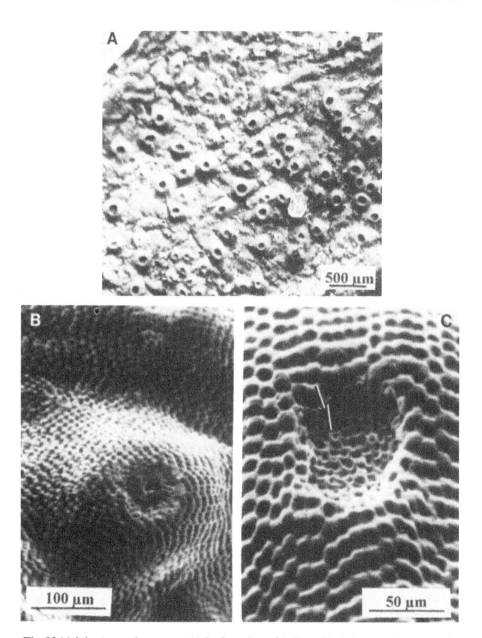

Fig. 93 *Melobesia membranacea*. (A) Surface view of thallus with dark centred conceptacles, on *Chondrus* (YMC 77/48). (B) Gametangial conceptacle (Wembury, Devon, on *Chondrus*, photo D.Garbary). (C) Tetrasporangial conceptacle; note pores (arrow) (collection as in B).

from central agglomeration of fusion cells; *tetrasporangial conceptacle* chambers low-domed, developing on basal filaments and shedding upper layers, 124–170µm diameter, 55–78µm high, roof 9–26µm thick, roof 2 cells thick, cells squarish with round lumen; *tetrasporangia* 52–62µm long × 22–52µm diameter, mostly zonate but very occasionally cruciate or bisporangial, to *c*.30 per conceptacle; true *bisporangial conceptacles* not known. *Germination disc* comprising up to about 56 small squarish cells.

Very common. Epiphytic, most abundantly on *Furcellaria* and *Laminaria*, also on many other algae and characteristically on *Chondrus* and *Mastocarpus* beneath *Titanoderma pustulatum* var. *pustulatum*, more rarely epilithic and epizoic. Littoral and sublittoral to at least 42m.

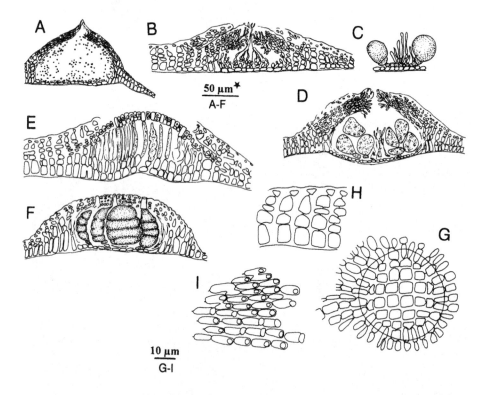

Fig. 94 Anatomical features of *Melobesia membranacea*. (A) VS spermatangial conceptacle (YMC 78/20). (B) VS carpogonial conceptacle (YMC 75/17). (C) VS gonimoblast filaments on fusion cells (YMC 79/63). (D) VS carposporangial conceptacle (YMC 75/13). (E) VS young tetrasporangial conceptacle (YMC 75/15). (F) VS mature tetrasporangial conceptacle (YMC 79/63). (G) Surface view of germination disc (YMC 86/173). (H) VS thallus (YMC 77/15). (I) Surface view of bistratose thallus (YMC 80/110).

Throughout the British Isles.

Norway (Hordaland) to Sénégal and Cape Verde Isles, Mediterranean, W. Baltic; Canada (Nova Scotia) to USA (Virginia); South Africa, Australia, New Zealand. Probably very widely distributed but not recorded from Japan.

Fertile throughout the year, tetrasporangial conceptacles much more abundant than gametangial ones. Cabioch (1972) found that plants in Brittany grew and reproduced very rapidly and were always fertile.

Undoubtedly the commonest epiphytic crustose coralline alga in the British Isles, it seems to be the epiphytic equivalent of the large epilithic species *Phymatolithon lenormandii*. Like that species it is overgrown by practically everything and yet survives such a wide range of conditions that it attains great abundance.

Underwater a host plant such as *Furcellaria* appears densely covered with substantial, pink, calcified thalli of *M. membranacea*, but the epiphyte becomes virtually invisible as soon as the host is removed from the water. All types of conceptacle show the instantly recognizable dark centre but it is not always easy to identify conceptacle type, especially in dried material.

Melobesia van-heurckii (Heydrich) Cabioch (1972), pp.177,179,272.
Figs 95; 96

Lectotype: AWH! (selected here). *H. Van Heurck*, St Brelade, Jersey, February 1903, on the surface of *Aglaosphenia* attached to *Halidrys*, drift. Isotypes: PC!, TRH (see Adey & Lebednik, 1967).

Lithothamnion van-heurckii Heydrich in Chalon (1905), p.207.
Epilithon van-heurckii (Heydrich) Heydrich in Van Heurck (1908), p.93.

Key characters: Thallus growing only on hydroids, uncalcified, cell fusions and secondary pit connections absent. Overgrowing characteristics unknown.

Illustrations: Heydrich (in Chalon, 1905), figs 3–5; Cabioch (1972), fig.14.

Thallus encrusting, closely adherent, uncalcified, very thin, to at least 350μm diameter, to 10μm thick, flat, lobed, becoming confluent, confluences flat; *margin* lobed, thin; *surface* smooth, colour pink; *bisporangial conceptacles* hemispherical, to 100μm diameter, with to *c*.12 pores in roof.

Basal filaments unistratose, cells in surface view 5–24μm long × 5–12μm diameter, in VS to 10μm high, contents very evenly stained, each with a shiny body (nucleus?), *"epithallial"* cells more or less triangular in VS, in surface view seen as lenticular cells at edge of basal cells, *erect filaments* absent; *trichocytes* unknown.

Gametangial and tetrasporangial plants unknown; *bisporangial conceptacle* chambers domed, *c*.20μm diameter, *c*.10μm high with roof *c*.5μm thick; *bisporangia c*.10μm long × *c*.5μm diameter. *Germination disc* comprising 4–6 cells; thallus usually developing from one part of the periphery.

Rare. Known only on the surface of hydroids in the lower littoral.

Jersey (type specimen only).
France (Golfe de St Malo).

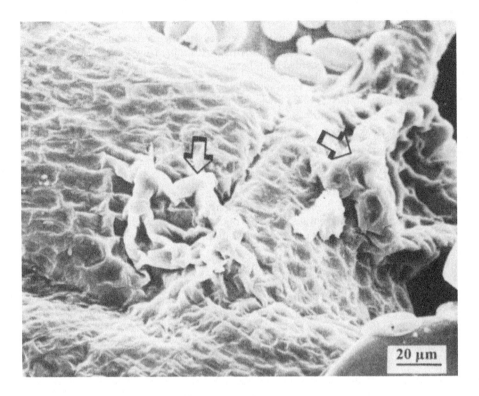

Fig. 95 Surface view of type of *Melobesia van-heurckii*; note two conceptacles (arrows) (AHW. *Chalon*, Jersey).

Recorded only between November and February; bisporangia present.

Cabioch (1972) found that the conceptacle was formed by special, coalescent filaments recalling those of *Rhodophysema*, and that spores germinated to form a 6-celled disc, each cell giving rise to a basal (hypothallial) filament. According to Lami (1940), this species is distinguished from *Audouinella spetsbergensis* (Kjellman) Woelkerling (as *Rhodochorton penicilliforme* (Kjellman) Rosenvinge), which often grows on the same host, by its thicker thallus and very refringent cell walls.

There remains considerable doubt as to whether this species should be referred to *Melobesia*, or even to the Corallinaceae, despite the presence of a type of epithallial cell and multiporate conceptacles. The texture of cell contents is unlike other corallines and it lacks calcification, cell fusions, secondary pit connections, and the characteristic *Melobesia* germination pattern, but we have retained it in this genus pending further investigations. According to Mme Lemoine (pers. comm.) it occurs abundantly in the Bay of Biscay area of France and is probably more abundant in the Channel Isles than the single record would indicate.

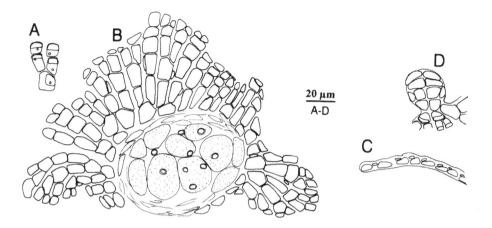

Fig. 96 Anatomical features of type of *Melobesia van-heurckii*. (A) Surface view of thallus cells; note shiny spots. (B) Surface view of thallus with multiporate 'conceptacle'. (C) VS thallus. (D) Surface view of germination disc.

MESOPHYLLUM Lemoine

MESOPHYLLUM Lemoine (1928), p.251.

Type species: *M. lichenoides* (Ellis) Lemoine (1928), p.252 (see Ishijima, 1942, p.174; Woelkerling & Irvine, 1986b, p.390).

Stereophyllum Heydrich (1904), p.198 nom. illeg.(ICBN Art.64.1) non *Stereophyllum* Mitten (1859), p.117.
Polyporolithon Mason (1953), p.316.

Plants nongeniculate, attached or unattached; varying from flat or lumpy crustose thalli composed either of thin, often superimposed foliose lamellae attached to substratum only centrally or by envelopment of host branches or of cylindrical to compressed protuberances; gametangial conceptacles uniporate, tetra/bisporangial conceptacles multiporate.

Thallus pseuodparenchymatous, monomerous, medulla coaxial with filaments of arching laterally aligned cells in RVS, lamellae dorsiventral, terminal epithallial cells with outermost walls rounded or flattened but not flared; cell elongation characteristics uncertain; cell fusions common, secondary pit connections unknown; trichocytes rare, single, terminal; spore germination characteristics poorly known.

Gametangial plants dioecious; spermatangial systems simple, arising from floor, walls and roof of conceptacle chamber; carposporangial conceptacles with central fusion cell producing gonimoblast filaments from margins; tetra/bisporangial conceptacles apparently arising from small groups of subepithallial initals, conceptacle roofs arising from filaments interspersed among sporocytes, tetra/bisporangia scattered across floor, each with apical plug blocking roof pore prior to sporangium release, columella absent.

For an account of the genus, see Woelkerling (1988), p.191. One species, *M. lichenoides*, in the British Isles.

Mesophyllum lichenoides (Ellis) Lemoine (1928), p.252.
Figs Front cover; Frontispiece; 27; 75; 97–99

Neotype: BM! (Box 1658); isoneotypes: BM!, LTB!. Cornwall (Hannafore Point, 12 May 1984, *Chamberlain & Woelkerling*).

Corallium lichenoides Ellis (1768), p.407, pl. 17, figs 9–11.
Lithophyllum lichenoides (Ellis) Rosanoff ex Hauck (1883), p.268, nom.illeg. (ICBN Art. 64.1) non
  Philippi (1837), p.389 (see Woelkerling 1983b).
Melobesia lichenoides (Ellis) Harvey (1849a), p.109, nom. illeg. (ICBN Art. 64.1) non *M.*
  lichenoides (Philippi) Endlicher (1843), p.49.
Lithothamnion lichenoides (Ellis) Foslie (1895), p.206 [reprint p.130].
?*Millepora agariciformis* Pallas (1766), p.269.
?*Lithothamnion agariciforme* (Pallas) Foslie (1897) p.5.
?*Mesophyllum lichenoides* f. *agariciforme* (Pallas) Hamel & Lemoine (1953), p.78.

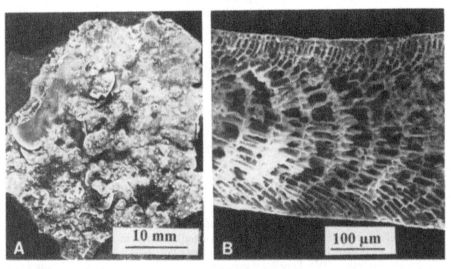

Fig. 97 *Mesophyllum lichenoides*. (A) Epilithic thalli (YMC 91/10). (B) Vertical fracture of lamella showing coaxial medulla (YMC 83/153).

Fig. 98 Vertical sections of *Mesophyllum lichenoides*. (A-C) Diagrammatic drawings to show position of conceptacles in thallus; note coaxial medulla. (A) Spermatangial conceptacle on epiphytic thallus (YMC 83/241). (B) Tetrasporangial conceptacle on epiphytic thallus (YMC 84/376). (C) Carpogonial conceptacle on epilithic thallus (YMC 87/17). (D) Carpogonial conceptacle drawn to same scale as* conceptacles; cf. Fig. 99 (YMC 83/96).

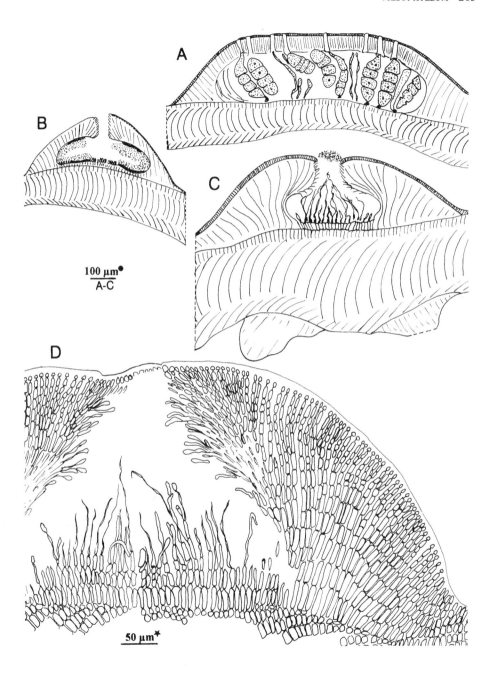

100 μm●
A-C

50 μm⋆

Key characters: Plants pinkish (in shade), brownish with yellowish loose margins (in strong light), lamellae to 20mm in diameter, often imbricating, surface usually smooth and shiny with concentric markings; usually epiphytic on *Corallina*; tetrasporangial conceptacles protruding, large (to 1mm). Overgrowing all other *Corallina* epiphytes; epilithic form overgrowing *Lithophyllum incrustans* and *Phymatolithon purpureum*.

Illustrations: Lemoine (1911), figs 59–61; Suneson (1937), figs 37–41; Cabioch & Giraud (1978a), figs 1–10; Woelkerling & Irvine (1986b), figs 1–27; Hiscock (1986) pl.4b; Cabioch *et al.* (1992), figs 147,235.

Thallus encrusting, usually consisting of thin, foliose, lobed or branched lamellae up to 620μm thick, 30mm long, 18mm broad, concave, convex or convoluted, attached centrally, populations of adherent crustose thalli also seen; *margins* free, sometimes enveloping host and more or less fusing with each other; secondary lobes and branches developing from margins and superficially; *surface* smooth or sometimes warty, with distinct concentric markings; colour brownish (Methuen: reddish brown to brownish grey) or pink in shade, with a pale yellowish border, texture quite glossy; *conceptacles* in scattered groups on dorsal surface, conspicuous, *carposporangial conceptacles* conical, to 1000μm in diameter; *spermatangial conceptacles* somewhat smaller; *tetrasporangial conceptacles* domed or flat-topped, up to 250μm high, 1250μm in diameter; *bisporangial conceptacles* not seen.

Structure monomerous; *medulla* coaxial in RVS with cells laterally aligned, 20–50μm long, 10–18μm in diameter; *cortical filaments* gradually becoming perpendicular to thallus surface with dorsal cells shorter than those in medulla and ventral cells often nearly as long, *epithallial cells* 1 layered, rectangular, up to 10μm long × 15μm diameter, marginal initials elongate and terminal when actively dividing, finally developing epithallial cells only when marginal extension ceases.

Gametangial conceptacles dioecious, *spermatangial conceptacles* with chambers 400–470μm in diameter, to 105μm high, canal to 90μm long, spermatangia simple, borne on floor, walls and roof; *carposporangial conceptacles* with chambers 430–600μm in diameter, to 220μm high and canal to 190μm long; *tetrasporangial conceptacles* protruding markedly with chambers 420–670μm in diameter, 150–370μm high and roof 60–130μm thick, pores surrounded by distinctive unbranched filaments with cells shorter than other roof cells and with denser contents.

Epiphytic, usually on *Corallina*, rarely epilithic in British Isles (Dorset, Channel Isles) but commonly so further south; lower littoral in areas fairly to very exposed to wave action, recorded sublittorally to 8m.

Northwards to Inverness, Orkney, eastwards to Dorset; in Ireland northwards to Leitrim, eastwards to Wexford. Records for Norfolk (Dashwood, 1853) and Yorkshire (James, 1923; Perkins, 1953) not confirmed.
British Isles to Mauritania; Mediterranean; Canary Isles.

Fig. 99 Vertical sections of conceptacles of *Mesophyllum lichenoides* (NOTE that these are too large to draw to same scale as * conceptacles; cf. *Leptophytum laeve* Fig. 83). (A) Spermatangial conceptacle; note vestigial carpogonia in centre of floor (YMC 83/241). (B) Carposporangial conceptacle with gonimoblast filaments (YMC 83/96). (C) Tetrasporangial conceptacle (YMC 84/376).

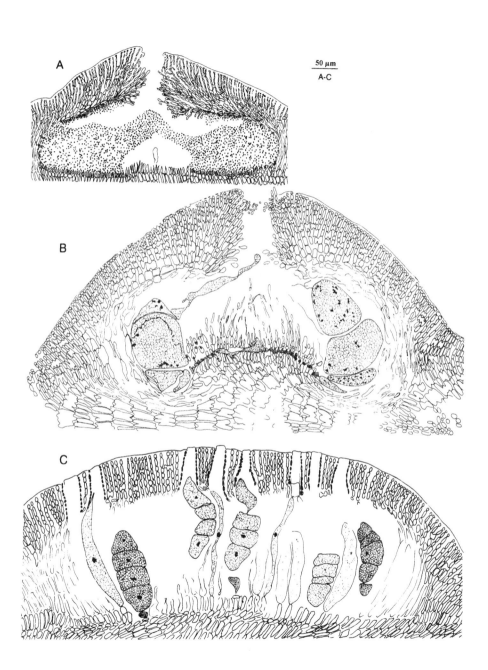

50 μm
A-C

Perennial; gametangial conceptacles recorded for February-April, tetrasporangial for April-December.

Populations of flat, adherent, epilithic plants recorded for Dorset (Kimmeridge) and Alderney.

Unattached balls were described for western Ireland (Roundstone, dredged) by Foslie (1897 p.5 as *Lithothamnion agariciforme* f. *hibernica* nom. nud., BM!) but have not been recorded this century. This form, illustrated by Johnston (1842), p.241, fig. 23); Harvey (1847) plate 73; Printz (1929), pl. 11, fig. 11 and Woelkerling (1988) fig. 222, has been considered to be related to *Lithophyllum dentatum*, *q.v.* (see Foslie, 1900a; Lemoine, 1913a). In *M. lichenoides* the lamellae remain thin and the convolutions are sometimes rolled into cones whereas those in *L. dentatum* are much thicker and fuse back to back forming protuberances with dimpled or grooved ends.

For further information on this species see Woelkerling & Irvine (1986b), who drew attention to the possible synonym *Millepora agariciformis* Pallas (1766) and noted that the taxonomic significance of the specialised pore cells in the tetrasporangial conceptacles was unknown (cf. *Leptophytum*).

PHYMATOLITHON Foslie

PHYMATOLITHON Foslie (1898a), p.4. nom. cons.

Apora Gunnerus (1768), p.72 nom. rejic.

Type species: *P. calcareum* (Pallas) Adey & McKibbin (1970), p.100 (see Woelkerling & Irvine, 1986a, Irvine & Woelkerling, 1986).

Eleutherospora Heydrich (1900), p.64. (type not seen).
Lithothamnium Philippi subgen. *Phymatolithon* (Foslie) Rosenvinge (1917),p.227.

Plants nongeniculate, attached or unattached; varying from flat, crustose thalli to bushy plants composed entirely of branches or protuberances; gametangial conceptacles uniporate, tetra/bisporangial conceptacles multiporate.

Thallus pseudoparenchymatous; structure dorsiventral and monomerous in crustose parts, more or less radial and monomerous in protuberances and branches; medulla of crustose thallus composed of varying numbers of filaments subparallel to substratum with cells of filaments usually elongate; cortex composed of distal portions of medullary filaments curving towards thallus surface with cells of filaments increasing in length downwards from subepithallial initials; epithallial cells with outer walls more or less domed, not flared; fusions between cells of contiguous filaments common, secondary pit connections unknown; cell surface predominantly *Phymatolithon*-type (SEM); trichocytes rare; spore germination characteristics poorly known.

Gametangial plants normally dioecious, sometimes monoecious, conceptacles sometimes containing both spermatangia and carpogonia; spermatangial systems dendroid on conceptacle floor, simple to arcuate on walls and roof; carposporangial conceptacles lacking fusion cells, gonimoblast filaments borne individually; tetrasporangial conceptacles multiporate, primordium developing from about three to many cells deep within cortex, conceptacles thus causing either a thin veil or a thick cortical disc to be shed; tetrasporangia zonate, borne across conceptacle floor, each terminating in an apical plug in conceptacle roof, lacking special cells surrounding plug; bisporangial conceptacles as tetrasporangial ones but producing bisporangia with two uninucleate bispores.

For an account of the genus see Woelkerling (1988), p.197. Six species are recorded for the British Isles; together with *Lithophyllum* spp., they make a major contribution to the coralline cover of the entire littoral zone, but unlike that genus *P. lamii* and *P. purpureum* also extend throughout the photic sublittoral, covering vast areas of bedrock (Adey & Adey, 1973, p.374; J. Clokie, pers. comm.). Possibly the most familiar species is *P. lenormandii,* a relatively thin, crustose species extending to quite high levels in the littoral. Other species are also common and widespread and *P. purpureum* forms a brightly coloured band of pink especially noticeable in winter at low tide level on rocky shores. *P. calcareum* appears to be restricted to maerl-rhodolith populations in the British Isles (see Introduction). For further information on the taxonomy and biology of these species, especially in the sublittoral, see Adey & Adey (1973) and Adey & McKibbin (1970). Lebednik (1977) reported several small fusion cells in the carposporangial conceptacles but this observation has not been confirmed. The name *Phymatolithon* Foslie (1898a) was proposed for conservation against *Apora* Gunnerus (1768) by Irvine & Woelkerling (1986).

KEY TO SPECIES

1　Plants comprising unattached, branched thalli .... *P. calcareum*
　　Plants epilithic, crustose, flat to knobbly 2
2　Tetra/bisporangial conceptacles with a conspicuous rim.............................. 3
　　Tetra/bisporangial conceptacles without a rim, raised or flat......................... 4
3　Plants becoming thick; old conceptacles buried in thallus, conceptacle rim to *c*.100μm wide.. *P. purpureum*
　　Plants rarely exceeding 1mm thick; old conceptacles not buried in thallus, conceptacle rim to *c*.50μm wide... *P. laevigatum*
4　Old conceptacles not buried in thallus .... *P. lenormandii*
　　Old conceptacles buried in thallus 5
5　Colour brown, conceptacles shedding a thin veil.............................. *P. brunneum*
　　Colour pink to violet, conceptacles shedding a thick disc .... *P. lamii*

Phymatolithon brunneum Y.Chamberlain sp. nov.
Figs 74; 75; 100–102

Holotype: BM! *Y.M.Chamberlain*, Lulworth, Dorset, rocks in lower littoral, immediately above *P. purpureum* zone, 17 February 1987, YMC 87/22.

Diagnosis: Thallus crustose, without protuberances, dull brown, winter fertile; tetrasporangial conceptacles domed, not shedding cortical discs, old conceptacles becoming buried in thallus. Differing from *P. calcareum* in not forming free-living plants; from *P. calcareum, P. lamii, P. laevigatum* and *P. purpureum* in not shedding cortical discs; from *P. lenormandii* in burying old conceptacles.

Thallus crustaceus, sine protuberationibus, fuscus, hieme fertilis; conceptacula tetra-sporangialia tholiformia, discos corticales non exeuntia, conceptacula vetera in thallo infossescentia. A *P. calcareo* differt plantis libere viventibus non faciendis; a *P. calcareo, P. lamii, P. laevigato* et *P. purpureo* discis corticalibus non exeundis; a *P. lenormandii* conceptaculis veteribus infossescentibus.

Key characters: Thallus brownish with conspicuously white edges to thallus and thallus lobes, tetra/bisporangial conceptacles shedding a thin veil, old conceptacles becoming buried, winter fertile. Overgrowing *P. lenormandii;* overgrown by *P. purpureum.*

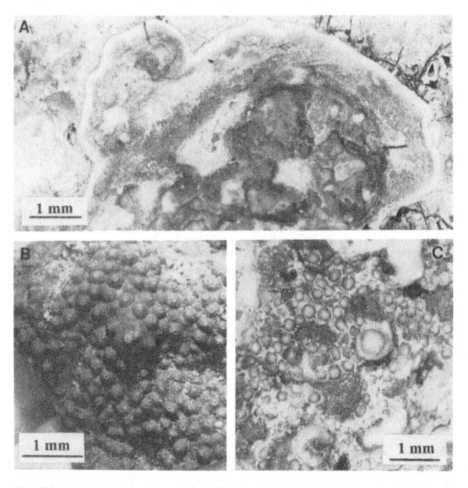

Fig. 100 *Phymatolithon brunneum* (YMC 87/22 – type). (A) Thallus edge; note white margin and irregular white patches on surface. (B) Thallus with carposporangial conceptacles. (C) Thallus with tetrasporangial conceptacles.

Thallus encrusting, adherent, to at least 6cm diameter, to 2mm thick, orbicular to spreading, becoming confluent with confluences more or less flat; *margin* entire, fairly thick, with or without orbital ridges, thallus flat, to lumpy, to irregularly protuberant, without branches; *surface* with frequent, white-edged lobes and excrescences, particularly encircling protuberances such as worm tubes, frequent patches of white thallus being shed, cell surface *Phymatolithon*-type (SEM), brownish (Methuen greyish brown, dark brown, dark violet, dark purple), edges white, grainy; all primordial conceptacles shedding thin white veils, *conceptacles* obscure, occurring in depressions between thallus protuberances; gametangial plants dioecious; *spermatangial conceptacles* low-domed to conical, adjacent conceptacles sometimes fusing, to *c.*300 μm. diameter; *carposporangial conceptacles* raised to conical, to 300μm diameter; *tetrasporangial*

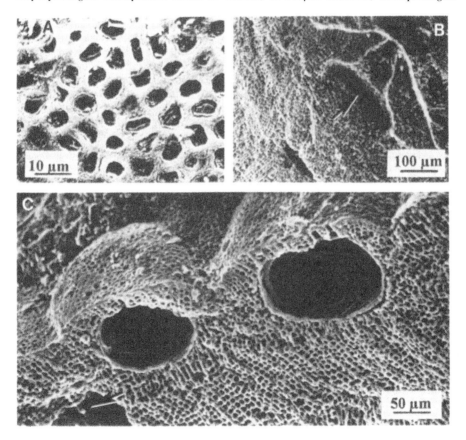

Fig. 101 *Phymatolithon brunneum.* (A) *Phymatolithon*-type surface cells (YMC 83/347). (B) Vertical fracture of carpogonial conceptacle (arrow) (YMC 87/22). (C) Vertical fracture of two tetrasporangial conceptacles at thallus surface; note also a buried conceptacle (arrow) (YMC 83/437).

conceptacles flush to hemispherical, without apparent rims, with *c.*20 pores, to 220μm diameter, adjacent conceptacles sometimes fusing; *bisporangial conceptacles* not known.

Medulla relatively thin and patchily distributed, to 25% thallus thickness; medullary filaments subparallel to substratum, often coaxial at thallus margin with cells elongate, 7–20μm long × 3–7μm diameter; *cortical filaments* becoming long with cells short near surface to elongate below and 3–10μm long × 3–7μm diameter; *epithallial cells* rectangular to domed, 1 layered; thallus surface layers often being shed, thallus regenerating to form lobes.

Gametangial plants dioecious; *spermatangial conceptacle* chambers flask-shaped, 156–208μm diameter, 52–78μm high with roof (neck) 26–52μm thick, dendroid spermatangial systems borne across floor with simpler systems interspersed and on walls and roof; *carposporangial conceptacle* chambers rounded to elliptical, 195–312μm diameter, 117–208μm high with roof 65–78μm thick and pore canal lined with papillae, fusion cell lacking, *gonimoblast filaments* borne individually; *tetrasporangial conceptacle* chambers elliptical, 117–156μm diameter, 78–94μm high with roof 26–36μm thick, pore plate very slightly concave, to 7 cells thick, cells flattened at surface and elongating towards conceptacle chamber; *tetrasporangia* 50–86μm long × 34–51μm diameter, occasional uninucleate *bisporangia* seen in tetrasporangial conceptacles; *bisporangial conceptacles* not known; old conceptacles of all types becoming buried in thallus and then noticeably rounded.

Rare. Epilithic on rocks and detritus, at present known only from the littoral, usually occurring just above the *P. purpureum* zone, sometimes higher up the shore.

Dorset, Devon, Pembroke, Channel Isles.

France (Normandy and Brittany). Probably more widely distributed.

Fertile from November to February.

Carpogonial conceptacles are flask-shaped and may have very long necks; after (presumed) fertilization the chamber swells upwards causing the roof to arch and become thinner. Buried conceptacles in carpogonial plants develop apparently normal, healthy carpogonia despite the presumed impossibility of fertilisation and further development.

This species is probably closely related to *P. lenormandii*. On the shore the brown colour and conspicuous white edges usually distinguish it from the more purple or pink tones of *P. lenormandii*. Anatomically the distinguishing features are the shallow medulla and buried conceptacles of *P. brunneum* as compared with the relatively thick medulla and non-burying conceptacles of *P. lenormandii*. In addition *P. lenormandii* bears conceptacles throughout the year whereas *P. brunneum* is almost exclusively winter-fertile.

Fig. 102 Vertical sections showing anatomical features of *Phymatolithon brunneum*. (A) Spermatangial conceptacle (YMC 87/6). (B) Carpogonial conceptacle (YMC 87/7). (C) Old carposporangial conceptacle (YMC 87/22). (D) Tetrasporangial conceptacle (YMC 87/6). (E) Carpogonial branches (YMC 87/7). (F) Diagrammatic drawing of carpogonial thallus; note medulla (m), cortex (c) and buried conceptacles (YMC 87/7).

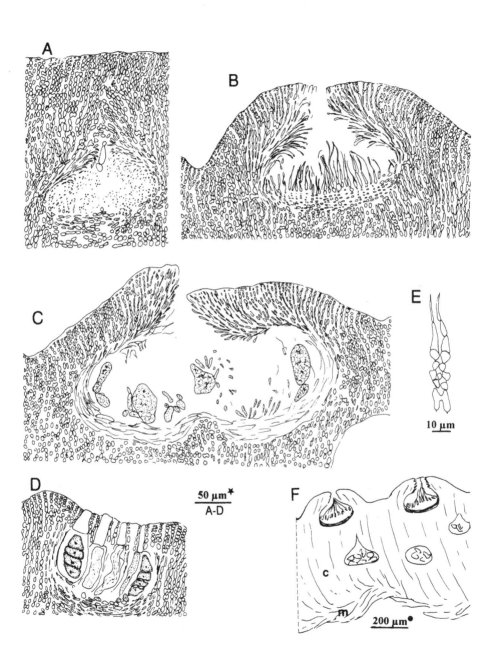

Phymatolithon calcareum (Pallas) Adey & McKibbin (1970), p.100.
Figs 8C; 74; 75; 103; 104; Table 4

Neotype: BM. (Box 1626, see Woelkerling & Irvine, 1986a, figs 1,15); isoneotypes: BM!,
LTB! Falmouth Harbour, Cornwall, *W. F. Farnham*, 11 Dec 1983.

Millepora calcarea Pallas (1766), p.265.
Millepora polymorpha Linnaeus (1767), p.1285, nom. superfl.
Melobesia calcarea (Pallas) Harvey (1849a), p.110.
Lithothamnion calcareum (Pallas) Areschoug (1852), p.523.
Lithothamnion polymorphum (Linnaeus) Areschoug (1852), p.524. nom. superfl.
Lithothamnion corallioides f. *subsimplex* Batters (1892), p.23 (BM!).
Phymatolithon polymorphum (Linnaeus) Foslie (1898a), p.4.
Lithothamnion calcareum f. *subsimplex* (Batters) Foslie (1905a), p.68.

Key characters: Plants mauvish-brown, composed of unattached, branched thalli, branches
to 6mm diameter, surface smooth or flaky, tetrasporangial conceptacles shedding thick discs.
Overgrowing characteristics of crustose form not known.

Illustrations: Lemoine (1910), pl. 1; Cabioch (1966b), pls I,A,IV, figs3,4; Cabioch (1970),
figs 1A,2,3A; Adey & McKibbin (1970), figs 1–3,6; Bosence (1976), text fig.10; Cabioch
& Giraud (1978b), figs 5,7–9,22–24,32; Woelkerling & Irvine (1986a); Cabioch *et al.*
(1992), fig.145.

Thallus comprising either attached crustose plant or unattached branching systems; no
crustose plants have been recorded from the British Isles but a summary of descriptions
given by Cabioch 1966b, 1970) and Cabioch & Giraud (1978b) is given below.
   Unattached plants varying from single branches to subglobular or flattened branching
systems, to about 7cm diameter, sparsely to densely branched, branches sometimes confluent,
slender to robust, terete to flattened, to 6mm diameter with apices rounded to flattened; *surface*
mainly rather smooth or with white flaky areas, cell surface *Phymatolithon*-type (SEM),
colour mauvish brown (Methuen violet brown), drying to mauvish (Methuen pale red, reddish
grey, purplish grey), texture matt; *tetrasporangial conceptacles* occurring in scattered and
sometimes dense groups mainly near branch apices, to 500µm diameter, coloured similarly
to thallus, usually somewhat raised and with distinct marginal rim, sometimes sunken below
thallus surface and lacking rim, shedding white cortical disc up to 10 cells thick (Woelkerling
& Irvine, 1986a), pore plate with 30–60 pores, pores surrounded by *c*.6 cells (SEM).
   Unattached plants secondarily monomerous, in LS thallus comprising a central core of
filaments (*medulla*) composed of elliptical cells 5–18µm long × 3–10µm diameter,
nonaligned and either empty or containing starch grains; *cortical filaments* radiating to the
periphery, subepithallial initials flattened, densely pigmented, inner cells gradually
elongating, extensively fused from about third cell down, cells 8–18µm long × 7–10µm
diameter, primary pit connections usually only occupying central part of end walls;
epithallial cells mainly dome shaped, sometimes absent or two to three cells deep.
   Gametangial and bisporangial plants not recorded; *tetrasporangial conceptacle* chambers
elliptical, 230–350µm diameter, 117–130µm high, with the roof 30–40µm thick;
tetrasporangia plump, 90–125µm long × 49–73µm diameter, one *bisporangium* with
uninucleate bispores seen; old conceptacles becoming buried in the thallus (Hamel &
Lemoine, 1953).

PHYMATOLITHON 213

Common. Unattached plants occur in sublittoral and occasionally lower littoral (e.g. Muckinish, Co.Clare) maerl beds, mostly in sheltered areas, usually together with *Lithothamnion corallioides* (*q.v.*) in England and Ireland, or *L. glaciale* (*q.v.*) in Scotland.

Widely distributed as maerl on southern and western coasts from Dorset to the Shetland Isles.

Norway (Hordaland) to N. Spain, W. Baltic, Mediterranean.

Attached crustose plants not known in British Isles. Crustose plants from France were described by Cabioch (1966a) as 80–150µm thick, with 5–6 layers of medullary cells 10–15µm long × 4–5µm diameter, containing many floridean starch grains, cell fusions absent; cortex composed of loose files of ovoid cells 5–10µm long × 3–5µm diameter, with numerous cell fusions narrower than those of *Lithothamnion corallioides* (see also Cabioch & Giraud, 1978b). Gametangial conceptacles are known from France (Cabioch, 1970) where they occur on crustose plants. Evidence regarding periodicity of tetrasporangial reproduction is somewhat conflicting. Woelkerling & Irvine (1986a) described all stages of conceptacle

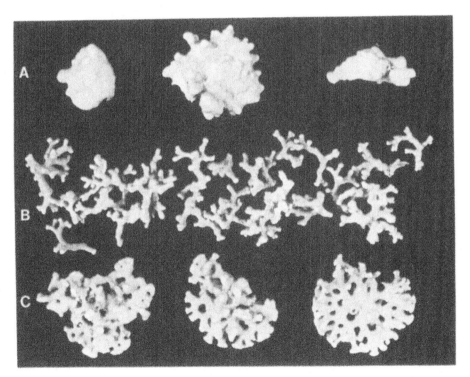

Fig. 103 *Phymatolithon calcareum.* (A) Crustose plants growing on pebbles and shells (*J.Cabioch*, Brittany, YMC 86/85) ×1. (B) Unattached, branched plants (W. Ireland, YMC 83/186) ×1. (C) Unattached, branched plants forming flattened discs (W. Ireland, YMC 83/168) ×1.

development in plants collected in December but mature conceptacles have been observed in plants collected in Dorset in July (YMC); according to C.Maggs (pers. comm.) some conceptacles with mature sporangia were found at all times of year in Galway maerl beds, but the numbers showed a peak at different times in different beds. Cabioch (1969b) found tetrasporangial conceptacles in winter in the Baie de Morlaix and suggested (Cabioch, 1969b, 1970) that phasic reproduction occurred, reaching a peak about every 6 years (see *L. corallioides*). In N. Spain Adey & McKibbin (1970) found developing tetrasporangial conceptacles in a small percentage of plants in March/April, whereas Suneson (1950) found them in July in Swedish plants. It seems, therefore, that reproduction may occur throughout the year - a feature seen in other more or less exclusively sublittoral species such as *L. sonderi*. As no crustose plants have been seen in the British Isles it must be assumed that unattached plants are propagated vegetatively; Cabioch (1969b, p.157, 1970) suggested, however, that in France recruitment is almost exclusively from branches cut off crustose plants by animal activity.

Field growth studies undertaken by Adey & McKibbin (1970) in Spain showed a

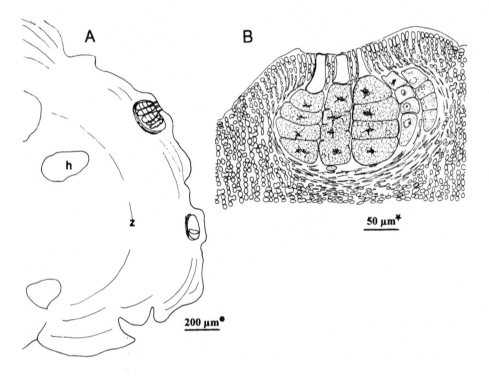

Fig. 104 Anatomical features of *Phymatolithon calcareum* (YMC 83/505). (A) Diagrammatic drawing of TS branch of tetrasporangial plant; note conceptacles at thallus surface, thallus zoning (z) and holes (h) in centre of thallus. (B) VS tetrasporangial conceptacle.

maximum growth rate of 5μm/day falling to zero in winter; this correlated with laboratory experiments giving 5μm/day at 13°C, 14h dark: 10h light daylength. According to their calculations, mean yearly growth rate in southwestern British Isles should be 15% higher, but this would still be less than 1mm/year. Böhm et al. (1978), however, calculated apical branch elongation of Baltic plants as 0.5–2.7mm/year. Adey & McKibbin (1970) also showed that at 12–13°C maximum growth occurred, but growth ceased if illumination fell below 170 lux at this temperature. Bosence (1976) found light to be the limiting factor for maerl growth in Mannin Bay, growth was best from 1–8m and ceased below 16m at 11.75–13.5°C.

This species is often confused with *Lithothamnion corallioides*, *q.v.* (see Table 4) and the comments regarding growth forms of *L. corallioides* apply equally to *P. calcareum*. Lemoine (1910) recorded a number of forms in Brittany and Cabioch (1966c) recorded the following formae from that area: f. *calcarea*, f. *simpliciuscula* Cabioch, f. *pseudosubvalida* Cabioch, f. *squarrulosa* Foslie, f. *major* Lemoine and f. *crassa* Lemoine. Cabioch (1969b) later concluded that difference in form is correlated with substratum. Cabioch (1966c) included f. *subsimplex* Batters (1892) in *Lithothamnion corallioides* but examination of the holotype (BM!) showed that, as concluded by Foslie (1905a), it pertains to *Phymatolithon calcareum*. Heydrich (1911b), in a study of crustose coralline algae from Roscoff, recorded a new form, f. *primiramosa*, of *Lithothamnion calcareum* and his drawing (p.II, fig.3) strongly resembles crustose plants of *P. calcareum* collected there recently.

Recent research by Woelkerling & Irvine (1986a) showed that *P. calcareum* must be regarded as the type species of the genus *Phymatolithon* and they selected neotype material collected by Dr W.F.Farnham from the Falmouth maerl beds.

Cabioch & Giraud (1978b) studied epithallial cells (TEM), concluding that they most closely resembled *P. (Lithothamnion) lenormandii*; they are characteristically domed (see Woelkerling & Irvine, 1986a) but occasionally flared as in *L. corallioides q.v.*.

Adey & McKibbin reported occasional staining bodies and large pit bodies.

Phymatolithon laevigatum (Foslie) Foslie (1898b), p.8.
Figs 74; 75; 105–107
Lectotype: TRH! *Kjellman*, Helgoland, undated. (see Foslie, 1895, pl.19, fig.23; Printz, 1929, pl.39, fig.14; Adey & Lebednik, 1967, p.91).

Lithothamnion laevigatum Foslie (1895), p.167 (reprint p.139).
Lithothamnion emboloides Heydrich (1900), p.74. (type not seen).

Key characters: Thallus in summer with coarse 'watermarks' (c.f. fine ones in *P. lamii*); medulla thin, tetra/bisporangial conceptacles shedding thick disc, pore plates silvery, slightly concave, with narrow slightly raised rims, conceptacles not becoming buried. Overgrowing *Pneophyllum* spp.; overgrown by *Lithophyllum crouanii*, *L. incrustans*, *L. orbiculatum*, *Phymatolithon purpureum*, *P. lamii*; about equal in competition with *P. lenormandii*.

Illustrations: Printz (1929), pl.XXXIX, fig. 14; Adey (1964), figs 21–25, 30–34, 45–50; Kornmann & Sahling (1977), pl. 119, figs A–E.

Thallus encrusting, closely adherent, to at least 5cm diameter, to 600(1000)μm thick, flat, without branches or protuberances, irregularly orbicular becoming confluent, confluences flat with adjacent thalli often separated by small channel; *margin* entire, thin, with or without orbital ridges; *surface* smooth when fertile (winter), variously 'watermarked' with low mounds, scales due to epithallial shedding, shallow depressions of old conceptacles etc. in summer, cell surface *Phymatolithon*-type (SEM); colour red-, to violet-, to grey-brown (Methuen violet brown, violet, greyish brown), texture matt; all primordial conceptacles shedding thick, white cortical discs, mature *spermatangial conceptacles* flush with thallus surface or slightly raised, appearing as white circles with conspicuous central pores, without rims, to 220µm diameter; *carposporangial conceptacles* as spermatangial ones but with slight rims, to c.250µm diameter; *tetra/bisporangial* conceptacles from slightly raised to slightly sunk, with narrow slightly raised rims to c.50µm wide, pore plates concave, silvery with up to c.90 pores, conceptacles 150–450µm diameter, size particularly variable and

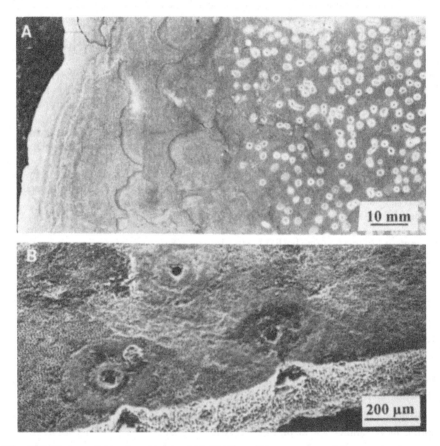

Fig. 105 *Phymatolithon laevigatum.* (A) Thallus with spermatangial conceptacles; note coarse thallus surface marking (YMC 87/159). (B) Uniporate carposporangial conceptacles; note thin thallus as seen in vertical fracture (YMC 88/10).

adjacent pore plates often fused; old conceptacles of all types shedding to leave shallow depressions that usually persist throughout summer.

Medulla relatively thin, to 55µm thick, comprising less than 20% thallus thickness, with filaments subparallel to substratum and cells elongate, (5)20–45µm long × (4)8–40µm diameter; *cortical filaments* moderately long with cells ovoid, (4)5–20µm long × (6)8–18µm diameter and pit bodies often conspicuous; *epithallial cells* rectangular, 0–1(to 3) cells thick.

Gametangial plants mainly dioecious, sometimes monoecious; *spermatangial conceptacle* chambers immersed, rounded to flask-shaped, 78–190µm diameter, (60)65–78(80)µm high with roof (neck) 20–34(50)µm thick, spermatangial systems dendroid, borne round most of conceptacle surface but with simpler ones at top; *carpogonial conceptacle* chambers flask-shaped to elliptical, with spermatangia occasionally occurring at top; *carposporangial conceptacles* with base composed of squashed cells, chambers rounded to elliptical, 104–234µm diameter, 90–120µm high and roof (neck) 26–55µm thick, composed of upwardly orientated filaments with some downwardly orientated papillae round base of pore, fusion cell lacking, *gonimoblast filaments* borne individually or in small groups; *tetrasporangial conceptacle* chambers sausage-shaped in VS with base lower than periphery, 135–169(260)µm diameter, 57–91(125)µm high with roof (20)21–31(35)µm thick, rim thickened and somewhat raised above pore plate, pore plate below, flush with, or slightly above thallus surface, up to *c.*5 cells thick with cells slightly to very elongate; *tetrasporangia* 60–78µm long × 33–51µm diameter; *bisporangial conceptacle* chambers similar in shape and structure to tetrasporangial ones, 166–260µm diameter, 78–130µm high with roof 26–39µm thick; *bisporangia* 91µm long × 65µm diameter.

Common. Epilithic on rocks and stones, very rarely on shells, mid to lower littoral, to 35m depth but mainly 0–10m. Occurring higher on shore in areas of strong wave action, characteristic of sheltered shores with somewhat lowered salinity (to 15 %.) and wide temperature variation. Occurring throughout the British Isles, mainly together with *P. lenormandii* on littoral rocks and stones and appearing redder than that species; very abundant locally in the south (e.g. Kimmeridge), occurring extensively in Scottish littoral (J. Clokie, pers. comm.).

Generally distributed throughout the British Isles.

Norway (Tromsø) to France (Brittany), W. Baltic; Arctic Canada to USA. (Connecticut), W. Greenland.

All reproductive types fertile from September (primordial) to February. Adey & Adey (1973) suggested that reproduction begins earlier in Scotland than the south.

The undated material selected as lectotype by Adey & Lebednik (1967) is part of that sent to Foslie by Kjellman from Helgoland. Much other material from the same shipment (coll. 1893) is present in TRH and elsewhere, having been distributed by Gjaervoll. It is mixed with *P. purpureum* and part of Foslie's (1895) original description refers to that species. Material sent to Foslie by Batters, and also mentioned in the protologue, is also *P. purpureum.*

At Porthleven, Cornwall, *P. laevigatum* and *P. lenormandii* are abundant on midlittoral rocks, but whereas *P. lenormandii* flourishes round a sewage outfall, *P. laevigatum* is diminished (YMC, obs.) in that area.

Adey & Adey (1973) suggested that temperatures below 3–4°C are required for

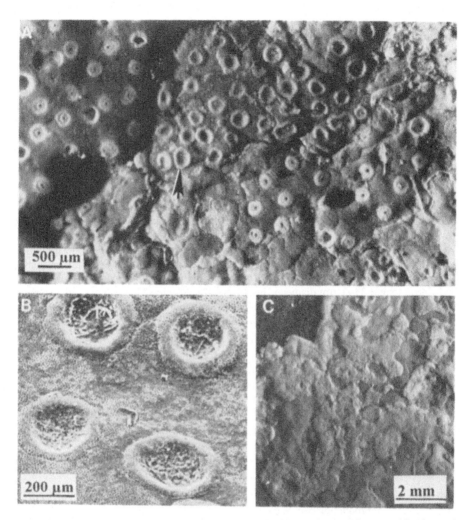

Fig. 106 *Phymatolithon laevigatum.* (A) Surface of thallus with multiporate, thin-rimmed tetrasporangial conceptacles (arrow) (cf. *P. purpureum* Fig. 115C) and uniporate conceptacles (YMC 88/10). (B) Detail of tetrasporangial conceptacle roofs with narrow raised rims (YMC 88/10). (C) Thallus surface in summer with coarse markings (cf. *P. lamii* Fig. 108C).

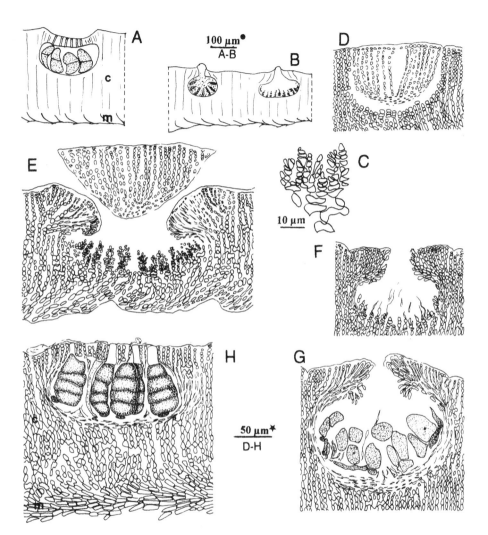

Fig. 107 Vertical sections showing anatomical features of *Phymatolithon laevigatum*. (A) Diagrammatic drawing of bisporangial conceptacle; note cortex (c) and thin medulla (m) (YMC 88/10). (B) Diagrammatic drawing of gametangial thallus with spermatangial conceptacle (left) and carpogonial conceptacle (right) (YMC 84/406). (C) Immature spermatangial system from conceptacle floor (YMC 83/380). (D) Young carpogonial conceptacle shedding cortical disc (YMC 84/406). (E) Immature spermatangial conceptacle shedding cortical disc (YMC 83/380). (F) Mature carpogonial conceptacle (YMC 84/406). (G) Carposporangial conceptacle (YMC 83/469). (H) Tetrasporangial conceptacle; note medulla (m) and cortex (c) (YMC 85/196).

reproduction and that the species is absent or infertile in southern England but we have not found this to be the case.

Walker (1984) recorded trichocytes (megacells) in *P. laevigatum* for the first time. The thin, flat thalli of *P. laevigatum* are not always easy to distinguish from *P. lenormandii* although when growing together the former are usually of a distinctly redder shade. *P. laevigatum* never has conceptacles in summer but they are present on *P. lenormandii* throughout the year. Primordial conceptacles of *P. laevigatum* shed white cortical discs, those of *P. lenormandii* shed white veils. Gametangial conceptacles of *P. laevigatum* are flat, those of *P. lenormandii* are conical, tetra/bisporangial conceptacles of *P. laevigatum* are silvery and concave and have distinct rims, those of *P. lenormandii* are usually raised and rimless.

It can also be difficult to distinguish *P. laevigatum* from *P. lamii*: the latter usually has smaller, deeper, rimless conceptacles (although there is some overlap) and thicker thalli and occurs mainly in the sublittoral (Adey & Adey, 1973, figs 35, 37). In summer *P. laevigatum* has coarser surface 'watermarks' than the delicate ones of *P. lamii*; both species are winter-fertile.

Phymatolithon lamii (Lemoine) Y.Chamberlain (1991a), p.224.

Figs Frontispiece; 74; 75; 108–110

Holotype: PC! (see Lemoine, 1931, p.13). *Lami*, Pointe de Cancaval, Rance, France, 1930, 8m.

Lithophyllum lamii Lemoine (1931), p.13.
Lithophyllum melobesioides P.& H.Crouan ex Lemoine in Hamel & Lemoine (1953), p.55, (CO!).
Phymatolithon rugulosum Adey (1964), p.381.

Key characters: Thallus in summer with delicate 'watermarks' (c.f. coarse ones in *P. laevigatum)*, medulla very thin, pore plates deeply sunken until mature, without rims, conceptacles sometimes burying, sometimes not. Overgrowing *P. lenormandii, P. laevigatum*; overgrown by *P. purpureum, Lithophyllum incrustans*; more or less equal in competition with *Lithothamnion sonderi*.

Illustrations: Adey (1964), figs 15–20, 27–29, 35, 36, 39–44, 51–64; Adey & Adey (1973), table VIII (as *P. rugulosum*); Chamberlain (1991a), figs 1–20.

Thallus encrusting, adherent but usually easily flaking off, to at least 6cm diameter, to 4mm thick, flat, without branches or protuberances, orbicular to spreading, becoming confluent with confluences slightly crested; *margin* entire, thick, without orbital ridges; *surface* smooth while fertile (winter), variously 'watermarked' with delicate ridges, mounds and frills in summer, cell surface *Phymatolithon*-type (SEM), uniform milky- pink to violet in winter (Methuen reddish lilac), patchy pink and white in summer, texture matt; all primordial conceptacles shedding thick white discs, then becoming deeply sunken, finally more or less flush with surface at maturity; *spermatangial conceptacles* appearing as white circles with central pore, to 55µm diameter; *carposporangial conceptacles* similar, to 120 µm; *tetra/bisporangial conceptacles* appearing initially as deep holes, finally as small, flat rimless pore plates, to 230µm with up to c.40 pores.

Medulla very thin, to c.60µm thick, comprising less than 10% thallus thickness, filaments subparallel to substratum, cells elongate, (5)11–40µm long × (2)3–7(10)µm diameter;

cortical filaments becoming very long with cells elongate ovoid, 2–11μm long × 2–7μm diameter and pit bodies often very conspicuous; *epithallial cells* high-domed, 1 layered; old degenerate thalli often resuming growth or overgown by surrounding areas, becoming thick and invaded by endophytic blue-green algae etc.; *trichocytes* rare.

Gametangial plants mainly dioecious, sometimes monoecious; *spermatangial conceptacle* chambers globular, (70)99–120μm diameter, (45)65–78μm high with roof 27–44μm thick, spermatangial systems dendroid, borne round entire surface; *carpogonial conceptacle* chambers flask-shaped with wide neck; *carposporangial conceptacles* with layer of squashed

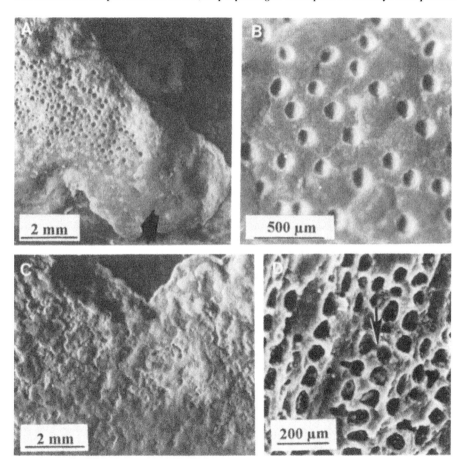

Fig. 108 *Phymatolithon lamii*. (A) Carposporangial thallus with conceptacle roofs that are either sunken or flush with thallus surface; note thick thallus margin lacking orbital ridges (arrow) (YMC 88/335). (B) Sunken tetrasporangial conceptacles (YMC 88/12). (C) Thallus surface in summer with delicate markings (cf. *P. laevigatum* Fig. 106C) (YMC 83/335). (D) Vertical fracture of cortical filaments; note narrow cell fusions (arrow) (YMC 84/408).

cells at base, chambers rounded-elliptical, 190–218μm diameter, 117–135μm high and roof (neck) 44–52μm thick, composed of upwardly orientated filaments, fusion cell lacking, *gonimoblast filaments* borne individually; *tetrasporangial conceptacle* chambers elliptical, 143–208μm diameter, 73–104μm high with roof 18–34μm thick, rim lacking, pore plate to 68μm below thallus surface, finally becoming flush with it, to *c*.6 cells thick, cells squarish; *tetrasporangia* 35–73μm long × 28–39μm diameter; *bisporangial conceptacle* chambers (one plant only) elliptical, 125–156μm diameter, 73–99μm high with roof 18–26μm thick, structure as in tetrasporangial conceptacles; *bisporangia* 59–68μm long × 33–43μm diameter. Old conceptacles of all types either shedding or becoming buried to many layers deep.

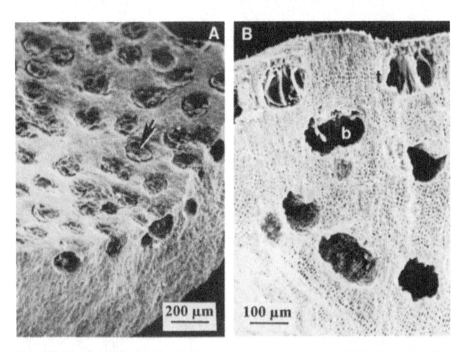

Fig. 109 *Phymatolithon lamii*. (A) Tetrasporangial conceptacles with thick cortical discs (arrow) (YMC 84/408). (B) Vertical fracture of tetrasporangial thallus showing conceptacles at the surface (t) and buried (b) (YMC 82/117).

Fig. 110 Vertical sections to show anatomical features of *Phymatolithon lamii*. (A) Diagrammatic drawing of gametangial thallus to show carpogonial (G) and carposporangial (S) conceptacles (YMC 83/179). (B) Spermatangial conceptacle (YMC 83/435). (C) Carpogonial conceptacle (YMC 84/342). (D) Carposporangial conceptacle (YMC 83/179). (E) Tetrasporangial conceptacle (YMC 84/92). (F) Thallus with mature bisporangial conceptacle and bisporangial initials with thick cortical disc *in situ* (YMC 83/494). (G) Carpogonial branches (YMC 86/16). (H) Diagrammatic drawing of tetrasporangial thallus (YMC 84/92).

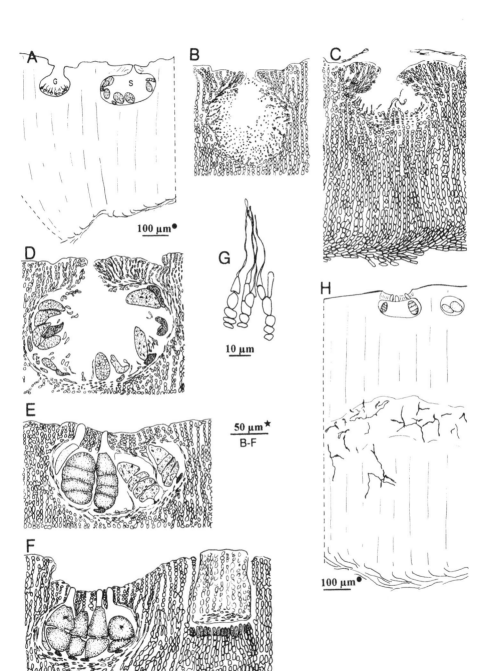

Quite common. Epilithic on rocks and stones, also on shells, occasionally on *Laminaria* holdfasts. On rocks at the bottom of low littoral pools, sublittorally to 90m (Clokie *et al.*, 1981), most abundantly from 3–21m on open and wave-exposed sites (Adey & Adey, 1973). *P. lamii* and *Lithophyllum crouanii* (as *orbiculatum*) replace *P. purpureum* (as *polymorphum*) in the shallow sublittoral of Shetland (Adey & Adey, 1973, p.367). In littoral pools it is usually accompanied by *P. purpureum* which has superficial, heavily rimmed tetra/bisporangial conceptacles compared with its deep to flush, rimless ones.

Probably throughout the British Isles but may be absent from SE coast.

Arctic Norway to northern Spain, Canada (Newfoundland) to USA (Massachusetts, Cape Cod).

All reproductive types fertile from September (primordial) to February/March. Adey (1964) computed upward growth (in Maine) as 50–200μm per year, based on annual reproductive cycle and conceptacle infilling. Although fastest growing in warmest months, the highest magnesium content is in rapidly infilling conceptacles, though this may not be biogenic (Moberley, 1968).

For the most part *P. lamii* shows little variation. However, particularly thin plants occur on stones in midlittoral channels at Bembridge and in Jersey that conform qualitatively with this species except in often having aligned cortical cells; it is possible that these plants remain small and thin because they are continually being abraded as the stones roll around. R. Hooper (pers. comm.) reports that in eastern Canada this species occurs only in niches that are not ice-scoured.

Recent examination (Chamberlain, 1991a) of the type of *Lithophyllum lamii* (PC) showed it to be conspecific with the taxon widely reported as *P. rugulosum*; the epithet *lamii* has priority.

Adey (1964) found trichocytes in this species, the first record for *Phymatolithon*. Adey & Adey (1973) described the cortical discs of all conceptacles as thin and fragmenting, in contrast to the thick ones of *P. laevigatum* and *P. purpureum*. We have found much variation in this feature, however.

Experiments with *P. lamii* and *P. laevigatum* in Nova Scotia (Johnson & Mann, 1986) showed that few fleshy algae were able to settle on crustose corallines in comparison with local granite substrata. This is due to continual epithallial shedding which also prevents fouling of the coralline surface in areas where grazing herbivores are absent.

In littoral situations the thick-looking, milky-pink thalli of this species are reasonably distinctive at the bottoms of lower littoral pools; we agree with R. Hooper (pers. comm.), that thalli detach quite easily like thick flakes of paint.

Adey & Adey (1973, p.375) devised a multivariate diagram to distinguish *P. lamii* (as *rugulosum*) from *P. purpureum* (as *polymorphum*). See *P. purpureum* for further discussion.

Lemoine (1913a, p.125) recorded *Lithothamnion* (now *Clathromorphum*) *compactum* from Clew Bay, Ireland. The specimen (BM!) on which this record was based has been examined (YMC, obs.) and reidentified as *Phymatolithon lamii*.

Phymatolithon lenormandii (Areschoug) Adey (1966b), p.325.
Figs Frontispiece; 20A; 74; 75; 111–114

Lectotype: LD! (selected by Woelkerling, 1988, p.219). Herb. Agardh, no.50673, Arromanches, France, September 1847. Isolectotype CN!

Melobesia lenormandii Areschoug (1852), p.514 (as *M. Lenormandi*).
Lithothamnion lenormandii (Areschoug) Foslie (1895), p.178 (reprint p.150) (as *L. lenormandi*).
Lithothamnion lenormandii f. *sublaeve* Foslie (1895), p.179 (reprint p.151).
Lithothamnion squamulosum Foslie (1895), p.183 (reprint p.155).

Key characters: Plants thin, margin mostly with strong orbital ridges, medulla to 50% thallus thickness, tetra/bisporangial conceptacles shedding thin veil, conceptacles present (not necessarily fertile) throughout the year, old conceptacles not becoming buried. Overgrown by most other large crustose coralline algae and *Peyssonnelia*, especially *P. dubyi* P.& H.Crouan; about equal in competition with *Phymatolithon laevigatum*; overgrowing *Pneophyllum* spp. and possibly *Corallina officinalis* basal crust.

Illustrations: Suneson (1943), figs 1, 2, pl. I, figs 1–4; Hamel & Lemoine (1953), pls XVI, XVII, XVIII, fig. 1; Adey (1966b), figs 25, 26, 43–50, 57. 91–95, 99–112; Kornmann & Sahling (1977), figs 118A-D; Rueness (1977), pl.VII, fig.3; Stegenga & Mol (1983), pl.74, figs 6–7; Cabioch *et al.* (1992), fig.144.

Thallus encrusting, closely adherent, to at least 8cm diameter, to 250μm thick, either entirely flat, or with a mosaic of low mounds, or partly or entirely squamulose, without branches or protuberances, orbicular becoming confluent, or overgrowing, confluences flat or slightly crested; *margin* entire, thin, with or without conspicuous orbital ridges; *surface* smooth, cell surface *Phymatolithon*-type (SEM); colour brownish, greyish, mauvish, pinkish, or reddish (Methuen, dark ruby, greyish ruby, dull violet, dull red etc.), often young parts reddish and old parts brownish on same plant, texture glossy, matt or grainy; all primordial conceptacles shedding thin, white veils; *spermatangial conceptacles* hemispherical, conical, or more or less flat, with rim of ruptured thallus at base, to 400μm diameter; *carposporangial* conceptacles similar, to 430μm diameter; *tetra/bisporangial* conceptacles orbicular to elliptical, either immersed to hemispherical and without rims, or raised with flat tops and slight rims, adjacent pore plates sometimes fused, 6–60 pores, to 430μm diameter, with rim of ruptured thallus at base. Conceptacles losing roofs and pore plates as they senesce, leaving high-rimmed craters which eventually become infilled, conceptacles never burying.
    Medulla relatively thick, to *c.*100 μm, comprising up to 50% thallus thickness, filaments subparallel to substratum with cells elongate, 8–18(28)μm long × 3–11(12)μm diameter; *cortical filaments* to *c.*30 cells long, cells flattened near surface, becoming ovate, 1.5–10(12)μm long × (2)3–10μm diameter, pit bodies common; *epithallial cells* domed, pale, often absent.
    Gametangial plants monoecious or dioecious, conceptacles sometimes with both spermatangia and carpogonia; *spermatangial conceptacle* chambers domed, 136–210μm diameter, 65–78(122)μm high, roof 39–73μm thick, spermatangial systems dendroid, borne across conceptacle floor, simpler ones round the walls and roof; *carpogonial conceptacles* similar in shape, sometimes containing spermatangia as well as carpogonia; *carposporangial conceptacles* with layers of squashed cells at base, chambers globular, elliptical or domed, (143)145–275(291)μm diameter, (76)95–160μm high with roof 52–78μm thick composed of upwardly orientated filaments giving rise to papillae lining pore canal, no central fusion cell, *gonimoblast filaments* borne individually; *tetrasporangial conceptacle* primordia developing just below subepithallial meristem, chambers narrowly to widely elliptical, 130–221 μm diameter, 70–130μm high with roof 21–55μm thick, rimless or with very slight rim, pore plate flush with or raised above thallus surface, to *c.*6 cells thick with cells squarish

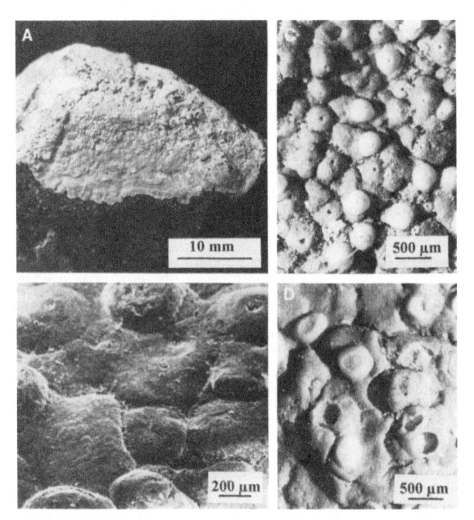

Fig. 111 *Phymatolithon lenormandii*. (A) Plant growing on a stone (BM. *LMI*, Guernsey, 13.iv.1969). (B) Surface of uniporate, squamulose thallus (YMC 84/241). (C) Carposporangial conceptacles (YMC 83/90). (D) Tetrasporangial conceptacles (YMC 83/90).

to elongate; *tetrasporangia* elliptical and plump to elongate and crescent-shaped, 70–109µm long × 26–65µm diameter; *bisporangial conceptacles* similar in shape and structure to tetrasporangial ones with chambers 174–333µm diameter, 65–130µm high and roof 21–60µm thick; *bisporangia* 47–114µm long × 31–57µm diameter. Tetra/bisporangial and carpogonial conceptacles sometimes on same plant.

Very common. Epilithic on rock and pebbles, also on worm tubes and shells including living *Mytilus*, occurring in damp places throughout the littoral and sublittoral to 30m depth, mainly 0–6m (Adey & Adey, 1973, fig.34), on both sheltered shores and those exposed to wave action. Characteristically a thin, mauve, encrusting plant in pools, and on rocks and stones in the midlittoral above the *Phymatolithon purpureum* zone. In competition with *Lithophyllum orbiculatum* as a colonizer in midlittoral pools, *L. orbiculatum* will usually be successful. Despite the fact that *P. lenormandii* is apparently a poor competitor, being overgrown by all but *P. laevigatum* and thin (e.g. *Pneophyllum* spp.) crustose corallines, it is a particularly widespread and successful pioneer species, being able to tolerate lowered salinity, (probably to about 15‰, Adey & Adey, 1973), sheltered, turbid estuarine and harbour waters, and unstable substrata such as soft sand- and lime-stone.

Throughout the British Isles.

Recorded in the literature from Arctic Norway to Mediterranean, Canary Isles, Arctic Canada to USA (Cape Cod), California, Mexico, Japan, Pacific Russia, Mauritius, southern south America. Geographical limits still uncertain owing to present taxonomic confusion, Adey & Adey (1973) did not record the species north of Newfoundland or in northern Norway.

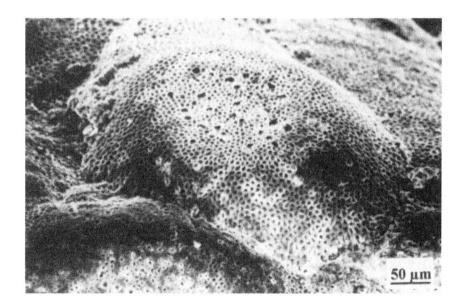

Fig. 112 *Phymatolithon lenormandii*, tetrasporangial conceptacle (YMC 84/241).

Conceptacles apparent throughout the year in the British Isles, possibly most abundantly reproductive in winter, but bisporangial conceptacles, for example, contained apparently normal sporangia in June.

Phymatolithon lenormandii shows wide variation in size, colour, texture and surface configuration. It is treated here as an aggregate species as it has not been possible to correlate this variation either with ecological conditions or with a taxonomic scheme at species level. Different growth forms are usually intermingled and the only obvious correlation with environmental factors has been the presence of smooth, glossy, dark brown plants on chalk. The following principal growth forms are recognisable : 1) thin, very strongly adherent,

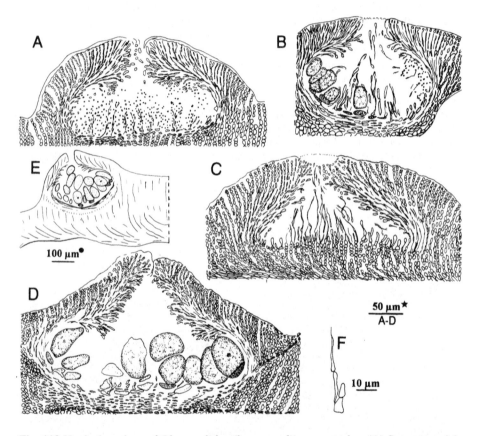

Fig. 113 Vertical sections of *Phymatolithon lenormandii* conceptacles. (A) Spermatangial conceptacle (YMC 83/445A). (B) Carposporangial conceptacle, red form (YMC 88/1). (C) Carpogonial conceptacle (YMC 84/418). (D) Carposporangial conceptacle, conical form (YMC 83/400). (E) Diagrammatic drawing of carposporangial conceptacle, red form (YMC 88/13A). (F) Carpogonial branch.

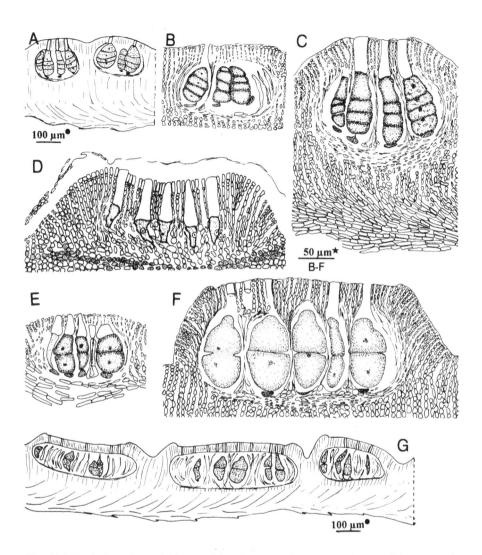

Fig. 114 Vertical sections of *Phymatolithon lenormandii* conceptacles. (A) Diagrammatic drawing of tetrasporangial thallus, squamulose form (YMC 84/421). (B) Tetrasporangial conceptacle, red form (YMC 83/344). (C) Tetrasporangial conceptacle, squamulose form (YMC 84/421). (D) Initiation of bisporangial conceptacle (YMC 87/8A). (E) Bisporangial conceptacle, red form (YMC 88/13A). (F) Bisporangial conceptacle (YMC 83/449). (G) Diagrammatic drawing of bisporangial thallus (YMC 83/449).

mauvish, matt plants with strong orbital ridges at margin and raised tetra/bisporangial conceptacles (c.f. isolectotype in CN!); 2) extensive (up to c.8cm diameter), smooth, brown-purple, glossy plants with orbital ridges at margin and raised, flat-topped tetra/bisporangial conceptacles, particularly characteristic of chalk substrata (f. *sublaevis* Foslie, 1895); 3) plants entirely composed of prolific squamules, grey-mauve, rough, with hemispherical conceptacles (f. *squamulosa* (Foslie) Foslie, 1901); 4) minute, dark, rough plants with small hemispherical conceptacles; and 5) smooth, flat, reddish, matt plants, without marginal ridges and with flat to slightly raised tetra/bisporangial conceptacles. Adey & Adey (1973, p.372) found that rough, squamulose plants may have smooth, glossy margins; these may eventually overgrow areas of the same plant and appear very different.

For details of spore germination see Cabioch (1966a) and spore adhesion see Walker & Moss (1984).

Giraud & Cabioch (1977) and Cabioch & Giraud (1978b) have made ultrastructural studies of this species. Further ultrastructural details and information about storage products are given in Millson & Moss (1985). Walker (1984) has observed trichocytes in *P. lenormandii*.

Shallow conceptacle initiation and flattened upper thallus cells are characters shared with *Leptophytum* spp. and forms with raised conceptacles closely resemble *L. elatum, q.v.*; the absence of a *Leptophytum*-type cell surface (SEM) provides the conclusive diagnostic character.

Plants with conical gametangial conceptacles are easily confused with forms of *Titanoderma pustulatum* (*q.v.*) and with '*Pneophyllum* sp'. (*q.v.*).

Phymatolithon purpureum (P.& H. Crouan) Woelkerling & L. Irvine (1986a), p.71.
Figs Frontispiece; 74; 75; 115; 116

Lectotype: CO! (see Woelkerling & Irvine, 1986a). Herb. P.-L. & H.-M. Crouan, Mingant, Brest, France, 19 February 1859.

Millepora polymorpha Linnaeus (1767), p.1285, pro parte, excl. typ.
Lithothamnion polymorphum (Linnaeus) Areschoug (1852), p.524, pro parte, excl. typ.
Lithothamnion purpureum P. & H. Crouan (1867), p.150. (CO!).
Phymatolithon polymorphum (Linnaeus) Foslie (1898a), p.4, pro parte, excl. typ.
Eleutherospora polymorpha (Linnaeus) Heydrich (1900), p.65, pro parte, excl. typ.

Key characters: Medulla relatively thick, tetra/bisporangial conceptacles with thick rims, shedding thick discs, becoming buried. Overgrowing *P.lenormandii, P. laevigatum, P. lamii*, and crustose base of *Corallina officinalis*; more or less equal in competition with *Lithophyllum incrustans*.

Illustrations: Hamel & Lemoine (1953), pl. XVIII, figs 3–6; Kornmann & Sahling (1977), pl.120, figs A-F; Rueness (1977), pl.VII, fig.4; Woelkerling & Irvine (1986a), figs 16–18.

Fig. 115 *Phymatolithon purpureum.* (A) Thallus showing thick, rolling, tyre-like margin and conceptacles shedding white discs (YMC 84/303). (B) Detail showing cortical discs popping off tetrasporangial conceptacles (arrow) and young stages of conceptacle roofs; scale as C (YMC 84/303). (C) Mature tetrasporangial conceptacles with thick, raised rims (cf. *P.laevigatum*, Fig. 106A)(YMC 83/181) (cf. Hamel & Lemoine, 1953, pl.XIX, figs 3,4, as *Lithothamnium bornetii*).

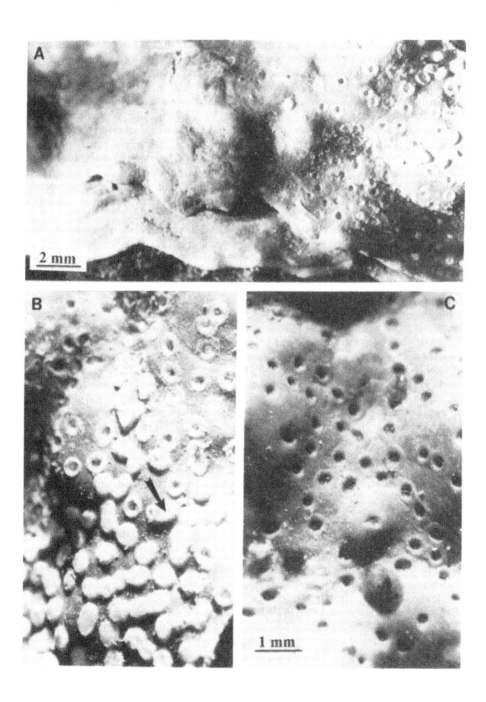

Thallus encrusting, usually closely adherent, sometimes easily detached, to at least 20cm diameter, to 4mm or more thick, flat, to undulating, to knobbly but not branched, orbicular to spreading, becoming confluent with confluences slightly to strongly crested, often many thalli superimposing; *margin* entire, thickish, with strong orbital ridges; *surface* smooth while fertile (winter), with pale flaky patches in summer, cell surface *Phymatolithon*-type (rarely *Leptophytum*-type)(SEM), colour glowing red-pink (Methuen greyish red) in winter, in summer dull pink with white patches and often small pimples marking superficial conceptacle remains, characteristically greenish when dried, texture glossy in winter, matt in summer, often grazed; all primordial conceptacles shedding thick, white cortical discs, *spermatangial conceptacles* from slightly below to slightly above thallus surface, appearing as pale rings and finally white discs with central pore, to 250µm diameter; *carposporangial conceptacles* similar to spermatangial ones but with slightly raised rims, to 300µm; *tetrasporangial conceptacles* to 350µm diameter, from slightly sunken to slightly raised, with thick (to 100µm wide), tyre-like, raised rims up to 100µm wide, pore plates concave, with up to c.70 pores, adjacent conceptacles often fusing to look like spectacles; *bisporangial conceptacles* not known.

Medulla thick, to 50% thallus thickness with filaments subparallel to substratum and cells elongate, 5–21 (29)µm long × 2–6 (13)µm diameter; *cortical filaments* becoming very long with cells short near surface to elongate below, 1.5–10 (13)µm long × 2–6 (8)µm diameter, lacunae between cortical filaments not uncommon in summer-collected plants; *epithallial cells* elliptical to domed, usually 1, or absent, or to 3 cells. Old thalli often invaded by polychaete worms.

Gametangial plants dioecious; *spermatangial conceptacle* chambers immersed, rounded, (160)164–182µm diameter, 60–130(140)µm diameter with roof (neck) 23–65µm thick, spermatangial systems dendroid, borne round lower parts of chamber, simpler ones above; *carpogonial conceptacles* flask-shaped; *carposporangial conceptacles* with base of layers of squashed cells, elliptical, 208–260(350)µm diameter, 130–143(225)µm high, with roof (neck) (35)57–78µm thick and composed of filaments of very thin, elongate cells bending inwards to form small papillae lining pore canal, central fusion cells not developing, *gonimoblast filaments* borne individually, conceptacle base formed of layers of squashed cells; *tetrasporangial conceptacle* chambers elliptical, (135)182–250µm diameter, (65)104–130(145)µm high, roof (20)23–42(45)µm thick, rim thickened and raised above pore plate, pore plate either below, or flush with, or above thallus surface, to c.6 cells thick, cells squarish to elongate; *tetrasporangia* 91–96µm long × 42–60µm diameter; *bisporangial conceptacles* not known. Old conceptacles of all types becoming buried in thallus to several layers deep.

Very common. Epilithic on rocks and stones, sometimes extending a short way up *Laminaria* holdfasts and stipes, also forming nodules, with or without protuberances, in maerl (Adey

---

Fig. 116 Vertical sections of *Phymatolithon purpureum* conceptacles. (A) Diagrammatic drawing of thallus with carposporangial conceptacles at surface and buried (YMC 84/44). (B) Initiation of tetrasporangial conceptacle with cortical disc *in situ* (YMC 83/483). (C) Carpogonial conceptacle (YMC 87/10). (D) Carposporangial conceptacle (YMC 84/8B). (E) Tetrasporangial conceptacle (YMC 83/477). (F) Spermatangial conceptacle (83/395). (G) Dendroid (left) and simple (right) spermatangial systems (YMC 84/21).

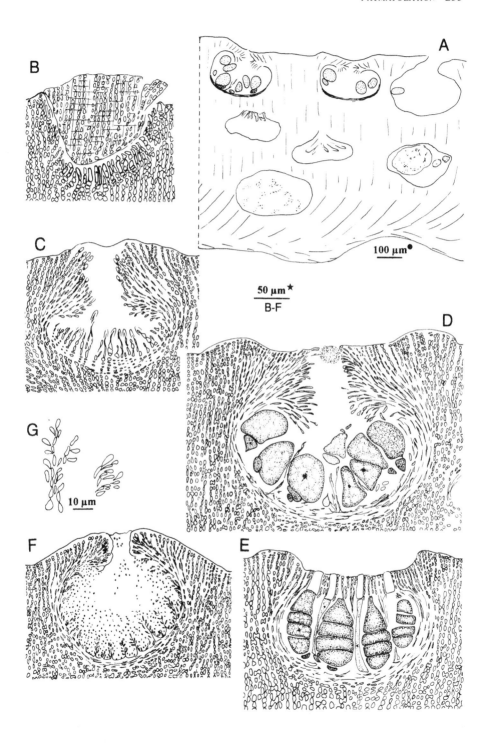

& Adey, 1973; Blunden *et al.*, 1981), to 200mm diameter in Norway (Adey & Adey, 1973). Low littoral and sublittoral to 45m, most abundant from 0–20m (Adey & Adey, 1973, fig. 36); occurring higher on shore in caves and areas of strong wave action as thin pale plants, resembling *Lithophyllum orbiculatum*, q.v. Abundant throughout the British Isles on open, rocky shores, forming a characteristically glowing pink (in winter) cover on low littoral rocks and in pools, plants often covered with conspicuous white discs in November and December. Only *Lithophyllum incrustans* reaches similarly massive proportions in the British Isles. Also common in sublittoral but apparently replaced by *P. lamii* and *Lithophyllum crouanii* in Shetland (Adey & Adey, 1973).

Throughout the British Isles.

Arctic Russia to Morocco, Iceland, Faroes, W. Baltic. Not known in NW Atlantic, previous records based on misidentifications (Adey, 1964).

All reproductive types mainly fertile from October (primordial) to March; but conceptacles sometimes seen in summer, and large apparently entirely infertile plants not uncommon in winter. Adey (in Adey & Vassar, 1975, p.62) suggested an average marginal growth rate of 0.3mm per month for plants kept in tanks in Norway.

The pale lumpy thallus seen in summer-collected plants appears to result from the development of lacunae in the cortex combined with renewed upward growth of groups of cortical filaments; these plants sometimes show overgrown conceptacles where the cover is incomplete, resembling a uniporate conceptacle. Older plants are commonly infested by polychaete worms; the wormholes can be mistaken for conceptacles which are similar in size.

Woelkerling & Irvine (1986a) showed that all combinations, including *Phymatolithon* (*Lithothamnion*) *polymorphum*, based on *Millepora polymorpha* Linnaeus refer nomenclaturally to *P. calcareum* (*q.v.*); the correct name for the entity known in literature mainly as *P.* (*L.*) *polymorphum* is *P. purpureum* (P.& H. Crouan) Woelkerling & L.Irvine.

Adey & Adey (1973) reported thalli up to 20cm thick but we have not seen such massive plants. Material described by Suneson (1943, as *Lithothamnion polymorphum*, C!, LD!) seems to be a mixture of this species and *P. laevigatum*, *q.v.*.

Walker & Moss (1984) gave detailed observations on spore adhesion.

Lemoine (1953) reported *P. purpureum* (as *polymorphum*), *P. laevigatum* and *P. lenormandii* as fossils on post-glacial fossil shells in Co.Antrim.

Chamberlain & Cooke (1984) described this species as forming a lining to cups of the sea-urchin, *Paracentrotus lividus*, (Frontispiece) on the shore at Fanore, Co.Clare.

The records of *Lithothamnion foecundum* Kjellman for the British Isles (Adams, 1908; Lemoine, 1913a) probably refer to this species.

HILDENBRANDIALES

by Linda M. Irvine and Curt M. Pueschel*

HILDENBRANDIALES Pueschel & Cole

HILDENBRANDIALES Pueschel & Cole (1982), p.718.

Thalli coriaceous, closely adherent to substratum, rhizoids absent, entirely crustose or with erect axes arising from crustose base; composed of cylindrical cells arranged in vertical files. Pit plugs with a cap membrane but lacking outer cap layer. Gametangial reproduction unknown; *tetrasporangia* zonate, produced in conceptacles that enlarge by continuous conversion of vegetative cells lining chamber into tetrasporocytes; spore germination unipolar, cytoplasmic contents of germinating spore evacuated into germ tube.

The single family, Hildenbrandiaceae, was until recently placed in the Cryptonemiales. Proposal of ordinal status by Pueschel & Cole (1982) was based on characters that distinguished the family from the Cryptonemiales - presence of conceptacles and the erosive manner by which they enlarge, and presence of a pit plug layer - as well as absence of characters indicating affinity with the Cryptonemiales or any other order. The presence of conceptacles in the Corallinales could be construed as evidence of affinity, and both orders apparently have a mechanism for formation of secondary pit connections without conjunctor cells (Cabioch & Giraud, 1982; Pueschel, 1988). Both vegetative structure and pit plug organization (Pueschel & Cole, 1982) are quite different in the two orders, however, and modes of conceptacle development are dissimilar, indicating that conceptacles are analogous, rather than homologous, structures. Furthermore, in the Corallinales tetrasporocytes are not produced continuously, are restricted to the floor of the conceptacles and are unique in showing cleavages which arise simultaneously (Guiry, 1978b; Guiry & Irvine, 1989).

*Department of Biological Sciences, State University of New York at Binghamton, USA

HILDENBRANDIACEAE Rabenhorst

HILDENBRANDIACEAE Rabenhorst (1868), p.408 [as Hildenbrandtiaceae].

Thallus entirely crustose or with simple or branched axes arising from crustose base, basal surface closely adherent to substratum, rhizoids absent; crustose portions composed solely of crowded small cylindrical cells, erect axes with similar cells forming a cortex surrounding elongate cylindrical cells, secondary pit connections common; gametangia unknown; tetrasporangia borne in conceptacles, transversely or obliquely zonate, but sometimes appearing obliquely cruciate; fungal hyphae commonly penetrating cell walls and into conceptacle chamber.

Two genera are known, *Hildenbrandia*, which is cosmopolitan, and *Apophlaea*, which is endemic to New Zealand and the subantarctic islands; they share a similar vegetative anatomy and a unique type of conceptacle. The only reproductive organs known are tetrasporangia; in culture, tetraspores grow into tetraspore-bearing thalli (DeCew & West, 1977; Fletcher, 1983). *Apophlaea* has been assigned to three different orders or considered *incertae sedis*. We follow Denizot (1968) who placed it in the Hildenbrandiaceae, a proposal supported by evidence from Hawkes (1983) and Pueschel (1989).

Erythroclathrus Liebman (1839) has been listed as a synonym of *Hildenbrandia* (see Kylin, 1956), based on *E. rivularis* (= *H. rivularis* (Liebman) J. Agardh). The original description of the genus included the type of *Cruoria* Fries (1836), however, and so the name *Erythroclathrus* is technically a later name for *Cruoria*. *H. rivularis* is a freshwater species widespread in the British Isles, usually in shaded, running water (West & Fritsch, 1927; Israelson, 1942). Its generic placement is uncertain because no conceptacles have been demonstrated; however, the anatomy of the thallus and the fine structure of pit plugs (Pueschel, 1989) are both indistinguishable from those of *Hildenbrandia*.

HILDENBRANDIA Nardo *orth. cons.*

HILDENBRANDIA Nardo (1834b), p.675 (as *Hildbrandtia*)

Type species: *H. prototypus* Nardo (1834b), p.675 (= *H. rubra* Sommerfelt) Meneghini (1841), p.426.

Hildenbrandia sect. *Haematophlaea* J. Agardh (1852), p.495.
Haematophlaea (J. Agardh) P.& H.Crouan (1858), p.73, pro parte, type only.

Thallus crustose, closely adherent to substratum by whole of undersurface, firm and coriaceous; composed of compact, coherent, unbranched or sparsely branched, erect filaments arising from a prostrate, unistratose basal layer, rhizoids absent; cells usually very small, cylindrical, occasionally elongate in direction of growth, especially on an uneven substratum, sometimes appearing horizontally stratified; secondary pit connections direct, abundant; cells uninucleate, chloroplasts single per cell; cytoplasmic contents of germinating spore evacuated into germ tube which divides to form disc directly or to form short filament that subsequently divides to form disc.

Gametangia unknown; tetrasporangia developing from both floor and walls of spherical or ovate conceptacles which enlarge by repeated conversion of vegetative cells into tetrasporocytes; tetrasporangia essentially zonate, cleavages not simultaneous, first cleavage transverse or oblique, second cleavage also transverse or oblique, sometimes parallel to the first, sometimes non-parallel and then resulting in an apparently cruciate or irregular arrangement; walls of discharged tetrasporangia often persistent, paraphyses absent.

The name *Rhodytapium* given by Kylin (1956) as a synonym of *Hildenbrandia* is invalid; it is a provisional name lacking a description (Silva, pers. comm.).

Many of the characters which have been used to distinguish species are questionable. The thickness of the thallus varies not only with age but also with frequent browsing by grazing animals; the perceived colour is affected by the colour and opacity of the substratum, as pointed out by Sommerfelt (1826), as well as by the thickness of the plant itself, thus accounting for the frequency of records of '*H. rosea*' on white quartz. It is also possible that the size and shape of the conceptacles varies with environmental conditions (Denizot, 1968). DeCew & West (1977) described differences in the mode of development of conceptacles in different species and suggested this may be a character of taxonomic importance. They found that, in the Pacific species *H. occidentalis* Setchell, tetrasporangia are initiated basipetally from the surface whereas in Pacific material of *H. rubra* (as *H. prototypus*) tetrasporangial initials are immersed and the overlying vegetative cells later disintegrate. Pueschel (1982 and LMI obs.) considered that the latter is a misinterpretation based on the position and angle of sections studied. Rosenvinge (1917) described development of tetrasporangia in both *H. rubra* (as *H. prototypus*) and *H. crouanii* as superficial, the conceptacle being formed as they are enveloped by the continuing growth of surrounding vegetative filaments (see also Denizot, 1968).

A stalk cell is sometimes reported to subtend the tetrasporangium. However, unlike regenerative stalk cells which divide to form a new stalk cell and tetrasporocyte after discharge of the tetrasporangium, this subtending cell becomes a tetrasporocyte directly and its subtending cell will appear as a stalk cell (Rosenvinge, 1917; Pueschel, 1982).

Hildenbrandia is one of the few genera in the Rhodophyta for which both cruciate and zonate tetrasporangia have been reported. We consider, however, that tetrasporangial cleavage is essentially zonate but that the plane of any cleavage can be either transverse or oblique (Fig.117). The arrangement of spores in a tetrasporangium is used here to distinguish species occurring in the British Isles. In *H. crouanii* the second cleavages are parallel to the first; the first is often oblique but, in such cases, the sporangia appear regularly zonate if revolved through 90° about the long axis. In *H. rubra* the cleavages are not parallel and the arrangement of spores usually forms a pattern which has been described as obliquely cruciate (Rosenvinge, 1917; Guiry, 1978b). Such sporangia present widely differing appearances when viewed at different angles, especially when the two second cleavages occur in different planes, and can even appear transversely zonately cleaved, but into three cells. (These differences can best be understood by constructing plasticine models.) The conceptacles of both species may thus appear to contain sporangia variously and/or irregularly cleaved. Careful interpretation is needed since the arrangement is not usually as 'irregular' as described in the literature (see Guiry & Irvine, 1989). Both types of spore arrangement were illustrated by Umezaki (1969) for *H. rubra* (as *H. prototypus*) in Japan.

The two species thus defined, *H. crouanii* and *H. rubra*, often occur side by side and a colour difference between them can be observed. Unfortunately, it does not seem possible to use colour as a character to identify a species growing alone. Most British Isles specimens can, however, be identified as *H. rubra* or *H. crouanii* as circumscribed above; apparently intermediate specimens have been found occasionally in eastern Canada as well as the British Isles (Hooper and LMI obs.)

KEY TO SPECIES

1 Thallus brownish red; tetrasporangia appearing zonate, planes of cleavage parallel, transverse or oblique, 4–11 × 20–32µm.. *H. crouanii*
   Thallus rosy to dark red; tetrasporangia appearing cruciate or irregular, planes of cleavage not parallel, 8–15 × 18–35µm.. *H. rubra*

Hildenbrandia crouanii J. Agardh (1852), p.496 (as *Crouani*).
Figs 117; 118

Lectotype: LD! (Herb. Alg. Agardh. 27613). France (Brest).

?Hildenbrandia dawsonii (Ardré) Hollenberg (1971), p.286.
?Hildenbrandia canariensis var.*dawsonii* Ardré (1959), p.230.
?Hildenbrandia kerguelensis (Askenasy) Y. Chamberlain (1962), p.372.
?Hildenbrandia prototypus var. *kerguelensis* Askenasy (1888), p.30.
Hildenbrandia canariensis Børgesen (1929), p.56, fig.26d.

Illustrations: Harvey (1849b), pl. CCL (as *H. rubra*); P.& H.Crouan (1867), pl. 19, figs 1–3 (as *H. rosea*); Rosenvinge (1917), fig. 126; Børgesen (1929) fig. 26d (as *H. canariensis*); Ardré (1959), pl. III, fig.6 (as *H. canariensis* var. *dawsonii*); Chamberlain (1962), pl. 82 (as *H. kerguelensis*); Denizot (1968), pp. 196–197.

Thallus encrusting, indefinitely large, sometimes very extensive, outline usually irregular, coriaceous, up to at least 150µm thick; *surface* smooth, interrupted only by groups of flush or slightly raised conceptacles, colour rose-red to brownish red; consisting of a *basal layer* of prostrate filaments giving rise to *closely compacted homogeneous* cells *c.* 5µm diameter, not obviously filamentous but often appearing in vertical files.

Gametangial plants unknown; *tetrasporangial conceptacles c.* 110µm deep, 80µm diameter, *tetrasporangia* lining firstly floor and then walls of conceptacle, appearing transversely or obliquely zonate, 20–32µm long × 5–11µm diameter.

On stable rock and loose stones; forming indefinite, sometimes very extensive patches; margins rarely extending slightly over coralline algae, otherwise not epiphytic. Apparently extending, with *H. rubra*, from upper littoral to at least 18m sublittorally; comparative distribution of the two species not known because of sampling difficulties.

Distribution records incomplete because of uncertain species limits. Specimens with zonate tetrasporangia have been reported for Norway to Mauritania; W. Baltic; Mediterranean; Canary Isles; Greenland, eastern Canada (Newfoundland) to USA (Mass.), and many other parts of the world including Kerguelen, California, Chile and Japan.

Tetrasporangia recorded for January to September, probably occurring throughout the year.

Thickness much affected by grazing; conceptacles usually flush with surface but sometimes protruding in thin thalli, chambers spherical but deepening with age.

Børgesen (1929) described *H. canariensis* as differing from *H. crouanii* by having transversely rather than obliquely cleaved tetrasporangia; the original description of the latter (Agardh, 1852) does not mention oblique cleavage, however, and a sketch by Agardh (Herb. Alg. Agardh. no. 276612) shows transversely zonate sporangia (see also Printz, 1926). Rosenvinge (1917) was apparently the first to report oblique cleavage. The layer of mucilage surrounding the tetraspores within the tetrasporangium has little inherent rigidity, as is apparent when spores are shed (CMP obs.). Therefore, the significance of small

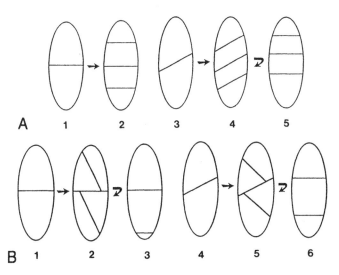

Fig. 117 Diagrams showing successive cleavages of tetrasporocytes. (A) *Hildenbrandia crouanii*. 1,2 show all divisions transverse and parallel; 3,4 show all divisions oblique and parallel; 5 shows appearance of 4 when rotated through 90°. (B) *Hildenbrandia rubra*. 1,2 show a transverse division followed by two oblique divisions; 4,5 show an oblique division followed by two further oblique divisions in a different plane from the first; 3,6 show appearance of 2 and 4 when rotated through 90°.

differences in orientation of cleavage planes is questionable and the possibility of distortion during diverse preservation procedures should not be overlooked. This has implications for whether *H. crouanii* is regarded as restricted to the North Atlantic or as cosmopolitan. Specimens with either obliquely or transversely zonate tetrasporangia are included here in *H. crouanii*.

Both this species and *H. rubra* can be distinguished from other red crusts by the conceptacles which are almost always present and visible with a handlens as minute craters on the surface. They are similar in vegetative structure to the *Porphyrodiscus simulans* stage of *Ahnfeltia plicata* (see Vol 1(1), p.229). The latter is more purplish in colour, with tetrasporangia in mucilaginous superficial sori.

Hildenbrandia rubra (Sommerfelt) Meneghini (1841), p.426.
Figs 117; 118

Lectotype: O! Norway (Saltdal).

Verrucaria rubra Sommerfelt (1826), p.140.
?*Hildenbrandia prototypus* Nardo (1834b), p.675.
?*Hildenbrandia rosea* Kützing (1843), p.384.
Hildenbrandia sanguinea Kützing (1843), p.384 (L!).
?*Rhododermis drummondii* Harvey (1844), p.27.

Illustrations: Kützing (1843), pl. 78, fig. V (as *H. sanguinea, H. rosea*); Rosenvinge (1917), figs 121–125 (as *H. prototypus*); Newton (1931), fig. 184; Chemin 1937, fig. 36 (as *H. prototypus*); Denizot (1968), p. 196 (as *H. prototypus*); Pueschel & Cole (1982) fig. 17; Stegenga & Mol 1983) pl. 76, figs.4–5.

Vegetative thallus as in *H. crouanii*, colour varying from rose red to dark red or nearly black, but not brownish.

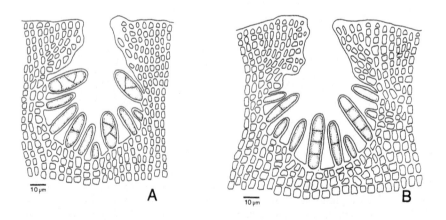

Fig. 118 VS mature tetrasporangial conceptacles. (A) *Hildenbrandia crouanii*. (B) *Hildenbrandia rubra*.

Gametangial plants unknown; *tetrasporangial conceptacles* ovate *c*.80–90μm diameter × 65–70μm high, *tetrasporangia* appearing obliquely cruciate or irregular, (13)22–30(35)μm long × (7)11–13(15)μm diameter.

Epilithic on stable rock and loose stones, forming indefinite, sometimes very extensive, patches; common, together with *H. crouanii*, in estuaries and harbours and often conspicuous on rocks and cave walls in upper littoral; recorded sublittorally to at least 12m.

Recorded throughout the British Isles.

Arctic Ocean to Ghana; Baltic; Mediterranean; Azores; Canary Isles; Arctic Canada to Brazil; Pacific, probably cosmopolitan.

Data on seasonal behaviour and form variation scanty because of sampling difficulties; tetrasporangia recorded throughout the year (Lewis, 1936; LMI obs.). Thickness much affected by grazing. Umezaki (1969) reported that, in Japan, all plants have a quiescent period from mid-February to late March during which spores are not developed and plants at higher levels die away; after 8 months in culture his plants became 5–10mm diameter and 70–100μm thick, but remained sterile.

H. prototypus, *H. rosea* and *Rhododermis drummondii* are included as doubtful synonyms. The type specimen of the first is missing from Nardo's collection (in Herb. Zanardini, Venice) and the kind of tetrasporangial cleavage is not specified in the original description; the type material of *H. rosea* is sterile. *R. drummondii* appears to be a thin form of *Hildenbrandia* with protruding conceptacles, but the original specimen lacked spores.

Glossary

The following definitions apply to their use in the context of this Flora; an extended glossary is available in Woelkerling (1988, p.225). See also glossaries in Irvine (1983) and Johansen (1981).

Adelphoparasite: a parasite that is closely related to its host; cf. alloparasite
Aligned: rows of adjacent cells with end walls at same level
Alloparasite: a parasite that is unrelated to its host; cf. adelphoparasite
Apiculate: ending abruptly in short point
Apomictic: reproducing asexually, without karyogamy and meiosis
Applanate: flattened in one plane
Appressed: pressed together without being united
Arcuate: bent like a bow; spermatangial systems in Melobesioideae
Axial conceptacle: a medullary conceptacle originating in line with and terminating the axis of a branch; in the Corallinoideae

Basal filaments: unistratose, basal layer of thallus filaments (Figs 1; 2A); in dimerous plants
Basionym: the original name of a taxon
Biogenic: originating from living matter
Bisporangium: a sporangium whose contents divide to form two spores; cf. tetrasporangium
Bistratose: having two layers; cf. unistratose
Buried conceptacle: old conceptacle that becomes covered over by continued thallus growth, but not infilled

Caducous: not persistent
Calcite: hexagonal, rhombohedral crystals of calcium carbonate in the walls of vegetative cells (Fig. 4)
Coaxial: with cells laterally aligned in curving rows (Fig.97B)
Columella: filamentous, sometimes calcified structure often occurring in the centre of the floor of a tetra/ bisporangial conceptacle
Concatenate: linked together like a string of beads
Concavity: depression on thallus or conceptacle surface formerly occupied by an epithallial cell (SEM)
Conceptacle: an enclosed chamber opening outwards, which contains reproductive structures (no homology with conceptacles in different algal groups is implied)
Conceptacle primordium: a group of meristematic cells from which a conceptacle develops

Confluence: where two adjacent thalli meet

Confluent: running together; cf. discrete

Contiguous: adjacent; used particularly to denote the proximity of different filaments whose cells are aligned or secondarily joined by fusions or secondary pit connections

Coplanar: in the same plane

Cortex: collective term for filaments derived from subepithallial initials, occurring externally to the medulla (Fig.2C) and running more or less at right angles to the thallus surface; in monomerous plants

Cortical disc: thick white disc shed by some conceptacles; in Melobesioideae; cf. veil

Corymbose: with reproductive bodies borne at same level (Fig.17C)

Dendroid: tree-like, as a result of lower branches not developing or falling away; in spermatangial systems

Dimerous: a type of thallus construction involving two sets of filaments (basal and erect) more or less at right angles to one another (Fig.2A); cf. monomerous

Discrete: individually distinct; cf. confluent

Divaricate: widely spreading

Dorsal: upper; cf. ventral

Elliptical: regularly oval

Endophytic: growing within a plant; cf. semi-endophytic; epiphytic

Endoplasmic reticulum: an extensive system of membranes in eukaryotic cells, dividing the cytoplasm into compartments and channels; those parts containing many ribosomes are called 'rough ER' and those lacking ribosomes are called 'smooth ER'

Epilithic: growing attached to an inorganic substratum (rock, stones etc.)

Epiphytic: growing on the surface of a plant; cf. endophytic

Epithallial cell: a cell formed outwardly from either a basal cell or a subepithallial initial

Epizoic: growing on animal substrata such as shells

Erect filament: filament derived from, and more or less at right angles to, a basal filament (Fig.2A); in dimerous plants

Excrescence: a general term used for outgrowths such as lamellae, protuberances, etc.

Fasciated: coalescent; e.g. of branch systems in *Lithothamnion corallioides*

Fastigiate: with conical or tapering outline

Filament: a row of cells joined by primary pit connections

Flared: a term used to describe the laterally extended periphery of the outermost wall of epithallial cells in *Lithothamnion, Exilicrusta* and members of Sporolithaceae, seen as two wings in VS (Fig.85C)

Flat thallus: a thallus of generally flattened appearance lacking protuberances etc.; cf. smooth thallus which applies to microscopic features

Fusion cell: a cell formed from amalgamated supporting cells, and sometimes additional cells, after presumed karyogamy, subsequently giving rise to gonimoblast filaments

Geniculum: a group of completely or partially noncalcified cells located below every intergeniculum; in geniculate corallines

Germination disc: pattern of cells developing within original spore wall

Glossy: thallus texture smooth and shiny; cf. grainy, matt

Gonimoblast filaments: filaments, usually produced from fusion cell, that produce carposporangia

Grainy: thallus texture rough but slightly shiny; cf. glossy, matt

Hermaphrodite: having male and female organs in the same conceptacle

Heterotrichous: thallus comprising a prostrate creeping system from which erect filaments project

Holotype: *either* the single specimen designated in the protologue of a species by the original author to serve as the nomenclatural type; *or* the single specimen used by an author in originally describing a species

Hypothallial filaments: name sometimes used for basal filaments or medulla

Imbricating thallus: thallus in which lobe-like parts of a thallus overgrow one another like tiles on a roof

Immersed conceptacle: conceptacle in which the roof is flush with the thallus surface

Intercalary: within a filament (not terminal)

Intergeniculum: a calcified segment many of which make up the frond of a geniculate coralline; separated from each other by noncalcified genicula

Initial: a meristematic cell, potentially destined to divide; cf. subepithallial, terminal initials

Isolectotype: a duplicate of a lectotype, *q.v.*

Lacunose: with many holes

Lamella: leaf-like thallus lobe

Lateral: borne from the side of an axis

Lateral conceptacle: conceptacle originating at the surface of an intergeniculum; in Corallinoideae but not found in British Isles species

Lectotype: a specimen or published illustration from an author's original collection selected by a subsequent author to serve as the nomenclatural type in the absence of a holotype

Lenticular: lens-shaped

Lobe: a forward projecting extension of an intergeniculum; in Corallinoideae

Maerl: unattached, branched corallines, living or dead; often occurring in extensive deposits (maerl beds or banks)

Margin: peripheral area of a dorsiventral, nongeniculate thallus, often composed of basal and epithallial cells only and not usually bearing conceptacles

Marginal conceptacle: conceptacle originating at the margin of a flattened intergeniculum; in the Corallinoideae

Matt: thallus surface smooth and dull, not shiny; cf. glossy, grainy

Medulla: collective term for filaments derived from terminal initials, occurring internally to the cortex and running more or less parallel to the thallus surface (Fig.2C); in a monomerous thallus

Monomerous: thallus type composed of medulla and cortex (Fig. 2C); cf dimerous

Multiaxial: a type of thallus construction in which each axis includes a core of principal, co-dominant filaments; cf. uniaxial

Neotenic: with features of early stages of development persisting in adult; plants sometimes reproducing in apparently juvenile state; see *Fosliella*-state of *Hydrolithon*

Orbital ridges: raised concentric rings round margin of thick thallus (e.g. *Phymatolithon lenormandii*)
Orbital rings: concentric markings round edge of bistratose thallus (e.g. *Fosliella*-state of *Hydrolithon*)
Orbicular: more or less circular

Palisade cell: cell of a basal filament which is relatively tall, narrow and obliquely orientated in VS; adjacent palisade cells have end walls at the same level
Papillae: small swollen cells sometimes lining the pore canal
Perithallial filaments: name sometimes used for erect filaments or cortex
Plug: mucilaginous substance blocking (usually multiporate) conceptacle pore
Pore: hole(s) in a conceptacle roof through which sporangia escape
Pore canal: passage between a pore and a conceptacle chamber, applicable when the conceptacle roof is thick
Pore-plate: roof of multiporate conceptacle that contains the pore plugs
Primary branching: branching which originates in or near apical meristems; in geniculate corallines; cf. secondary branching
Primary pit connection: connection formed between two successive cells of a filament during cell division; cf. secondary pit connection
Primordium: group of initials, especially those initiating a conceptacle
Prominent: protruding markedly and abruptly from the thallus surface; used of conceptacles; cf. raised
Protologue: all the information associated with the original publication of a taxonomic name
Protuberance: knob-like outgrowth from crustose thallus
Pseudodichotomous branching: appearing dichotomous but resulting from two successive cell divisions (Fig.1)
Pseudolateral conceptacle: conceptacle originating in apex of aborted secondary branch and protruding more than lateral conceptacles; in *Corallina*
Pseudoparenchyma: aggregated filaments having the appearance of a parenchyma but analogous not homologous
Pulvinate: cushion-like

Raised: (used of conceptacles) protruding gradually somewhat above the thallus surface; cf. prominent
Rhodolith: a term sometimes used for coralline nodules with a nonalgal core; cf. maerl (see Introduction 4.1)
Ribosome: cytoplasmic particle carrying out protein synthesis; made up of RNA and protein; free in cytoplasm or attached to endoplasmic reticulum
Row: laterally aligned contiguous cells of erect or cortical filaments

Secondary branching: branching originating well below branch apices in intergenicular cortex; in geniculate corallines; cf. primary branching
Secondary pit connection (direct): pit connection formed secondarily and directly between two mature cells of contiguous thallus filaments; cf. primary pit connection (other red algae form secondary pit connections indirectly)
Semi-endophytic: growing partly within a plant, e.g. *Choreonema*
Siderophilic: having an affinity for fatty substances

Simple: unbranched (e.g. of spermatangial systems)
Sinuate: with a wavy outline, used of basal cells and rows of erect filament cells (mainly in *Titanoderma*)
Smooth thallus: thallus surface lacking microscopic protrusions, flakes etc. and therefore not roughened; cf. flat thallus which applies to macroscopic features
Spout: tube-like extension of spermatangial (rarely tetrasporangial) pore
Squamulose: with small, slightly overlapping thallus lobes
Squarish: more or less isodiametric
Subepithallial initial: a subterminal initial on an erect or cortical filament, cutting off epithallial cells outwardly and erect filament or cortical cells inwardly
Substratum: any surface upon which plant grows
Supporting cell: a cell bearing one or more carpogonial branches and constituting a layer across the floor of a carpogonial conceptacle
Sympatric: with areas of distribution that coincide or overlap

Terminal initial: an initial terminating a filament; cf. subepithallial initial
Tetra/ bisporangial: term used when data apply to both types of conceptacle
Tetrasporangium: a sporangium whose contents divide meiotically (presumably) and (in Corallinaceae) simultaneously to form four zonately arranged haploid (presumably) tetraspores; cf.bisporangium
Tetrasporocyte: contents of a sporangium destined to divide into four spores
Tier: stratum of aligned vegetative cells (Fig.12)
Trichocyte: a complex of one or more surface or near surface cells potentially hair-bearing and persisting for varying lengths of time
Trichocyte field: group of trichocytes on thallus surface interspersed with normal vegetative cells, e.g. *Hydrolithon cruciatum*

Unconsolidated: thallus in which basal filaments are not contiguous
Unistratose: composed of single cell layer; cf. bistratose

Veil: thin layer of white cells shed by some conceptacles; in Melobesiodeae (see *Phymatolithon lenormandii*); cf. cortical disc
Ventral: lower; cf. dorsal
Verticillate: whorled

Wing: a flat, lateral protrusion of an intergeniculum; in Corallinoideae

Zonate: linear arrangement of four spores in tetrasporangium
Zoned thallus: see Introduction 8.4

References

ABBOTT I.A. & HOLLENBERG G.J. 1976. *Marine Algae of California*. Stanford University Press, Stanford, California.

ADAMS J. 1908. A synopsis of Irish algae, freshwater and marine. *Proceedings of the Royal Irish Academy*, Sec.B **27**(2):11–60.

ADEY W.H. 1964. The genus *Phymatolithon* in the Gulf of Maine. *Hydrobiologia* **24**:377–420.

ADEY W.H. 1965. The genus *Clathromorphum* (Corallinaceae) in the Gulf of Maine. *Hydrobiologia* **26**:539–573.

ADEY W.H. 1966a. The genus *Pseudolithophyllum* (Corallinaceae) in the Gulf of Maine. *Hydrobiologia* **27**:479–497.

ADEY W.H. 1966b. The genera *Lithothamnium, Leptophytum* (nov. gen.) and *Phymatolithon* in the Gulf of Maine. *Hydrobiologia* **28**:321–370.

ADEY W.H. 1966c. Distribution of saxicolous crustose corallines in the northwestern North Atlantic. *Journal of Phycology* **2**:49–54.

ADEY W.H. 1968. The distribution of crustose corallines on the Icelandic coast. *Scientia Islandica* (Anniversary Vol.) **1968**:16–25.

ADEY W.H. 1970a. A revision of the Foslie crustose coralline herbarium. *Det Kongelige Norske Videnskabers Selskabs Skrifter* **1970**(1):1–46.

ADEY W.H. 1970b. The effects of light and temperature on growth rates in boreal-subarctic crustose corallines. *Journal of Phycology* **6**:269–276.

ADEY W.H. 1970c. Some relationships between crustose corallines and their substrate. *Scientia Islandica* **2**:21–25.

ADEY W.H. 1970d. The crustose corallines of the northwestern North Atlantic, including *Lithothamnium lemoineae* n.sp. *Journal of Phycology* **6**:225–229.

ADEY W.H. 1971. The sublittoral distribution of crustose corallines on the Norwegian coast. *Sarsia* **46**:41–58.

ADEY W.H. 1979. Crustose coralline algae as microenvironmental indicators for the Tertiary. In: *Historical Biogeography, Plate Tectonics, and the Changing Environment* (Ed. by J.Gray and A.J.Boucot) pp.459–464. Oregon State Univ. Press, Corvallis, Oregon.

ADEY W.H. & ADEY P.J. 1973. Studies on the biosystematics and ecology of the epilithic crustose Corallinaceae of the British Isles. *British Phycological Journal* **8**:343–407.

ADEY W.H. & JOHANSEN H.W. 1972. Morphology and taxonomy of Corallinaceae with special reference to *Clathromorphum, Mesophyllum*, and *Neopolyporolithon* gen.nov. (Rhodophyceae, Cryptonemiales). *Phycologia* **11**:159–180.

ADEY W.H. & LEBEDNIK P.A. 1967. *Catalog of the Foslie Herbarium*. Det Kongelige Norske Videnskabers Selskab Museet, Trondheim, Norway. 92pp.

ADEY W.H. & MacINTYRE I.G. 1973. Crustose coralline algae: a re-evaluation in the geological sciences. *Geological Society of America Bulletin* **84**:883–904.

ADEY W.H. & McKIBBIN D. 1970. Studies on the maerl species *Phymatolithon calcareum* (Pallas) nov.comb. and *Lithothamnion corallioides* Crouan in the Ria de Vigo. *Botanica Marina* **13**:100–106.

ADEY W.H., MASAKI T. & AKIOKA H. 1974. *Ezo epiyessoense*, a new parasitic genus and species of Corallinaceae (Rhodophyta, Cryptonemiales). *Phycologia* **13**:329–344.

ADEY W.H., MASAKI T. & AKIOKA H. 1976. The distribution of crustose corallines in Eastern Hokkaido and the biogeographic relationships of the flora. *Bulletin of the Faculty of Fisheries, Hokkaido University* **26**:303–313.

ADEY W.H. & SPERAPANI C.P. 1971. The biology of *Kvaleya epilaeve*, a new parasitic genus and species of Corallinaceae. *Phycologia* **10**:29–42.

ADEY W.H., TOWNSEND R.A. & BOYKINS W.T. 1982. The crustose coralline algae (Rhodophyta, Corallinaceae) of the Hawaiian Islands. *Smithsonian Contributions to Marine Science* **15**:i-iv, 1–74.

ADEY W.H. & VASSAR J.M. 1975. Colonization, succession and growth rates of tropical crustose coralline algae (Rhodophyta, Cryptonemiales). *Phycologia* **14**:55–69.

AFONSO-CARRILLO J. 1984. Estudios en las algas Corallinaceae (Rhodophyta) de las Islas Canarias. II. Notas taxonomicas. *Vieraea* **13**:127–144.

AFONSO-CARRILLO J. 1988. Structure and reproduction of *Spongites wildpretii* sp.nov. (Corallinaceae, Rhodophyta) from the Canary Islands, with observations and comments on *Spongites absimile* comb.nov. *British Phycological Journal* **23**:89–102.

AFONSO-CARRILLO J., LOSADA-LIMA A. & LEON-ARRENCIBA M.C. 1986. Sobre la posición sistemática de *Choreonema* Schmitz (Corallinaceae, Rhodophyta). *Vierea* **16**:207–210.

AGARDH, J.G. 1852. *Species genera et ordines algarum* **2**(2). Lundae.

ANDRAKE W. & JOHANSEN H.W. 1980. Alizarin red dye as a marker for measuring growth in *Corallina officinalis* L. (Corallinaceae, Rhodophyta). *Journal of Phycology* **16**:620–622.

ARDISSONE F. 1883. Phycologia mediterranea. Part 1. Floridee. *Memorie della Società Crittogamologica Italiana* **1**:1–516.

ARDRÉ F. 1959. Un intéressant *Hildenbrandtia* du Portugal. *Revue Algologique* N.S. **4**:227–237.

ARDRÉ F. 1970. Contribution à l'étude des algues marines du Portugal. I. La flore. *Portugaliae Acta Biologica*, ser.B. **10**:1–423.

ARDRÉ F. & CABIOCH J. 1985. Marie Lemoine (1887–1984). *Cryptogamie: Algologie* **6**:61–70.

ARESCHOUG J.E. 1852. Ordo XII. Corallineae. In: *Species, Genera, et Ordines Algarum* (by J.G.Agardh) Vol.2, Part 2, pp.506–576. C.W.K. Gleerup, Lund.

ARESCHOUG J.E. 1875. Corallineae. Observationes phycologicae III. *Nova acta Regiae Societatis Scientarum Upsaliensis*, ser.3, **10**:1–6.

ÅSEN A. 1976. *Jania rubens* (L.) Lamour. (Rhodophyta, Cryptonemiales) in Norway. *Norwegian Journal of Botany* **23**:195–199.

ASKENASY E. 1888. Algen. In Engler,A., ed., *Forschungsreise S.M.S.Gazelle* **IV**. Botanik. Berlin.

ATHANASIADIS A. 1989. North Aegean marine algae. III Structure and development of the encrusting coralline *Titanoderma cystoseirae* (Rhodophyta, Lithophylloideae). *Nordic Journal of Botany* **9**:435–441.

BABA M., JOHANSEN H.W. & MASAKI T. 1988. The segregation of three species of *Corallina* (Corallinales, Rhodophyta) based on morphology and seasonality in northern Japan. *Botanica Marina* **31**:15–22.

BALAKRISHNAN M.S. 1947. The morphology and cytology of *Melobesia farinosa* Lamour. *Journal of the Indian Botanical Society* **1946** (M.O.P. Iyengar Commemoration Volume) :305–319.

BATTERS E.A.L. 1890. A list of the marine algae of Berwick-on-Tweed. *Transactions of the Berwickshire Naturalists' Club* **12**:221–392.

BATTERS E.A.L. 1892. New or critical British algae. *Grevillea* **21**:13–23, 49–58, 97–109.

BATTERS E.A.L. 1893. New or critical British algae. *Grevillea* **22**:20–24.

BATTERS E.A.L. 1895. On some new British algae. *Annals of Botany* **9**:168–169, 307–321.

BATTERS E.A.L. 1896. Some new British marine algae. *Journal of Botany, British and Foreign* **34**:6–11; 384–390.

BATTERS E.A.L. 1897. New or critical British marine algae. *Journal of Botany, British and Foreign* **35**:433–440.

BATTERS E.A.L. 1902. A catalogue of the British marine algae. *Journal of Botany, British and Foreign* **40**(Suppl):1–101.

BERNER L. 1979. Note sur le *Corallina officinalis* L. 1761. *Nova Hedwigia* **31**:977–991.

BIZZOZERO G. 1885. *Flora Veneta Crittogamica.* Parte II. Seminario, Padova. [i], [1]–255pp.

BLACKLER H. 1951. An algal survey of Lough Foyle, North Ireland. *Proceedings of the Royal Irish Academy* **54** (Sect. B):97–139.

BLACKLER H. 1956. The phenology of certain algae at St. Andrews, Fife. *Transactions and Proceedings of the Botanical Society of Edinburgh* **37**:61–78.

BLUNDEN G., FARNHAM W.F., JEPHSON N., BARWELL C.J., FENN R.H., & PLUNKETT B.A. 1981. The composition of maerl beds of economic interest in northern Brittany, Cornwall, and Ireland. *Proceedings of the International Seaweed Symposium* **10**:651–656.

BÖHM, L., SCHRAMM, W. & RABSCH, U. 1978. Ecological and physiological aspects of some coralline algae from the Western Baltic: calcium uptake and skeletal formation in *Phymatolithon calcareum. Kieler Meeresforschung* **4**:282–288.

BØRGESEN F. 1903. Marine Algae. In: Warming, E., Botany of the Faroes, Part II, pp.339–532. Copenhagen. [preprint 1902].

BØRGESEN F. 1929. Marine algae from the Canary Islands. III Rhodophyceae. Part II. Cryptonemiales, Gigartinales, and Rhodymeniales. *Det Kongelige Danske Videnskabernes Selskabs Biologiske Meddelelser* **8**(1):1–97.

BOROWITZKA M.A. 1977. Algal calcification. *Oceanography and Marine Biology Annual Review* **15**:189–223.

BOROWITZKA M.A. & VESK M. 1978. Ultrastructure of the Corallinaceae. I. The vegetative cells of *Corallina officinalis* and *C. cuvierii. Marine Biology* **46**:295–304.

BOROWITZKA M.A. & VESK M. 1979. Ultrastructure of the Corallinaceae (Rhodophyta) II. Vegetative cells of *Lithothrix aspergillum. Journal of Phycology* **15**:146–153.

BORY J.B. 1832. Notice sur les polypiers de la Grèce. *Expédition Scientifique de Morée.* Sciences physiques **3**:204–209.

BOSELLINI A. & GINSBERG R.N. 1971. Form and internal structure of recent algal nodules (Rhodolites) from Bermuda. *Journal of Geology* **79**:669–682.

BOSENCE D.W.J. 1976. Ecological studies on two unattached coralline algae from Western Ireland. *Palaeontology* **19**:365–395.

BOSENCE D.W.J. 1979. Live and dead faunas from coralline algal gravels, Co. Galway. *Palaeontology* **22**:449–478.

BOSENCE D.W.J. 1980. Sedimentary facies, production rates and facies models for recent coralline algal gravels, Co. Galway, Ireland. *Geological Journal* **15**:91–111.

BOSENCE D.W.J. 1983a. Description and classification of rhodoliths (rhodoids, rhodolites). In: *Coated Grains* (Ed. by T.M. Peryt) pp.217–224. Springer-Verlag, Berlin.

BOSENCE D.W.J. 1983b. The occurrence and ecology of recent rhodoliths – a review. In: *Coated Grains* (Ed. by T.M. Peryt) pp.225–242. Springer-Verlag, Berlin.

BOSENCE D.W.J. 1984. Construction and preservation of two modern coralline algal reefs, St. Croix, Caribbean. *Palaeontology* **27**:549–574.

BRAGA, J.C., BOSENCE, D. & STENECK, R.S. 1993. New anatomical characters in fossil coralline algae and their taxonomic implications. *Palaeontology* **36**:535–547.

BRESSAN G. 1974. Rodoficee calcaree dei mari Italiani. *Bollettino della Società Adriatica di Scienze* **59**:1–132.

BRESSAN G. 1980. La polarité cellulaire dans la progénèse des *Fosliella* (Corallinacées). *Giornale Botanico Italiano* **114**:15–24.

BRESSAN G. & BENES M. 1978. Individuazione di caratteri quantitativi diacritici in *Corallina mediterranea* Areschoug e *C. officinalis* Linneo (Corallinaceae-Rhodophyceae). *Bollettino della Società Adriatica di Scienze* **61**:1–10.

BRESSAN G., GHIRARDELLI L.A. & BELLEMO A. 1981. Research on the genus *Fosliella* (Rhodophyta, Corallinaceae): Structure, ultrastructure, and function of transfer cells. *Botanica Marina* **24**:503–508.

BRESSAN G., MINIATI-RADIN D. & SMUNDIN L. 1977. Su una nuova Corallinacea: *Fosliella cruciata* sp. nov. *Giornale Botanico Italiano* **110**:438.

BRESSAN G. & TOMINI I. 1980. Individuazione di ritmi di acrescimento nelle alghe: problemi metodologici. *Informatore Botanico Italiano* **12**:357–370.

BRESSAN G. & TOMINI I. 1981. Quelques observations sur la croissance des algues rouges calcaires du genre *Fosliella* (Rhodophycophyta, Corallinaceae). *Vie Milieu* **31**:283–291.

BROWN V., DUCKER, S.C. & ROWAN K.S. 1977. The effect of orthophosphate concentration on the growth of articulated coralline algae (Rhodophyta). *Phycologia* **16**:125–131.

BUFFHAM, T.H. 1888. On the reproductive organs, especially the antheridia, of some of the Florideae. *Journal of the Quekett Microscopical Club*, ser.2, **3**:257–266.

BURROWS, E.M. 1958. Sublittoral algal population in Port Erin Bay, Isle of Man. *Journal of the marine Biological Association of the United Kingdom* **37**:187–703.

CABIOCH J. 1966a. Sur le mode de développement des spores chez les Corallinacées. *Compte Rendu Hebdomadaire des Séances de l'Académie des Sciences* Paris **262**(D):2025–2028.

CABIOCH J. 1966b. Sur le mode de formation du thalle articulé chez quelques Corallinacées. *Compte Rendu Hebdomadaire des Séances de l'Académie des Sciences* Paris **263**(D):339–342.

CABIOCH J. 1966c. Contribution à l'étude morphologique, anatomique et systématique de deux Mélobésiées: *Lithothamnion calcareum* (Pallas) Areschoug et *Lithothamnion coralloides* Crouan. *Botanica Marina* **9**:33–53.

CABIOCH J. 1968. Quelques particularités anatomiques du *Lithophyllum fasciculatum* (Lamarck) Foslie. *Bulletin de la Société Botanique de France* **115**:173–186.

CABIOCH J. 1969a. Persistance de stades juveniles et possibilité d'une néotène chez le *Lithophyllum incrustans*. *Compte Rendu Hebdomadaire des Séances de l'Académie des Sciences* Paris **268**(D):497–500.

CABIOCH J. 1969b. Les fonds de maerl de la Baie de Morlaix et leur peuplement végétal. *Cahiers de Biologie Marine* **10**:139–161.

CABIOCH J. 1970. Le maërl des côtes de Bretagne et le problème de sa survie. *Penn ar Bed* **7**:421–429.

CABIOCH J. 1971a. Étude sur les Corallinacées. I. Caractères généraux de la Cytologie. *Cahiers de Biologie Marine* **12**:121–186.

CABIOCH J. 1971b. Essai d'une nouvelle classification des Corallinacées actuelles. *Compte Rendu Hebdomadaire des Séances de l'Académie des Sciences* Paris **272**(D):1616–1619.

CABIOCH J. 1972. Étude sur les Corallinacées. II. La morphogenèse; conséquences systématiques et phylogénétiques. *Cahiers de Biologie Marine* **13**:137–288.

CABIOCH J. 1979. Sur un nouveau cas de semi-parasitisme chez les Corallinacées (Rhodophycées): *Dermatolithon corallinae* (Crouan) Foslie. *Compte Rendu Hebdomadaire des Séances de l'Académie des Sciences* Paris **288**(D):1533–1535.

CABIOCH J. 1980. Le parasitisme du *Choreonema thuretii* (Bornet) Schmitz (Rhodophycées, Corallinacées) et son interprétation. *Compte Rendu Hebdomadaire des Séances de l'Académie des Sciences* Paris **290**(D): 707–710.

CABIOCH J. 1988. Morphogenesis and generic concepts in coralline algae – a reappraisal. *Helgoländer Meeresuntersuchungen* **42**:493–509.

CABIOCH J., FLOC'H, J-Y, LE TOQUIN, A, BOUDOURESQUE, C-F, MEINESZ, A. & VERLAQUE, M. 1992. *Guide des Algues des Mers d'Europe.* Delachaux et Niestlé, Paris. 231pp.

CABIOCH J. & GIRAUD G. 1978a. Comportement cellulaire au cours de la régénération directe chez le *Mesophyllum lichenoides* (Ellis) Lemoine (Rhodophycées, Corallinacées). *Compte Rendu Hebdomadaire des Séances de l'Académie des Sciences* Paris **286**(D):1783–1785.

CABIOCH J. & GIRAUD G. 1978b. Apport de la microscopie électronique à la comparaison de quelques espèces de *Lithothamnium* Philippi. *Phycologia* **17**:369–381.

CABIOCH J. & GIRAUD G. 1982. La structure hildenbrandioïde, stratégie adaptive chez les Floridées. *Phycologia* **21**:307–315.

CABIOCH J. & GIRAUD G. 1986. Structural aspects of Biomineralization in the coralline algae. In: Leadbeater, B. & Riding, R., eds., *Biomineralization in lower plants and animals.* Systematics Association Special Volume **30**:141–156.

CAMPBELL S.J. & WOELKERLING W.J. 1990. Are *Titanoderma* and *Lithophyllum* (Corallinaceae, Rhodophyta) distinct genera ? *Phycologia* **29**:114–125.

CARDINAL A., CABIOCH J., & GENDRON L. 1979. Les Corallinacées (Rhodophytes-Cryptonemiales) des côtes du Québec. II. *Lithothamnium* Philippi emend Adey. *Cahiers de Biologie Marine* **20**:171–179.

CHALON J. 1905. *Liste des Algues Marines observées jusqu'a ce jour entre l'Embouchure de l'Escaut et la Corogne (incl. Iles Anglo-Normandes).* J. Buschmann, Anvers. 259pp.

CHAMBERLAIN Y.M. 1962. Notes on two species of *Hildenbrandia. Nova Hedwigia* **4**:371–373.

CHAMBERLAIN Y.M. 1977a. The occurrence of *Fosliella limitata* (Foslie) Ganesan (a new British record) and *F. lejolisii* (Rosanoff) Howe (Rhodophyta, Corallinaceae) on the Isle of Wight. *British Phycological Journal* **12**:67–81.

CHAMBERLAIN Y.M. 1977b. Observations on *Fosliella farinosa* (Lamour.) Howe (Rhodophyta, Corallinaceae) in the British Isles. *British Phycological Journal* **12**:343–358.

CHAMBERLAIN Y.M. 1978. *Dermatolithon litorale* (Suneson) Hamel and Lemoine (Rhodophyta, Corallinaceae) in the British Isles. *Phycologia* **17**:396–402.

CHAMBERLAIN Y.M. 1983. Studies in the Corallinaceae with special reference to *Fosliella* and *Pneophyllum* in the British Isles. *Bulletin of the British Museum (Natural History),* Botany Series **11**:291–463.

CHAMBERLAIN Y.M. 1984. Spore size and germination in *Fosliella, Pneophyllum* and *Melobesia* (Rhodophyta, Corallinaceae). *Phycologia* **23**:433–442.

CHAMBERLAIN Y.M. 1985a. The typification of *Melobesia membranacea* (Esper) Lamouroux (Rhodophyta, Corallinaceae). *Taxon* **34**:673–677.

CHAMBERLAIN Y.M. 1985b. Trichocyte occurrence and phenology in four species of *Pneophyllum* (Rhodophyta, Corallinaceae) from the British Isles. *British Phycological Journal* **20**:375–379.

CHAMBERLAIN Y.M. 1985c. In memoriam: Mme Paul Lemoine (née Marie Dujardin-Beaumetz) 29 December 1887–29 December 1984. *Phycologia* **24**:369–373.

CHAMBERLAIN Y.M. 1986. A reassessment of the type specimens of *Titanoderma verrucatum* and *T. macrocarpum* (Rhodophyta, Corallinaceae). *Cryptogamie: Algologie* **7**:193–213.

CHAMBERLAIN Y.M. 1987. Conceptacle production and life history in four species of *Pneophyllum* (Rhodophyta, Corallinaceae) from the British Isles. *British Phycological Journal* **22**:43–48.

CHAMBERLAIN Y.M. 1988. Observations on an autogenous fertile outgrowth of *Lithophyllum crouanii* (Corallinales, Rhodophyta) from southern England. *Phycologia* **27**:378–386.

CHAMBERLAIN Y.M. 1990. The genus *Leptophytum* (Rhodophyta, Corallinaceae) in the British Isles with descriptions of *Leptophytum bornetii, L. elatum* sp. nov. and *L. laeve. British Phycological Journal* **25**:179–199.

CHAMBERLAIN Y.M. 1991a. Observations on *Phymatolithon lamii* (Lemoine) Y.Chamberlain comb. nov. (Rhodophyta, Corallinales) in the British Isles with an assessment of its relationship to *P. rugulosum, Lithophyllum lamii* and *L. melobesioides. British Phycological Journal* **26**:219–233.

CHAMBERLAIN Y.M. 1991b. Historical and taxonomic studies in the genus *Titanoderma* (Rhodophyta, Corallinales) in the British Isles. *Bulletin of the British Museum (Natural History)*, Botany Series **21**:1–80.

CHAMBERLAIN Y.M. 1992. Observations on two melobesioid crustose coralline red algal species from the British Isles: *Exilicrusta parva*, a new genus and species, and *Lithothamnion sonderi* Hauck. *British Phycological Journal* **27**:185–201.

CHAMBERLAIN Y.M. 1993. Observations on the crustose coralline red alga *Spongites yendoi* (Foslie) comb. nov. in South Africa and its relationship to *S. decipiens* (Foslie) comb. nov. and *Lithophyllum natalense* Foslie. *Phycologia* **32**:100–115.

CHAMBERLAIN Y.M. & COOKE P.J. 1984. Crustose corallines growing in sea-urchin cups at Fanore, Co. Clare. *British Phycological Journal* **19**:190–191. Abstract.

CHAMBERLAIN Y.M., IRVINE L.M. & WALKER R. 1988. A redescription of *Lithophyllum crouanii* (Rhodophyta, Corallinales) in the British Isles with an assessment of its relationships to *L. orbiculatum*. *British Phycological Journal* **23**:177–192.

CHAMBERLAIN Y.M., IRVINE L.M. & WALKER R. 1991. A redescription of *Lithophyllum orbiculatum* (Rhodophyta, Corallinales) in the British Isles and a reassessment of generic delimitation in the Lithophylloideae. *British Phycological Journal* **26**:149–167.

CHAPMAN V.J. & PARKINSON P.G. 1974. *The Marine Algae of New Zealand. Part III: Rhodophyceae. Issue 3: Cryptonemiales* J.Cramer, Lehre. pp.155–278.

CHEMIN É. 1937. Le developpement des spores chez les Rhodophycées. *Revue Générale de Botanique* **49**:205–234, 300–327, 353–374, 424–428, 478–536.

CHIHARA M. 1972. Reproductive cycles and spore germination of the Corallinaceae and their possible relevance in the systematics. (1) *Amphiroa, Marginosporum*, and *Lithothrix*. *The Journal of Japanese Botany* **47**:15–25.

CHIHARA M. 1973a. The significance of reproductive and spore germination characteristics in the systematics of the Corallinaceae: articulated coralline algae. *The Japanese Journal of Botany* **20**:369–379.

CHIHARA M. 1973b. Reproductive cycles and spore germination of the Corallinaceae and their possible relevance in the systematics. (3). *Corallina, Jania*, and their related genera. *The Journal of Japanese Botany* **48**:13–19.

CHIHARA M. 1973c. Reproductive cycles and spore germination of the Corallinaceae and their possible relevance in the systematics. 4. *Lithophyllum, Lithothamnion* and their related genera. *The Journal of Japanese Botany* **48**:345–352.

CHIHARA M. 1974a. Reproductive cycles and spore germination of the Corallinaceae and their possible relevance in the systematics. (5). Five species of *Fosliella*. *The Journal of Japanese Botany* **49**:89–96.

CHIHARA M. 1974b. The significance of reproductive and spore germination characteristics to the systematics of the Corallinaceae: nonarticulated coralline algae. *Journal of Phycology* **10**:266–274.

CLOKIE J.J.P., BONEY A.D. & FARROW G. 1979. The significance of *Conchocelis* as an indicator organism: data from the Firth of Clyde and NW Shelf. *British Phycological Journal* **14**:120–121.

CLOKIE J.J.P., SCOFFIN T.P. & BONEY A.D. 1981. Depth maxima of *Conchocelis* and *Phymatolithon rugulosum* on the NW Shelf and Rockall platform. *Marine Progress Series* **4**:131–133.

COLTHART B.J. & JOHANSEN H.W. 1973. Growth rates of *Corallina officinalis* (Rhodophyta) at different temperatures. *Marine Biology* **18**:46–49.

CONOVER J.T. 1958. Seasonal growth of benthic marine plants as related to environmental factors in an estuary. *Institute of Marine Sciences* **5**:97–147.

COTTON A.D. 1911. *Lithophyllum* in the British Isles. *The Journal of Botany, British and Foreign* **49**:115–117.

COTTON A.D. 1912. Clare Island survey. Part 15. Marine algae. *Proceedings of the Royal Irish Academy* **31**:1–178.

COULL B.C. & WELLS J.B.J. 1983. Refuges from fish predation: experiments with phytal meiofauna from the New Zealand rocky intertidal. *Ecology* **64**:1599–1609.

CROUAN H.-M. & CROUAN P.-L. 1852. *Algues Marines du Finistère. Recueillies et publiées par Crouan Frères, pharmaciens à Brest.* Vol.2. (Issued as an exsiccata).

CROUAN P.-L. & CROUAN H.-M. 1858. Note sur quelques algues marines nouvelles de la rade de Brest. *Annales des Sciences Naturelles,* Botanique, sér. 4, **9**:69–75.

CROUAN H.-M. & CROUAN P.-L. 1859. Notice sur le genre *Hapalidium*. *Annales des Sciences Naturelles* Ser. 4., Bot. **12**:284–287.

CROUAN H.-M. & CROUAN P.-L. 1860. Liste des algues marines découvertes dans le Finistère depuis la publication des algues de ce département en 1852. *Bulletin de la Société Botanique de France* **7**:367–373.

CROUAN H.-M. & CROUAN P.-L. 1867. *Florule du Finistère*. F. Klincksieck, Paris. x+262pp.

DASHWOOD C.H. 1853. Notes on the marine botany of the coast of Norfolk. *Naturalist, Morris,* **3**:12–14; 75–78.

DAVIS A.N. & WILCE R.T. 1987. Floristics, phenology, and ecology of the sublittoral marine algae in an unstable cobble habitat (Plum Cove, Cape Ann, Massachusetts, USA). *Phycologia* **26**:23–34.

DAWSON E.Y. 1953. Marine red algae of Pacific Mexico. Part I. Bangiales to Corallinaceae subf. Corallinoideae. *Allan Hancock Pacific Expeditions* **17**(1):1–171.

DAWSON E.Y. 1956. Some marine algae of the southern Marshall Islands. *Pacific Science* **10**:25–66.

DAWSON E.Y. 1960. Marine red algae of Pacific Mexico. Part 3. Cryptonemiales, Corallinaceae, subf. Melobesioideae. *Pacific Naturalist* **2**:3–125.

DECAISNE J. 1842. Mémoire sur les Corallines ou Polypiers calcifères. *Annales des Sciences Naturelles (Botanique),* Ser.2 **18**:96–128. Note: Also appears as pp.85–117 in *Essais sur une Classification des Algues et des Polypiers Calcifères – Mémoire sur les Corallines.* Paul Renouard, Paris. 120pp.

DE CEW T.C. & WEST J.A. 1977. Culture studies on the marine red algae *Hildenbrandia occidentalis* and *H. prototypus* (Cryptonemiales, Hildenbrandiaceae). *Bulletin of the Japanese Society of Phycology* **25** (Supplement: Yamada Memorial Issue):31–41.

DENIZOT M. 1968. *Les Algues Floridées Encroûtantes (à l'exclusion des Corallinacées).* Privately published, Paris. 310pp.

DICKINSON C.I. 1963. *British Seaweeds.* The Kew Series, Eyre & Spottiswoode, Frome & London.

DIGBY P.S.B. 1977a. Growth and calcification in the coralline algae, *Clathromorphum circumscriptum* and *Corallina officinalis,* and the significance of pH in relation to precipitation. *Journal of the Marine Biological Association of the United Kingdom* **57**:1095–1109.

DIGBY P.S.B. 1977b. Photosynthesis and respiration in the coralline algae, *Clathromorphum circumscriptum* and *Corallina officinalis* and the metabolic basis of calcification. *Journal of the Marine Biological Association of the United Kingdom* **57**:1111–1124.

DIGBY P.S.B. 1979. Reducing activity and the formation of base in the coralline algae: an electrochemical model. *Journal of the Marine Biological Association of the United Kingdom* **59**:455–477.

DIXON P.S. 1973. *Biology of the Rhodophyta.* Oliver & Boyd, Edinburgh. xiii+285 pp.

DIXON P.S. & IRVINE L.M. 1976. Rhodophyta. pp.532–541 in Parke, M. & Dixon, P.S. eds, Checklist of British marine algae – third revision. *Journal of the Marine Biological Association of the United Kingdom* **56**:527–594.

DIXON P.S. & IRVINE L.M. 1977. *Seaweeds of the British Isles. Vol.1. Rhodophyta. Part 1. Introduction, Nemaliales, Gigartinales.* British Museum [Natural History], London. xi+252pp.

DIZERBO A.H. 1969. Les limites géographiques de quelques algues marines du Massif Armoricain. *Proceedings of the International Seaweed Symposium* **6**:141–149.

DOMMASNES A. 1968. Variations in the meiofauna of *Corallina officinalis* L. with wave exposure. *Sarsia* **34**:117–124.

DOMMASNES A. 1969. On the fauna of *Corallina officinalis* L. in western Norway. *Sarsia* **38**:71–86.

DUERDEN R.C. & JONES W.E. 1981. The host specificity of *Jania rubens* (L.) Lamour. in British waters. *Proceedings of the International Seaweed Symposium* **8**:313–319.

EDYVEAN R.G.J. & FORD H. 1984a. Population biology of the crustose red alga *Lithophyllum incrustans* Phil. 2. A comparison of populations from three areas of Britain. *Biological Journal of the Linnean Society* **23**:353–363.

EDYVEAN R.G.J. & FORD H. 1984b. Population biology of the crustose red alga *Lithophyllum incrustans* Phil. 3. The effects of local environmental variables. *Biological Journal of the Linnean Society* **23**:365–374.

EDYVEAN R.G.J. & FORD H. 1986a. Spore production by *Lithophyllum incrustans* (Corallinales, Rhodophyta) in the British Isles. *British Phycological Journal* **21**:255–261.

EDYVEAN R.G.J. & FORD H. 1986b. Population structure of *Lithophyllum incrustans* (Philippi) (Corallinales, Rhodophyta) from south-west Wales. *Field Studies* **6**:397–405.

EDYVEAN R.G.J., FORD H. & HARDY F.G. 1983. Population biology and reproduction of *Lithophyllum incrustans* (Rhodophyta: Lithophyllaceae) on north east coast of Britain. *British Phycological Journal* **18**:202.

EDYVEAN R.G.J. & MOSS B.L. 1984. Conceptacle development in *Lithophyllum incrustans* Philippi (Rhodophyta, Corallinaceae). *Botanica Marina* **27**:391–400.

ELLIOTT G.F. 1970. *Pseudaethesolithon*, a calcareous alga from the fars (Persian Miocene). *Geologica Romana* **9**:31–46.

ELLIS J. 1755. *An Essay towards a Natural History of the Corallines.* Privately published, London. xxviii+103pp.

ELLIS J. 1768. Extract of a letter from John Ellis Esquire, F.R.S. to Dr. Linnaeus of Upsala, F.R.S. on the animal nature of the genus of zoophytes, called *Corallina. Philosophical Transactions of the Royal Society of London* **57**(1):404–427.

ELLIS J. & SOLANDER D. 1786. *The Natural History of many Curious and Uncommon Zoophytes.* B.White and Son, London. xii+208pp.

ENDLICHER S.L. 1843. *Genera Plantarum Secundum Ordines Naturales Disposita. Suppl. 3. Mantissa Botanica Altera.* F.Beck, Wien. vi+111pp.

ERWIN, D. & PICTON, B. 1987. *Guide to inshore marine life.* Marine Conservation Society, Immel Publishing.

ESPER E.G.C. 1806. *Fortsetzungen der Pflanzenthiere.* Vol.2, Part 10 (pages 25–48). Raspe, Nurnberg.

EVANS R.G. 1957. The intertidal ecology of some localities on the Atlantic coast of France. *Journal of Ecology* **45**:245–271.

FARLOW W.G. 1881. The marine algae of New England. *Report of U.S. Fish Commission* **1879**: Appendix A–1, 1–210.

FARNHAM W.F. & BISHOP G.M. 1985. Survey of the Fal estuary, Cornwall. *Progress in Underwater Science* **10**:53–63.

FARROW G., SCOFFIN T., BROWN B. & CUCCI M. 1979. An underwater survey of facies variation on the inner Scottish shelf between Colonsay, Islay and Jura. *Scottish Journal of Geology* **15**:13–29.

FELDMANN J. 1937. Recherches sur la végétation marine de la Mediterranée. La côte des Albères. *Revue Algologique* **10**:1–339.

FELDMANN, J. ['1939'] 1942. Les algues marines de la côte des Albères. IV. Rhodophycées. *Revue algologique* 11:247–330.

FIGUEIREDO M., KAIN J.M. & NORTON T.A. 1992. Crustose coralline algae responses to epiphytic cover. *British Phycological Journal* 27:89.

FLETCHER R.L. 1983. Studies on *Hildenbrandia* Nardo from the south coast of England. *British Phycological Journal* 18:203–204.

FLETCHER R.L. 1987. *Seaweeds of the British Isles*. Volume 3, Part 1. British Museum (Natural History), London.

FORD H., HARDY F.G. & EDYVEAN R.G.J. 1983. Population biology of the crustose red alga *Lithophyllum incrustans* Phil. Three populations on the east coast of Britain. *Biological Journal of the Linnean Society of London* 19:211–220.

FOSLIE M. 1887. Nye havsalger. *Tromsø Museums Aarshefter* 10:175–195.

FOSLIE M. 1891. Contribution to knowledge of the marine algae of Norway. II. Species from different tracts. *Tromsø Museums Aarshefter* 14:36–58. [Also issued as an independently paginated reprint, title page, pp.1–23.]

FOSLIE M. 1895. The Norwegian forms of *Lithothamnion*. *Det Kongelige Norske Videnskabers Selskabs Skrifter* 1894:29–208. [Also issued as an independently paginated reprint (title page, pp.1–180.]

FOSLIE M. 1896. New or critical lithothamnia. *Det Kongelige Norske Videnskabers Selskabs Skrifter* 1895(2):1–10.

FOSLIE M. 1897. On some Lithothamnia. *Det Kongelige Norske Videnskabers Selskabs Skrifter* 1897(1):1–20.

FOSLIE M. 1898a. Systematic survey of the lithothamnia. *Det Kongelige Norske Videnskabers Selskabs Skrifter* 1898(2):1–7.

FOSLIE M. 1898b. List of species of the lithothamnia. *Det Kongelige Norske Videnskabers Selskabs Skrifter* 1898(3):1–11.

FOSLIE M. 1898c. Some new or critical lithothamnia. *Det Kongelige Norske Videnskabers Selskabs Skrifter* 1898(6):1–19.

FOSLIE M. 1900a. New or critical calcareous algae. *Det Kongelige Norske Videnskabers Selskabs Skrifter* 1899(5):1–34

FOSLIE M. 1900b. Remarks on Melobesieae in Herbarium Crouan. *Det Kongelige Norske Videnskabers Selskabs Skrifter* 1899(7):1–16.

FOSLIE M. 1900c. Revised systematical survey of the Melobesieae. *Det Kongelige Norske Videnskabers Selskabs Skrifter* 1900(5):1–22.

FOSLIE M. 1901. New melobesieae. *Det Kongelige Norske Videnskabers Selskabs Skrifter* 1900(6):1–24.

FOSLIE M. 1904a. Algologiske notiser. *Det Kongelige Norske Videnskabers Selskabs Skrifter* 1904(2):1–9.

FOSLIE M. 1904b. Die Lithothamnien des Adriatischen Meeres und Marokkos. *Wissenschaftliche Meeresuntersuchungen Abteiling Helgoland* Neue Folge 7(1):1–40. (issued as a preprint without change in pagination in 1904; journal version was published in 1905.)

FOSLIE M. 1905a. Remarks on northern lithothamnia. *Det Kongelige Norske Videnskabers Selskabs Skrifter* 1905(3):1–138.

FOSLIE M. 1905b. New lithothamnia and systematical remarks. *Det Kongelige Norske Videnskabers Selskabs Skrifter* 1905(5):1–9.

FOSLIE M. 1906. Algologiske notiser II. *Det Kongelige Norske Videnskabers Selskabs Skrifter* 1906(2):1–28.

FOSLIE M. 1908a. Die lithothamnien der deutschen südpolar- expedition 1901–1903. *Deutsche Südpolar-Expedition 1901–1903* 8:205–219.

FOSLIE M. 1908b. Algologiske Notiser V. *Det Kongelige Norske Videnskabers Selskabs Skrifter* 1908(7):1–20.

FOSLIE M. 1908c. *Pliostroma*, a new subgenus of *Melobesia*. *Det Kongelige Norske Videnskabers Selskabs Skrifter* **1908**(11):1–7.

FOSLIE M. 1909. Algologiske Notiser, VI. *Det Kongelige Norske Videnskabers Selskabs Skrifter* **1909**(2):1–63.

FRIES E.M. 1836. *Corpus florarum provincialium sueciae. I. Floram scanicam.* Upsaliae.

FUJITA D. 1988. Seasonal changes of photosynthetic and respiratory rates of *Lithophyllum yessoense* Foslie (Corallinaceae). *Suisanzoshoku* **36**:7–10.

FURNARI F. & SCAMMACCA B. 1971. Prime osservazioni sulla flora algale di Capo Passero e isolette vicine. *Bollettino delle Sedute Dell'Accademia Gioenia di Scienze Naturali in Catania,.* ser.4 **10**(8):679–688.

GANESAN E.K. 1963. Notes on Indian red algae III. *Fosliella minutula* (Foslie) comb. nov. *Phykos* **2**:38–44. (Reprint dated October 1963, journal issued in April 1964).

GANESAN E.K. 1971. Studies on the morphology, anatomy, reproduction, and taxonomy of some Caribbean crustose Corallinaceae. In: *Symposium on Investigations and Resources of the Caribbean Sea and Adjacent Regions* (Ed. not stated) pp.411–416. Unesco, Paris.

GARBARY D.J. 1978. An introduction to the scanning electron microscopy of red algae. In: *Modern approaches to the Taxonomy of Red and Brown Algae* (ed. by D.E.G. Irvine & J.H. Price) pp.205–222. Academic Press, London.

GARBARY D.J. & JOHANSEN H.W. 1982. Scanning electron microscopy of *Corallina* and *Haliptilon* (Corallinaceae, Rhodophyta): surface features and their taxonomic implications. *Journal of Phycology* **18**:211–219.

GARBARY D. & SCAGEL R.F. 1979. Scanning electron microscopy of *Clathromorphum* spp. (Corallinaceae, Rhodophyta) *Proceedings of the Microscopical Society of Canada* **6**:14–15.

GARBARY D. & VELTKAMP C.J. 1980. Observations on *Mesophyllum lichenoides* (Corallinaceae, Rhodophyta) with the scanning electron microscope. *Phycologia* **19**:49–53.

GAYRAL P. 1958. *Algues de la Côte Atlantique Marocaine.* Société des Sciences naturelles et Physiques du Maroc, Rabat. 523pp.

GAYRAL P. 1966. *Les Algues des Côtes Françaises.* Deren & Cie, Paris. 632pp.

GIACONNE G. 1969. Raccolte di fitobenthos sulla banchina continentale italiana. *Giornale Botanico Italiano* **103**:485–514.

GIRAUD G. & CABIOCH J. 1976. Étude ultrastructurale de l'activité des cellules superficielles du thalle des Corallinacées (Rhodophycées). *Phycologia* **15**:405–414.

GIRAUD G. & CABIOCH J. 1977. Caractères généraux de l'ultrastructure des Corallinacées (Rhodophycées). *Revue Algologique* N.S. **12**:45–60.

GIRAUD G. & CABIOCH J. 1979. Ultrastructure and elaboration of calcified cell-walls in the coralline algae (Rhodophyta, Cryptonemiales). *Biologie Cellulaire* **36**:81–86.

GIRAUD G. & CABIOCH J. 1983. Inclusions cytoplasmiques remarquables chez les Corallinacées (Rhodophytes, Cryptonemiales). *Annales des Sciences Naturelles, Botanique, Paris,* ser.13, **5**:29–43.

GOSS-CUSTARD S., JONES J., KITCHING J.A. & NORTON T.A. 1979. Tide pools of Carrigathorna and Barloge Creek. *Philosophical Transactions of the Royal Society of London, B. Biological Sciences* **287**:1–44.

GREUTER W., chm. 1988. *International Code of Botanical Nomenclature adopted by the Fourteenth International Botanical Congress, Berlin, July – August 1987.* Koeltz Scientific Books, Königstein, Germany. xiv+328pp. (*Regnum Vegetabile* Vol.**118**).

GUIRY M.D. 1978a. A Concensus and Bibliography of Irish Seaweeds. *Bibliotheca Botanica* **44**:1–287.

GUIRY M.D. 1978b. The importance of sporangia in the classification of the Florideophyceae. In: *Modern Approaches to the Taxonomy of Red and Brown Algae* (Ed. by D.E.G.Irvine & J.H.Price) pp.111–144. Academic Press, London.

GUIRY M.D. & IRVINE L.M. 1989. Sporangial form and function in the Nemaliophycidae, (Rhodophyta). In Kumar,H.D. ed., *Phycotalk* **1**:155–184. Meerut.

GUNNERUS J.E. 1768. Om nogle Norske Coraller. *Det Kongelige Norske Videnskabers Selskabs Skrifter* **4**:38–73.

HAAS P. & HILL T.G. 1933. The metabolism of calcareous algae. I. *The Biochemical Journal* **27**:1801–1804.

HAAS P., HILL T.G. & KARSTENS W.K.H. 1935. The metabolism of calcareous algae. II. The seasonal variation in certain metabolic products of *Corallina squamata* Ellis. *Annals of Botany* **49**:609–619.

HAGEN N.T. 1983. Destructive grazing of kelp beds by sea urchins in Vestfjorden, Northern Norway. *Sarsia* **68**:177–190.

HAMEL G. & LAMI R. 1930. Liste préliminaire des algues recoltées dans la région de Saint-Servan. *Bulletin du Laboratoire Maritime du Muséum d'Histoire Naturelle à Saint-Servan* **6**:1–34.

HAMEL G. & LEMOINE M. 1953. Corallinacées de France et d'Afrique du Nord. *Archives du Museum National D'Histoire Naturelle*, Ser.7, **1**:15–136.

HARDY F.G. [`1992'] 1993. The marine algae of Berwickshire: a detailed checklist. *Botanical Journal of Scotland* **46**:199–232.

HARLIN M.M. & LINDBERGH J.M. 1977. Selection of substrata by seaweeds: optimal surface relief. *Marine Biology* **40**:33–40.

HARLIN M.M., WOELKERLING W.J. & WALKER D.I. 1985. Effects of a hypersalinity gradient on epiphytic Corallinaceae (Rhodophyta) in Shark Bay, Western Australia. *Phycologia* **24**:389–402.

HARVEY W.H. 1844. Description of a minute alga from the coast of Ireland. *Annals and Magazine of Natural History* **14**:27–28.

HARVEY W.H. 1847. *Phycologia Britannica*, plates 73–78. Reeve & Benham, London.

HARVEY W.H. 1848. *Phycologia Britannica*, plates 147–216. Reeve & Benham, London.

HARVEY W.H. 1849a. *A Manual of the British Marine Algae*. ed.2.John van Voorst, London.

HARVEY W.H. 1849b. *Phycologia Britannica*, plates 217–294. Reeve & Benham, London.

HARVEY W.H. 1849c. *Nereis Australis*. Part II. Pages 65–124 (plates 26–50). Reeve, London. [For data on publication date, see *Taxon* **17**: 82, 725 (1968).]

HARVEY W.H. 1850. *Phycologia Britannica*, plates 259–354. Reeve & Benham, London. For further dating information on Harvey's *Phycologia Britannica*, see Price, J.H. 1984.

HAUCK F. 1883. *Die Meeresalgen Deutschlands und Österreichs*. Part 5 (pages 225–272); Part 6 (pages 273–320). E.Kummer, Leipzig.

HAWKES M.W. 1983. Anatomy of *Apophlaea sinclairii* – an enigmatic red alga endemic to New Zealand. *Japanese Journal of Phycology* **31**:55–64.

HEYDRICH F. 1897a. Corallinaceae, inbesondere Melobesieae. *Berichte der Deutschen Botanischen Gesellschaft* **15**:34–71.

HEYDRICH F. 1897b. Melobesieae. *Berichte der Deutschen Botanischen Gesellschaft* **15**:403–420.

HEYDRICH F. 1899. Einige neue Melobesien des Mittelmeeres. *Berichte der Deutschen Botanischen Gesellschaft* **17**:221–227.

HEYDRICH, F. 1900. Die Lithothamnien von Helgoland. *Wissenschaftliche Meeresuntersuchungen* N.F.(Abt. Helgoland) **4**:63–82.

HEYDRICH F. 1901. Die Lithothamnien des Museum d'Histoire Naturelle in Paris. *Botanische Jahrbücher für Systematik, Pflanzengeschichte und Pflanzengeographie* **28**:529–545.

HEYDRICH F. 1904. Stereophyllum, ein neues Genus der Corallinaceen. *Berichte der Deutschen Botanischen Gesellschaft* **22**:196–199.

HEYDRICH F. 1905. Note sur l'Epilithon van heurckii *Heyd*. [offprint from Chalon, 1905 q.v.].

HEYDRICH F. 1911a. *Lithophyllum incrustans* Phil. mit einem Nachtrag über *Paraspora fruticulosa* (Ktz.) Heydrich. *Bibliotheca Botanica* **18**(75):1–24.

HEYDRICH F. 1911b. Die Lithothamnien von Roscoff. *Berichte der Deutschen Botanischen Gesellschaft* **29**:26–32.

HISCOCK, S. 1986. A field guide to the British red seaweeds (Rhodophyta). *Field Studies Council Occasional Publications*, **13**.

HOEK C. VAN DEN 1982a. The distribution of benthic marine algae in relation to the temperature regulation of their life histories. *Biological Journal of the Linnean Society* **18**:81–144.

HOEK C. VAN DEN 1982b. Phytogeographic distribution groups of benthic marine algae in the north Atlantic Ocean. *Helgoländer Meeresuntersuchungen* **35**:153–214.

HOEK C. VAN DEN 1984. World-wide latitudinal and longitudinal seaweed distribution patterns and their possible causes as illustrated by the distribution of Rhodophytan genera. *Helgoländer Meeresuntersuchungen* **38**:227–257.

HOLLENBERG G.J. 1971. Phycological notes. VI New records, new combinations, and noteworthy observations concerning marine algae of California. *Phycologia* **10**:281–289.

HOLMES E.M. 1883–1910. *Algae Britannicae rariores exsiccatae*, Fasc.I-XII, nos.1–312. London.

HOLMGREN P., HOLMGREN N. H. & BARTLETT L. C. 1990. *Index Herbariorum, Pt.1. The Herbaria of the World*. Ed.8. Koeltz Scientific Books, Königstein. x+693pp. [*Regnum Vegetabile* Vol.120].

HOOPER R.G. & WHITTICK A. 1984. The benthic marine algae of the Kaipokok Bay and Big River Bay region of the central Labrador coast. *Le Naturaliste Canadien* **111**:131–138.

HOWE M.A. 1920. Class 2. Algae. In: *The Bahama Flora* (by N.L. Britton & C.F. Millspaugh) pp.553–631. Privately published, New York.

HUVÉ H. 1956. Contribution à l'étude des fonds à *Lithothamnium* (?) *solutum* Foslie (= *Lithophyllum solutum* (Foslie) Lemoine) de la région de Marseille. *Recueil des Travaux de la Station Marine d'Endoûme* **18**:105–133.

HUVÉ H. 1962. Taxonomie, écologie et distribution d'une Mélobésiée Méditerranéenne: *Lithophyllum papillosum* (Zanardini) comb.nov., non *Lithophyllum (Dermatolithon) papillosum* (Zanard.) Foslie *Botanica Marina* **4**:219–240.

INTERNATIONAL Code of Botanical Nomenclature – see Greuter, 1988.

IRVINE, L.M. 1983. *Seaweeds of the British Isles*. Volume 1, Part 2A. British Museum (Natural History) London.

IRVINE L.M. & WOELKERLING W.J. 1986. Proposal to conserve *Phymatolithon* against *Apora* (Rhodophyta: Corallinaceae). *Taxon* **35**:731–733.

ISHIJIMA W. 1942. The first find of *Mesophyllum* from the Tertiary of Japan. *Journal of the Geological Society of Japan* **49**:174–176. (Concurrently published with *Transactions and Proceedings of the Palaeontological Society of Japan* **149**:153–155; both sets of page numbers appear in the paper).

ISRAELSON G. 1942. The freshwater Florideae of Sweden. *Symbolae Botanicae Upsalienses* **6**:1–81.

JAASUND E. 1965. Aspects of the marine algal vegetation of North Norway. *Botanica Gothoburgensis* **4**:1–174.

JAMES C.H. 1923. Marine algae. In Pearsall, W.H. & Mason, F.A., eds, Yorkshire Naturalists at Bridlington, pp.209–210. *Naturalist, Hull* **797**:205–212.

JOHANSEN H.W. 1969. Morphology and systematics of coralline algae with special reference to *Calliarthron. University of California Publications in Botany* **49**:1–78.

JOHANSEN H.W. 1970. The diagnostic value of reproductive organs in some genera of articulated coralline red algae. *British Phycological Journal* **5**:79–86.

JOHANSEN H.W. 1976. Current status of generic concepts in coralline algae (Rhodophyta). *Phycologia* **15**:221–244.

JOHANSEN H.W. 1981. *Coralline Algae, A First Synthesis*. CRC Press, Boca Raton, Florida. [vii]+239pp.

JOHANSEN H.W. & COLTHART B. 1975. Variability in articulated coralline algae (Rhodophyta). *Nova Hedwigia* **26**:135–149.

JOHANSEN H.W., IRVINE L.M. & WEBSTER A.M. 1973. *Haliptylon squamatum* (L.) comb. nov., a poorly known British coralline alga. *British Phycological Journal* **8**:212.

JOHANSEN H.W. & SILVA P.C. 1978. Janieae and Lithotricheae: two new tribes of articulated Corallinaceae (Rhodophyta). *Phycologia* **17**:413–417.

JOHN D.M., HAWKINS S.J. & PRICE J.H. 1992. *Plant-animal interactions in the marine benthos.* Systematics Association Special Volume **46**. Oxford.

JOHNSON C.R. & MANN H. 1986. The crustose coralline alga, *Phymatolithon* Foslie, inhibits the overgrowth of seaweeds without relying on herbivores. *Journal of Experimental Marine Biology and Ecology* **96**:127–146.

JOHNSON T. & HENSMAN R. 1899. A list of Irish Corallinaceae. *The Scientific Proceedings of the Royal Dublin Society*, N.S.**9**:22–30.

JOHNSTON G. 1842. *A History of British Sponges and Lithophytes.* W.H. Lizars, Edinburgh. xii+264pp.

JOHNSTONE W.G. & CROALL A. 1859. *The Nature Printed British Sea-Weeds. Vol.1. Rhodospermae. Fam I-IX.* Bradbury, Evans & Co., London. xv+188pp.

JONES P.L. & WOELKERLING W.J. 1983. Some effects of light and temperature on growth and conceptacle production in *Fosliella cruciata* Bressan (Corallinaceae, Rhodophyta). *Phycologia* **22**:449–452.

JONES P.L. & WOELKERLING W.J. 1984. An analysis of trichocyte and spore germination attributes as taxonomic characters in the *Pneophyllum-Fosliella* complex (Corallinacae, Rhodophyta). *Phycologia* **23**:183–194.

JONES W.E. & MOORJANI S.A. 1973. The attachment and early development of the tetraspores of some coralline red algae. *Special Publications of the Marine Biological Association of India* **1973**:293–304.

KASPERK C., EWERS R., SIMONS B. & KASPERK R. 1988. Knochenersatz material aus Algen. *Deutsch Zahnarztl.* **243**:116–119.

KEATS D.W. & CHAMBERLAIN Y.M. 1994. Three species of *Hydrolithon* (Rhodophyta, Corallinaceae) from South Africa: *Hydrolithon onkodes* (Foslie) Penrose & Woelkerling, *H. superficiale* sp. nov. and *H. samoënse* (Foslie) comb. nov. *Journal of South African Botany* **60**:8–21.

KEATS D.W., STEELE D.H. & SOUTH G.R. 1983. Food relations and short term aquaculture potential of the green sea urchin (*Strongylocentrotus droebachiensis*) in Newfoundland. *M.S.R.L. Technical Reports* **24**:1–24.

KJELLMAN F.R. 1883. Norra Ishafvets Algflora. *Vega-expeditionens Vetenskapliga Iaktagelser* **3**:1–431.

KLEEN E. 1875. Om Nordlandens hörge hafsalger. *Öfversigt af Kongliga Vetenskaps-Akademiens Förhandlingar* **1874**(9):3–46.

KLOCZCOVA N.G. 1987. A new genus and species of coralline algae (Corallinales, Rhodophyta) from the seas of the Far East of USSR. *Botanical Journal [USSR]*. **72**:100–105.

KLOCZCOVA N.G. & DEMESHKINA Z.V. 1987. Algae Corallinaceae (Rhodophyta) marium orientis extremi URSS. *Pneophyllum* Kütz. *Novitates Systematicae Plantarum non Vascularum* **24**:34–39.

KNIGHT M. & PARKE M.W. 1931. Manx Algae. An algal survey of the south end of the Isle of Man. *Proceedings and Transactions of the Liverpool Biological Society* **45** (Appendix II):vii+1–155.

KOOISTRA W.H.C.F., JOOSTEN A.M.T. & HOEK C.VAN DEN 1989. Zonation patterns in intertidal pools and their possible causes: a multivariate approach. *Botanica Marina* **32**:9–26.

KORNERUP A. & WANSCHER J.H. 1978. *Methuen Handbook of Colour.* 3rd Edition, English translation. Methuen, London.

KORNMANN P. & SAHLING P.H. 1977. Meeresalgen von Helgoland. Benthische Grün-, Braun-und Rotalgen. *Helgoländer wissenschaftliche Meeresuntersuchungen* **29**:1–289.

KÜTZING F.T. 1841. *Uber die "Polypiers Calcifères" de Lamouroux*. F.Thiele, Nordhausen. 34pp.

KÜTZING F.T. 1843. *Phycologia Generalis*. F.A.Brockhaus, Leipzig. xxxii+458pp.

KÜTZING F.T. 1847. Diagnosen und Bemerkungen zu Neuen oder Kritischen algen. *Botanische Zeitung* **5**:1–5,22–25,33–38,52–55,164–167,177–180,193–198,219–223.

KÜTZING F.T. 1849. *Species Algarum*. F.A.Brockhaus, Lipsiae. vi+922pp.

KÜTZING F.T. 1858. *Tabulae Phycologicae*. Vol.**8**. Privately published, Nordhausen. ii+48pp.

KÜTZING F.T. 1869. *Tabulae Phycologicae*. Vol.**19**. Privately published, Nordhausen. iv+36pp.

KYLIN H. 1928. Entwicklungsgeschichtliche Florideenstudien. *Lunds Universitets Årsskrift,* N.F.Avd.2, **24**(4):1–127.

KYLIN H. 1944. Die Rhodophyceen der Schwedischen Westküste. *Lunds Universitets Årsskrift,* N.F.Avd.2, **40**(2):1–104.

KYLIN H. 1956. *Die Gattungen der Rhodophyceen*. CWK Gleerups, Lund. xv+673pp.

LAMARCK J.B. 1815. Sur les polypiers corticifères. *Memoires du Muséum Nationale d'Histoire Naturelle de Paris* **2**:227–240.

LAMARCK J.B. 1816. *Histoire Naturelle des Animaux sans Vertèbres*. Vol.**2**. Verdière, Paris. 568pp.

LAMI R. 1931. Complément à la liste préliminaire des algues récoltées dans la région de Saint-Servan. *Bulletin du Laboratoire Maritime du Muséum d'Histoire Naturelle de Saint-Servan* **6**:25–28.

LAMI R. 1932. Micro-atolls et micro-récifs frangeants de *Lithophyllum incrustans*. *Revue Algologique* **6**:227–230.

LAMI R. 1940. Sur les epiphytes hivernaux des stipes de Laminaires et sur deux *Rhodochorton* qui s'y observent dans la région de Dinard. *Bulletin du Laboratoire maritime de Dinard* **22**:47–60.

LAMOUROUX J.V.F. 1812. Extrait d'un mémoire sur la classification des polypiers coralligènes non entièrement pierreux. *Nouveau Bulletin des Sciences, par la Société Philomatique de Paris* **3**:181–188.

LAMOUROUX J.V.F. 1816. *Histoire des Polypiers Coralligènes Flexibles, Vulgairement Nommés Zoophytes*. F.Poisson, Caen. lxxxiv+chart+559pp.

LARSON B.R., VADAS R.L. & KESER M. 1980. Feeding and nutritional ecology of the sea urchin *Strongylocentrotus drobachiensis* in Maine, USA. *Marine Biology (Berlin)* **59**:49–62.

LEBEDNIK P.A. 1977. Postfertilization development in *Clathromorphum, Melobesia,* and *Mesophyllum* with comments on the evolution of the Corallinaceae and the Cryptonemiales (Rhodophyta). *Phycologia* **16**:379–406. Addendum: *Phycologia* **17**:358 (1978).

LEBEDNIK P.A. 1978. Development of male conceptacles in *Mesophyllum* Lemoine and other genera of the Corallinaceae (Rhodophyta). *Phycologia* **17**:388–395.

LE JOLIS A.F. 1863. *Liste des Algues Marines de Cherbourg*. J.Baillière, Paris. 168pp.

LEMOINE M. (Mme P.) 1909. Répartition du *Lithothamnium calcareum* (maërl) et de ses variétés dans la région de Concarneau. *Bulletin du Museum National d'Histoire Naturelle*, Paris **15**:552–555.

LEMOINE M. (Mme P.) 1910a. Essai de classification des mélobésiées basée sur la structure anatomique. *Bulletin de la Société Botanique de France* **57**:323–331, 367–372.

LEMOINE M. (Mme P.) 1910b. Répartition et mode de vie du Maërl aux environs de Concarneau. *Annales de l'Institut Océanographique* **1**(3):1–28.

LEMOINE M. (Mme P.) 1911. Structure anatomique des mélobésiées. Application à la classification. *Annales de L'Institut Océanographique*, Monaco **2**(2):1–213. [Also issued in thesis form with identical pagination but a different title page. A summary appears in *Bulletin de la Société Botanique de France* **58**:394–397. (P. Hariot, 1911).]

LEMOINE M. (Mme P.) 1913a. Mélobésiées de l'ouest de l'Irlande (Clew Bay). *Nouvelles Archives du Muséum d'Historie Naturelle*, Paris Ser.5, **5**:121–145.

LEMOINE M. (Mme P.) 1913b. Mélobésiées. Revision des mélobésiées Antarctiques. In: *Deuxième Expédition Antarctique Française (1908–1910) Commandée par le Dr Jean Charcot* (Editor not stated). Sciences Naturelles, Vol.1. Botanique. 67pp. Masson et Cie, Paris.

LEMOINE M. (Mme P.) 1913c. Quelques expériences sur la croissance des algues marines à Roscoff. *Bulletin du Musée Océanographique de Monaco* **277**:1–19.

LEMOINE M. (Mme P.) 1915. Calcareous algae. *Report on the Danish Oceanographical Expeditions 1908–1910 to the Mediterranean and Adjacent Seas* **2**:1–30.

LEMOINE M. (Mme P.) 1924. Corallinacées du Maroc. *Bulletin de la Société des Sciences Naturelles du Maroc* **4**:113–134.

LEMOINE M. (Mme P.) 1928. Un nouveau genre de Mélobésiées: *Mesophyllum. Bulletin de la Société Botanique de France* **75**:251–254.

LEMOINE M. (Mme P.) 1929a. Sur la présence de *Lithophyllum orbiculatum* Fosl. dans la Manche et son attribution au genre *Pseudolithophyllum. Revue Algologique* **4**:1–6.

LEMOINE M. (Mme P.) 1929b. Family 6 Corallinaceae. Subfamily 1. Melobesieae. In: 'Marine algae from the Canary Islands'. III. Rhodophyceae. Part II. Cryptonemiales, Gigartinales and Rhodymeniales (by F. Børgesen). *Det Kongelige Danske Videnskabernes Selskab, Biologiske Meddelelser* **8** (1):1–97.

LEMOINE M. (Mme P.) 1931. Les algues mélobésiées de la région de Saint-Servan. *Bulletin du Laboratoire maritime du Muséum d'Histoire Naturelle de Saint-Servan* **7**:1–20.

LEMOINE M. (Mme P.) 1940. Les algues calcaires de la zone néritique. *Mémoires de la Société de Biogéographie* **7**:75–138.

LEMOINE M. (Mme P.) 1953. Report on the calcareous Algae. *Proceedings of the Royal Irish Academy* **56**(C):166–168.

LEMOINE M. 1965. *Pseudolithophyllum expansum* (Philippi) Lemoine, existe-t-il dans la Manche? *Revue Algologique*, N.S. **8**:46–49.

LEMOINE M. 1978. Typification du genre *Pseudolithophyllum* Lemoine. *Revue Algologique*, N.S. **13**:177.

LEVRING T. 1937. Zur Kenntnis der Algenflora der Norwegischen Westküste. *Lunds Universitets Årsskrift*, N.F., Avd.2. **33**(8):1–147.

LEWIN J.C. 1962. Calcification. In Lewin,R., ed., *Physiology and biochemistry of algae*, pp.457–465. New York & London.

LEWIS E.A. 1936. An investigation of the seaweeds within a marked zone of the shore at Aberystwyth, during the year 1933–34. *Journal of the Marine Biological Association of the United Kingdom.* **20**:615–620.

LIEBMAN F. 1839. Om et nyt genus *Erythroclathrus* af algernas familie. *Naturhistorik Tidskrift* 2. Kjöbenhaven.

LINDLEY J. 1846. *The Vegetable Kingdom.* Bradbury & Evans, London. lxvii+911pp.

LINNAEUS C. 1758. *Systema Naturae.* 10 ed. Vol.1. L. Salvii, Stockholm. ii+824pp.

LINNAEUS C. 1767. *Systema naturae.* 12 ed. Vol.1, Part 2. Salvii, Holmiae. pp.533–1327[36].

LITTLER M.M. 1976. Calcification and its role among the macroalgae. *Micronesica* **12**:27–41.

LITTLER M.M. & KAUKER B.J. 1984. Heterotrichy and survival strategies in the red alga *Corallina officinalis* L. *Botanica Marina* **27**:37–44.

LITTLER M.M. & LITTLER D.S. 1984. Relationships between macroalgal functional form groups and substrata stability in a subtropical rocky-intertidal system. *Journal of Experimental Marine Biology and Ecology* **74**:13–34.

LYLE L. 1937. Additions to the marine flora of Sark. *Journal of Botany, London* **75**:18–22.

MAGGS C.A. 1986. Scottish marine macroalgae: a distributional checklist, biogeographical analysis and literature abstract. *Nature Conservancy Council Report* 635.

MAGGS C.A., FREAMHAINN M.T. & GUIRY M.D. 1983. A study of the marine algae of subtidal cliffs in Lough Hyne (Ine), Co. Cork. *Proceedings of the Royal Irish Academy, B,* **83**:251–266.

MAGGS C.A. & HOMMERSAND M. 1993. *Seaweeds of the British Isles.* vol.1, Rhodophyta Part 3A Ceramiales. HMSO & Natural History Museum, London.

MAGNE F. 1964. Recherches caryologiques chez les Floridées (Rhodophycées). *Cahiers de Biologie Marine* **5**:461–671.

MANZA A.V. 1937. The genera of articulated corallines. *Proceedings of the national Academy of Sciences, Washington* **23**(2):44–48.

MANZA A.V. 1940. A revision of the genera of articulated corallines. *The Philippine Journal of Science* **71**(3):239–316

MASAKI T. 1968. Studies on the Melobesioideae of Japan. *Memoirs of the Faculty of Fisheries, Hokkaido University* **16**:1–80.

MASAKI T. & TOKIDA J. 1960a. Studies on the Melobesioideae of Japan. II. *Bulletin of the Faculty of Fisheries, Hokkaido University* **10**:285–290.

MASAKI T. & TOKIDA J. 1960b. Studies on the Melobesioideae of Japan. III. *Bulletin of the Faculty of Fisheries, Hokkaido University* **11**:37–42.

MASAKI T. & TOKIDA J. 1963. Studies on the Melobesioideae of Japan. VI. *Bulletin of the Faculty of Fisheries Hokkaido University* **14**:1–6.

MASON L.R. 1953. The crustaceous coralline algae of the Pacific Coast of the United States, Canada and Alaska. *University of California Publications in Botany* **26**:313–390.

MATTY P.J. & JOHANSEN H.W. 1981. A histochemical study of *Corallina officinalis* (Rhodophyta, Corallinaceae). *Phycologia* **20**:46–55.

MENDOZA M. 1976. Estudio de las variaciones morfologicas externas, internas and citologicas de las Corallineae (Rhodophyta) de la Argentina. *Physis* (Sect. A) **35**:15–25.

MENDOZA M. & CABIOCH J. 1984. Redefinition comparée de deux espèces de Corallinacées d'Argentine: *Pseudolithophyllum fuegianum* (Heydrich) comb.nov. et *Hydrolithon discoideum* (Foslie) comb.nov. *Cryptogamie: Algologie* **4**:141–154.

MENDOZA M.L. & CABIOCH J. 1986. Le genre *Hydrolithon* (Rhodophyta, Corallinaceae) sur les côtes subantarctiques et antarctiques d'Argentine et de quelques régions voisines. *Cahiers Biologiques Marines* **27**:163–191.

MENEGHINI G. 1841. Algologia dalmatica. *Atti del terza Riunione degli Scienziati Italiani tenuta in Firenze* **3**:424–431.

MESLIN R.E. 1976. Sur la position et la valeur taxonomique de *Corallina elegans* Lenormand (Rhodophycées, Cryptonemiales). *Phycologia* **15**:415–419.

MILLSON C. & MOSS B.L. 1985. Ultrastructure of the vegetative thallus of *Phymatolithon lenormandii* (Aresch. in J.Ag.) Adey. *Botanica Marina* **28**:123–132.

MINDER F. 1910. *Die Fruchtentwicklung von* Choreonema thureti. Oberh. Ztg. Genossenschaft, Bad-Nauheim. 32pp.

MITTEN W. 1859. Musci Indiae Orientalis: an enumeration of the mosses of the East Indies. *The Journal of the Linnean Society, Botany* **3**(suppl.):1–117.

MOBERLEY R. 1968. Composition of magnesian calcites of algae and Pelecypods by electron microprobe analysis. *Sedimentology* **11**:61–82.

MORSE A.N.C. & MORSE D.E. 1984. Recruitment and metamorphosis of *Haliotis* larvae induced by molecules uniquely available at the surface of crustose red algae. *Journal of Experimental Biology and Ecology* **75**:191–215.

MORTON O. & CHAMBERLAIN Y.M. 1985. Records of some epiphytic coralline algae in the north-east of Ireland. *Irish Naturalists Journal* **21**:436–440.

MORTON O. & CHAMBERLAIN Y.M. 1989. Further records of encrusting coralline algae on the north-east coast of Ireland. *Irish Naturalists Journal* **23**:102–106.

MOSS B. & WALKER R. 1984. Mode of attachment of six epilithic crustose Corallinaceae (Rhodophyta). *Phycologia* **23**:321–329.

MUNDA I. 1977. A comparison of the north and south European associations of *Corallina officinalis* L. *Hydrobiologia* **52**:73–87.

MURATA K. & MASAKI T. 1978. Studies of reproductive organs in articulated coralline algae of Japan. *Phycologia* **17**:403–412.

NÄGELI C. 1858. Die Stärkekörner. In Nägeli C. & Kramer, C. *Pflanzenphysiologische Untersuchungen*. Friedrich Schulthess, Zürich. x+624pp.

NARDO G.D. 1834a. De corallinis ac nulliporis auct. *Isis van Oken* **1834**(1):673–675.

NARDO G.D. 1834b. De novo genere algarum cui nomen est *Hildbrandtia prototypus*. *Isis van Oken* **1834**(1):675.

NELSON R.J. & DUNCAN, Prof. 1876. On some points in the histology of certain species of Corallinaceae. *The Transactions of the Linnean Society of London*, Ser.2. Bot. **1**:197–209.

NEWTON L. 1931. *A Handbook of the British Seaweeds*. British Museum (Natural History), London. xiii+478pp.

NORO T., MASAKI T. & AKIOKA H. 1983. Sublittoral distribution and reproductive periodicity of crustose coralline algae (Rhodophyta, Cryptonemiales) in southern Hokkaido, Japan. *Bulletin of the Faculty of Fisheries, Hokkaido University* **34**:1–10.

NORRIS J.N. & TOWNSEND R.A. 1984. Proposals to revise ICBN: Algal names and Article 39: Figuratively speaking. *Taxon* **30**:497–499.

NORTON T.A. 1972. The marine algae of Lewis and Harris in the outer Hebrides. *British Phycological Journal* **7**:373–385.

NORTON T.A., ed., 1985. *A provisional atlas of the marine algae of Britain and Ireland*. Biological Records Centre, Institute of Terrestrial Ecology and Nature Conservancy Council.

NOTOYA M. 1974. Spore germination in crustose coralline *Tenarea corallinae*, *T. dispar*, and *T. tumidula*. *The Bulletin of Japanese Society of Phycology* **22**:47–51.

NOTOYA M. 1976a. Spore germination in several species of crustose corallines (Corallinaceae, Rhodophyta). *Bulletin of the Faculty of Fisheries, Hokkaido University* **26**(4):314–320.

NOTOYA M. 1976b. On the influence of various culture conditions on the early development of spore germination in three species of the crustose corallines (Rhodophyta) (Preliminary Report). *The Bulletin of Japanese Society of Phycology* **24**:137–142.

NOVACZEK I. & McLACHLAN J. 1986. Recolonization by algae of the sublittoral habitat of Halifax County, Nova Scotia, following the demise of sea urchins. *Botanica Marina* **29**:69–73.

OKAZAKI M., FURUYA K., TSUKAYAMA K. & NISIZAWA K. 1982. Isolation and identificatioin of alginic acid from a calcareous red alga *Serraticardia maxima*. *Botanica Marina* **25**:123–131.

PAINE R.T. 1984. Ecological determinism in the competition for space. *Ecology* **65**:1339–1348.

PALLAS P.S. 1766. *Elenchus Zoophytorum* P.van Cleef, Hague. [28]+451pp.

PARKE M. & DIXON P.S. 1976. Check-list of British marine algae – third revision. *Journal of the Marine Biological Association of the United Kingdom* **56**:527–594.

PEEL M.C. & DUCKETT J.G. 1975. Studies of spermatogenesis in the Rhodophyta. *Biological Journal of the Linnean Society* **7**(Supplement):1–13.

PEEL M.C., LUCAS I.A.N., DUCKETT J.G. & GREENWOOD A.D. 1973. Studies of sporogenesis in the Rhodophyta. I. An association of the nuclei with endoplasmic reticulum in post-meiotic tetraspore mother cells of *Corallina officinalis* L. *Zeitschrift für Zellforschung und Mikroskopische Anatomie* **147**:59–74.

PENROSE D. 1991. The genus *Spongites* (Corallinaceae, Rhodophyta): *S. fruticulosus*, the type species, in southern Australia. *Phycologia* **30**:438–448.

PENROSE D. 1992a. *Hydrolithon cymodoceae* (Corallinaceae, Rhodophyta) in southern Australia and its relationships to *Fosliella*. *Phycologia* **31**:89–100.

PENROSE D. 1992b. *Neogoniolithon fosliei* (Corallinaceae, Rhodophyta), the type species of *Neogoniolithon*, in southern Australia. *Phycologia* **31**:338–350.

PENROSE D. & CHAMBERLAIN Y.M. 1993. *Hydrolithon farinosum* (Lamouroux) comb.nov: implications for generic concepts in the Mastophoroideae (Corallinaceae, Rhodophyta). *Phycologia* **32**:295–303.

PENROSE D. & WOELKERLING W.J. 1988. A taxonomic reassessment of *Hydrolithon* Foslie, *Porolithon* Foslie and *Pseudolithophyllum* Lemoine emend. Adey (Corallinaceae, Rhodophyta) and their relationships to *Spongites* Kützing. *Phycologia* **26**:159–176

PENROSE D. & WOELKERLING W.J. 1991. *Pneophyllum fragile* in southern Australia: implications for generic concepts in the Mastophoroideae (Corallinaceae, Rhodophyta). *Phycologia* **30**:495–506.

PENROSE D. & WOELKERLING W.J. 1992. A reappraisal of *Hydrolithon* and its relationship to *Spongites* (Corallinaceae, Rhodophyta). *Phycologia* **31**:81–88.

PENTECOST A. 1978. Calcification and photosynthesis in *Corallina offinalis* L. using the $^{14}CO_2$ method. *British Phycological Journal* **13**:383–390.

PERKINS R.A. 1953. Marine algae – seaweeds. In Walsh,G.B., Rimington,F.C.,eds., *The natural history of the Scarborough District, 1 Geology and botany*, pp.67–80. Scarborough.

PHILIPPI R. 1837. Beweis, dass die Nulliporen Pflanzen sind. *Archiv für Naturgeschichte* **3**:387–393.

PRICE J.H. 1978. Seaweeds on shore: data sources for Britain. *Natural History Book Reviews* **3**:3–14.

PRICE J.H. 1984. Bibliographic notes on works concerning the Algae IV. Publication dates of parts of the *Phycologia Britannica*...(1846–1851) of William Henry Harvey. *Archives of Natural History* **11**:431–442.

PRICE J.H., JOHN D.M. & LAWSON G.W. 1986. Seaweeds of the western coast of tropical Africa and adjacent islands: a critical assessment. IV. Rhodophyta (Florideae) 1. Genera A-F. *Bulletin of the British Museum (Natural History)*, Botany Series **15**:1–122.

PRINTZ H. 1926. Die algenvegetation des Trondhjemsfjordes. *Skrifter Utgitt av Det Norske Videnskaps-Akademi i Oslo*, Matem-Naturvid. Kl. **1926**(5):1–273.

PRINTZ H., ed., 1929. *M. Foslie – 'Contributions to a Monograph of the Lithothamnia'*. K. Norske Vidensk. Selsk. Museet, Trondhjem. 60pp.

PUESCHEL C.M. 1982. Ultrastructural observations of tetrasporangia and conceptacles in *Hildenbrandia* (Rhodophyta: Hildenbrandiales). *British Phycological Journal* **17**:333–341.

PUESCHEL C.M. 1987. Absence of cap membranes as a characteristic of pit plugs of some red algal orders. *Journal of Phycology* **23**:150–156.

PUESCHEL C.M. 1988. Secondary pit connections in *Hildenbrandia* (Rhodophyta, Hildenbrandiales). *British Phycological Journal* **23**:25–32.

PUESCHEL C.M. 1989. An expanded survey of the ultrastructure of red algal pit plugs. *Journal of Phycology* **25**:625–636.

PUESCHEL C.M. 1992. An ultrastructural survey of the diversity of crystalline, proteinaceous inclusions in red algal cells. *Phycologia* **31**:489–499.

PUESCHEL C.M. & COLE K.M. 1982. Rhodophycean pit plugs: an ultrastructural survey with taxonomic implications. *American Journal of Botany* **69**:703–720.

PUESCHEL C.M. & TRICK H.N. 1991. Unusual morphological and cytochemical features of pit plugs in *Clathromorphum circumscriptum* (Rhodophyta, Corallinales). *British Phycological Journal* **26**:335–342.

RABENHORST L. 1868. *Flora europaea algarum aquae dulcis et submarinae*, Vol.3. Leipzig.

RAY J. 1690. *Synopsis Methodica Stirpium Britannicarum*. Ed.I. G. & J. Innys, London.

ROSANOFF S. 1866. Recherches anatomiques sur les Mélobésiées. *Mémoires de la Société Impériale des Sciences Naturelles et Mathématique de Cherbourg* **12**:5–112.

ROSENVINGE L.K. 1893. Grönlands Havalger. *Meddelelser om Grönland* **3**:765–981.

ROSENVINGE L.K. 1898. Deuxième memoire sur les algues marine du Groenland. *Meddelelser om Grönland* **20**:1–125. (Also issued as a preprint with the same pagination in 1898).
ROSENVINGE L.K. 1917. The marine algae of Denmark. Part I. Rhodophyceae II. (Cryptonemiales). *Kongelige Danske Videnskabernes Selskabs Skrifter*, Ser.7 (Natur. og Math. Afd.), **7**:155–283.
RUENESS J. 1977. *Norsk Algenflora.* Universitetsflorlaget, Oslo. [iv]+266pp.

SCHNEIDER C.W. & SEARLES R.B. 1991. *Seaweeds of the southeastern United States.* Durham, USA.
SCHNETTER R. & RICHTER U. 1979. Systematische Stellung und Vorkommen einer Corallinoidee (Corallinaceae, Cryptonemiales, Rhodophyceae) aus der Karibischen See: *Corallina panizzoi* nom.nov. et stat.nov. *Berichte der Deutschen Botanischen Gesellschaft* **92**:455–466.
SETCHELL W.A. 1943. *Mastophora* and the Mastophoreae: Genus and subfamily of Corallinaceae. *Proceedings of the National Academy of Sciences*, Washington **29**:127–135.
SILVA P.C. & JOHANSEN H.W. 1986. A reappraisal of the order Corallinales (Rhodophyta). *British Phycological Journal* **21**:245–254.
SOLMS-LAUBACH H. 1881. Die Corallinenalgen des Golfes von Neapel und der angrenzenden Meeres-abschnitte. *Fauna und Flora des Golfes von Neapel* **4**:1–64.
SOMMERFELT C. 1826. *Supplementum florae lapponicae.* Christianiae.
SOUTH G.R. & TITTLEY I. 1986. *A Checklist and Distributional Index of the Benthic Marine Algae of the North Atlantic Ocean.* Huntsman Marine Laboratory [St. Andrews, New Brunswick, Canada] and British Museum (Natural History) [London]. 76pp.
STEGENGA H. & MOL I. 1983. *Flora van de Nederlandse Zeewieren.* Koninklijke Nederlandse Natuurhistorische Vereniging nr. 33.
STENECK R.S. 1983. Escalating herbivory and resulting adaptive trends in calcareous algal crusts. *Paleobiology* **9**:44–61.
STENECK R.S. 1985. Adaptations of crustose Coralline algae to herbivory: Patterns in Space and Time. In: *Palaeoalgology Contemporary Research and Applications.* (Ed. by D.F.Toomey & M.H.Nitecki) pp.352–366. Springer-Verlag, Berlin.
STENECK R.S. & ADEY W.H. 1976. The role of environment in control of morphology in *Lithophyllum congestum*, a Caribbean algal ridge builder. *Botanica Marina* **19**:197–215.
STOSCH H.A. VON 1969. Observations on *Corallina, Jania* and other red algae in culture. *Proceedings of the International Seaweed Symposium* **6**:389–399.
STRÖMFELT H.F.G. 1886. *Om Algevegetationen vid Islands Kuster.* D.F.Bonniers, Götegorg, Sweden. 89pp.
SUNESON S. 1937. Studien über die Entwicklungsgeschichte der Corallinaceen. *Lunds Universitets Årsskrift*, N.F., Avd 2, **33**(2):1–101.
SUNESON S. 1943. The structure, life-history, and taxonomy of the Swedish Corallinaceae. *Lunds Universitets Årsskrift*, N.F., Avd.2, **39**(9):1–66.
SUNESON S. 1945. On the anatomy, cytology and reproduction of *Mastophora. Kungliga Fysiografiska Sällskapets i Lund Förhandlingar* **15**(26):251–264. [Also issued as an independently paginated reprint (pp.1–14). Both sets of page numbers were printed in the journal version with those of the reprint located near the inside margin and preceded or followed by a single bracket.]
SUNESON S. 1950. The cytology of the bispore formation in two species of *Lithophyllum* and the significance of the bispores in the Corallinaceae. *Botaniska Notiser* **1950**:429–450.
SUNESON S. 1982. The culture of bisporangial plants of *Dermatolithon litorale* (Suneson) Hamel et Lemoine (Corallinaceae, Rhodophyta). *British Phycological Journal* **17**:107–116.

TAYLOR W.R. 1960. *Marine Algae of the Eastern Tropical and Subtropical Coasts of the Americas.* University of Michigan Press, Ann Arbor, Michigan. ix+860pp.
THURET G. & BORNET E. 1878. *Études Phycologiques.* G.Masson, Paris. iii+105pp.

TITTLEY I., FARNHAM W.F., HOOPER R.G. & SOUTH G.R. 1989. Sublittoral seaweed assemblages, 2. A transatlantic comparison. *Progress in Underwater Science.* **13**:185–205.

TITTLEY I., IRVINE L.M. & KARTAWICK T. 1984. *Catalogue of type Specimens and Geographical Index to the Collections of Rhodophyta (Red Algae) at the British Museum (Natural History).* Part 1. Corallinales. British Museum (Natural History), London. 64pp. (Issued as a microfiche).

TRAILL G.W. 1890. The marine algae of the Orkney Islands. *Transactions and Proceedings of the Botanical Society of Edinburgh* **18**:302–342.

UMEZAKI I. 1969. The germination of tetraspores of *Hildenbrandia prototypus* Nardo and its life history. *Journal of Japanese Botany* **44**:17–28.

VADAS R.L. & STENECK R.S. 1988. Zonation of deep water benthic algae in the Gulf of Maine. *Journal of Phycology* **24**:338–346.

VAN HEURCK H. 1908. *Prodrome de la Flore des Algues Marines des Iles Anglo-Normandes et des Côtes Nord-Ouest de la France.* Labey & Blampied, Jersey. xii+120pp.

VERHEIJ E. 1993. The genus *Sporolithon* (Sporolithaceae fam.nov., Corallinales, Rhodophyta) from the Spermonde Archipelago, Indonesia. *Phycologia* **32**:184–196.

VOIGT E. 1981. Erster fossiler Nachweis des Algen-Genus *Fosliella* Howe, 1920 (Corallinaceae; Rhodophyceae) in der Maastrichter und Kunrader Kreide (Maastrichtium, Oberkreide). *Facies* **5**:265–282.

WALKER R. 1984. Trichocytes and megacells in cultured crusts in three British species of *Lithothamnium* and *Phymatolithon* (Corallinaceae, Rhodophyta). *Botanica Marina* **27**:161–168.

WALKER R. & MOSS B. 1984. Mode of attachment of six epilithic crustose Corallinaceae (Rhodophyta). *Phycologia* **23**:321–329.

WEST G.S. & FRITSCH F.E. 1927. *A treatise on the British freshwater algae.* Cambridge.

WIGHAM G.D. 1975. The biology and ecology of *Rissoa parva* (Da Costa) (Gasteropoda, Prosobranchia). *Journal of the Marine Biological Association of the United Kingdom* **55**:45–67.

WILKS K.M. & WOELKERLING W.J. 1991. Southern Australian species of *Melobesia* (Corallinaceae, Rhodophyta). *Phycologia* **30**:507–533.

WOELKERLING W.J. 1983a. A taxonomic reassessment of *Lithothamnium* Philippi (Corallinaceae, Rhodophyta) based on studies of R.A. Philippi's original collections. *British Phycological Journal* **18**:165–197.

WOELKERLING W.J. 1983b. A taxonomic reassessment of *Lithophyllum* Philippi (Corallinaceae, Rhodophyta) based on studes of R.A. Philippi's original collections. *British Phycological Journal* **18**:299–328.

WOELKERLING W.J. 1984. *M.H.Foslie and the Corallinaceae: an Analysis and Indexes.* J.Cramer, Vaduz. 142pp. (Bibliotheca Phycologica Vol.69).

WOELKERLING W.J. 1985a. A taxonomic reassessment of *Spongites* (Corallinaceae, Rhodophyta) based on studies of Kützing's original collections. *British Phycological Journal* **20**:123–153.

WOELKERLING W.J. 1985b. Proposal to conserve *Lithothamnion* against *Lithothamnium* (Rhodophyta: Corallinaceae). *Taxon* **34**:302–303.

WOELKERLING W.J. 1986. The genus *Litholepis* (Corallinaceae, Rhodophyta): taxonomic status and disposition. *Phycologia* **25**:253–261.

WOELKERLING W.J. 1987a. The genus *Choreonema* in southern Australia and its subfamilial classification within the Rhodophyta. *Phycologia* **26**:111–127

WOELKERLING W.J. 1987b. The disposition of *Chaetolithon* and its type species, *C. deformans* (Corallinaceae, Rhodophyta). *Phycologia* **26**:277–280.

WOELKERLING W.J. 1988. *The Coralline Red Algae: An Analysis of the Genera and Subfamilies of Nongeniculate Corallinaceae.* British Museum (Natural History), London and Oxford University Press, Oxford. xi+268pp.

WOELKERLING W.J. 1993. Type collections of Corallinales (Rhodophyta) in the Foslie Herbarium (TRH). *Gunneria* **67**:1–289.

WOELKERLING W.J. & CAMPBELL S.J. 1992. An account of southern Australian species of *Lithophyllum* (Corallinaceae, Rhodophyta). *Bulletin of the British Museum (Natural History)*, Botany Series **22**:1–107.

WOELKERLING W.J., CHAMBERLAIN Y.M. & SILVA P.C. 1985. A taxonomic and nomenclatural reassessment of *Tenarea*, *Titanoderma* and *Dermatolithon* (Corallinaceae, Rhodophyta) based on studies of type and other critical specimens. *Phycologia* **24**:317–337.

WOELKERLING W.J. & IRVINE L.M. 1982. The genus *Schmitziella* Bornet et Batters (Rhodophyta): Corallinaceae or Acrochaetiaceae? *British Phycological Journal* **17**:275–295.

WOELKERLING W.J. & IRVINE L.M. 1986a. The neotypification and status of *Phymatolithon* (Corallinaceae, Rhodophyta). *British Phycological Journal* **21**:55–80.

WOELKERLING W.J. & IRVINE L.M. 1986b. The neotypification and status of *Mesophyllum* (Corallinaceae, Rhodophyta). *Phycologia* **25**:379–396.

WOELKERLING W.J. & IRVINE L.M. 1988a. General characteristics of nongeniculate Corallinaceae. In: *The Coralline Red Algae: An Analysis of the Genera and Subfamilies of Nongeniculate Corallinaceae* (by W.J. Woelkerling), pp.4–28. British Museum (Natural History), London and Oxford University Press, Oxford.

WOELKERLING W.J. & IRVINE L.M. 1988b. The terms primigenous and postigenous. In: *The Coralline Red Algae: An Analysis of the Genera and Subfamilies of Nongeniculate Corallinaceae* (by W.J. Woelkerling), pp.259–260. British Museum (Natural History), London and Oxford University Press, Oxford.

WOELKERLING W.J., IRVINE L.M. & HARVEY A. 1993. Growth-forms in non-geniculate coralline red algae (Corallinales, Rhodophyta). *Australian Systematic Botany* **6**: 277–293.

WOELKERLING W.J. & PENROSE D. 1988. *Spongites* Kützing. In: *The Coralline Red Algae: An Analysis of the Genera and Subfamilies of Nongeniculate Corallinaceae* (by W.J. Woelkerling) pp.150–157. British Museum (Natural History), London and Oxford University Press, Oxford.

WOELKERLING W.J., SPENCER K.G. & WEST J.A. 1983. Studies on selected Corallinaceae (Rhodophyta) and other algae in a defined marine culture medium. *Journal of Experimental Marine Biology and Ecology* **67**:61–77.

WRAY J.L. 1977. *Calcareous algae*. Elsevier, Amsterdam. xiv+185pp.

YAMANOUCHI S. 1921. The life-history of *Corallina officinalis* var. *mediterranea*. *The Botanical Gazette* **72**:90–96.

YENDO K. 1905. A revised list of Corallinae. *The Journal of the College of Science, Imperial University of Tokyo* **20**(12):1–46.

ZANARDINI G. 1844. *Sulle Corallinee* (polipaj calciferi di Lamouroux) *rivista*. G.Tasso, Venice. 38pp. [Estratto dall'*Enciclopedia italiana* fasc. **106**].

ZANARDINI G. 1878. Phyceae Papuanae novae vel minus cognitae a cl. Beccari in itinere ad Novam Guineam annis 1872–75 collectae. *Nuovo Giornale Botanica Italiana* **10**:34–40.

Index

Suprageneric taxa shown in CAPITALS; genera, species and varieties covered in this volume are in Roman type; names of genera etc. considered to be synonyms or not to occur in the British Isles shown in *italics*. Page numbers of main entries of genera etc. shown in **bold**.

The distribution of each species is given on the basis of the county system in use before 1974 with a few subdivisions as necessary.

CPSIA information can be obtained at www.ICGtesting.com
Printed in the USA
BVOW06s1121011115

424877BV00005B/12/P